Outcomes of post-2000 Fast Track Land Reform in Zimbabwe

The struggle over land has been a central issue in Zimbabwe ever since white settlers began to carve out large farms over a century ago. Their monopolisation of the better-watered half of the land was the key focus of the African war of liberation, and was partially modified following Independence in 1980. A dramatic further episode in this history was launched at the start of the last decade with the occupation of many farms by groups of African veterans of the liberation struggle and their supporters, which was then institutionalised by legislation to take over most of the large commercial farms for sub-division.

Sustained fieldwork over the intervening years, by teams of scholars and experts, and by individual researchers is now generating an array of evidence-based findings of the outcomes: how land was acquired and disposed of; how it has been used; how far new farmers have carved out new livelihoods and viable new communities; the major political and economic problems they and other stakeholders such as former farm-workers, commercial farmers, and the overall rural society now face.

This book will be an essential starting place for analysts, policy-makers, historians and activists seeking to understand what has happened and to spotlight the key issues for the next decade.

This book was published as a special issue of the *Journal of Peasant Studies*.

Lionel Cliffe is Emeritus Professor of Politics at the Centre for African Studies (LUCAS), University of Leeds, England.

Jocelyn Alexander is Professor of Commonwealth Studies at the Department of International Development, University of Oxford, England.

Ben Cousins is DST/URF Research Chair in Poverty, Land and Agrarian Studies, University of Western Cape, South Africa.

Rudo Gaidzanwa is Professor of Sociology and Dean of the Faculty of Social Studies at the University of Zimbabwe.

Critical Agrarian Studies
Series Editor: Saturnino M. Borras Jr.

Critical Agrarian Studies is the new accompanying book series to the *Journal of Peasant Studies*. It publishes selected special issues of the journal and, occasionally, books that offer major contributions in the field of critical agrarian studies. The book series builds on the long and rich history of the journal and its former accompanying book series, the Library of Peasant Studies (1973-2008) which had published several important monographs and special-issues-as-books.

Critical Perspectives in Rural Development Studies
Edited by Saturnino M Borras Jr.

The Politics of Biofuels, Land and Agrarian Change
Edited by Saturnino M. Borras Jr., Philip McMichael and Ian Scoones

New Frontiers of Land Control
Edited by Nancy Lee Peluso and Christian Lund

Outcomes of Post-2000 Fast Track Land Reform in Zimbabwe
Edited by Lionel Cliffe, Jocelyn Alexander, Ben Cousins and Rudo Gaidzanwa

Green Grabbing: A New Appropriation of Nature
Edited by James Fairhead, Melissa Leach and Ian Scoones

Outcomes of post-2000 Fast Track Land Reform in Zimbabwe

Edited by

Lionel Cliffe, Jocelyn Alexander, Ben Cousins and Rudo Gaidzanwa

Routledge
Taylor & Francis Group

LONDON AND NEW YORK

First published 2013
by Routledge
2 Park Square, Milton Park, Abingdon, Oxon, OX14 4RN

Simultaneously published in the USA and Canada
by Routledge
711 Third Avenue, New York, NY 10017

First issued in paperback 2017

Routledge is an imprint of the Taylor & Francis Group, an informa business

This book is a reproduction of *Journal of Peasant Studies*, vol. 38, issue 5. The Publisher requests to those authors who may be citing this book to state, also, the bibliographical details of the special issue on which the book was based.

British Library Cataloguing in Publication Data
A catalogue record for this book is available from the British Library

Typeset in Times New Roman
by Taylor & Francis Books

Publisher's Note
The publisher would like to make readers aware that the chapters in this book may be referred to as articles as they are identical to the articles published in the special issue. The publisher accepts responsibility for any inconsistencies that may have arisen in the course of preparing this volume for print.

ISBN 13: 978-1-138-10865-3 (pbk)
ISBN 13: 978-0-415-62791-7 (hbk)

Contents

Citation Information

The chapters in this book were originally published in the *Journal of Peasant Studies*, volume 38, issue 5 (December 2011). When citing this material, please use the original page numbering for each article, as follows:

Chapter 1
An overview of Fast Track Land Reform in Zimbabwe: editorial introduction
Lionel Cliffe, Jocelyn Alexander, Ben Cousins and Rudo Gaidzanwa
Journal of Peasant Studies, volume 38, issue 5 (December 2011) pp. 907-938

Chapter 2
Changing agrarian relations after redistributive land reform in Zimbabwe
Sam Moyo
Journal of Peasant Studies, volume 38, issue 5 (December 2011) pp. 939-966

Chapter 3
Zimbabwe's land reform: challenging the myths
Ian Scoones, Nelson Marongwe, Blasio Mavedzenge, Felix Murimbarimba, Jacob Mahenehene and Chrispen Sukume
Journal of Peasant Studies, volume 38, issue 5 (December 2011) pp. 967-994

Chapter 4
Contextualizing Zimbabwe's land reform: long-term observations from the first generation
Marleen Dekker and Bill Kinsey
Journal of Peasant Studies, volume 38, issue 5 (December 2011) pp. 995-1020

Chapter 5
Women's struggles to access and control land and livelihoods after fast track land reform in Mwenezi District, Zimbabwe
Patience Mutopo
Journal of Peasant Studies, volume 38, issue 5 (December 2011) pp. 1021-1046

Chapter 6
Restructuring of agrarian labour relations after Fast Track Land Reform in Zimbabwe
Walter Chambati
Journal of Peasant Studies, volume 38, issue 5 (December 2011) pp. 1047-1068

Chapter 7

Who was allocated Fast Track land, and what did they do with it? Selection of A2 farmers in Goromonzi District, Zimbabwe and its impacts on agricultural production
Nelson Marongwe
Journal of Peasant Studies, volume 38, issue 5 (December 2011) pp. 1069-1092

Chapter 8

A synopsis of land and agrarian change in Chipinge district, Zimbabwe
Phillan Zamchiya
Journal of Peasant Studies, volume 38, issue 5 (December 2011) pp. 1093-1122

Chapter 9

Land, graves and belonging: land reform and the politics of belonging in newly resettled farms in Gutu, 2000–2009
Joseph Mujere
Journal of Peasant Studies, volume 38, issue 5 (December 2011) pp. 1123-1144

Chapter 10

Local farmer groups and collective action within fast track land reform in Zimbabwe
Tendai Murisa
Journal of Peasant Studies, volume 38, issue 5 (December 2011) pp. 1145-1166

An overview of Fast Track Land Reform in Zimbabwe: editorial introduction

Lionel Cliffe, Jocelyn Alexander, Ben Cousins and Rudo Gaidzanwa

Events in the last decade around the land question in Zimbabwe and the broader political context in which they have played out have been dramatic and transformational. Sparked by land occupations (locally referred to as *jambanja* meaning 'violence' or 'angry argument'), and involving contested land expropriation and violent episodes, the process has not surprisingly proved contentious among policy-makers and commentators, nationally and internationally, and among all those who have sought to explain, justify or criticise it. With few exceptions, those who have engaged in writing or political rhetoric have tended to take positions on one or other end of the spectrum in what has been a highly polarised debate, between welcoming a reversal of a racial distribution of land – some of them bewailing the manner of implementation and its distorting of the state – and those who condemn the end, in principle, as well as the means. The fervour surrounding these dramatic events and their explanation was vastly heightened, as well as being framed by, a massively debilitating economic crisis. This was marked by a world record hyper-inflation, for the moment resolved, and by a vast shrinkage in GDP. Debate continues as to what extent the overall economic meltdown was caused by or generated declines in post-land reform production or whether and how these processes interacted (see Davies 2005; Mamdani 2008; Scoones *et al.* 2010; UNDP 2008 for different positions in this debate). The political context was no less dramatic and transformational. A nationalist party, Zimbabwe African National Union-Patriotic Front (ZANU-PF), dominant for 20 years was seriously challenged for the first time by a new party, the Movement for Democratic Change (MDC), and reacted with a string of repressive laws and actions. These events also arguably shaped and were shaped by the land reform; one view explored below (Alexander 2006) is that repressive mechanisms were a requirement of enacting FTLRTP.

The very intensity of the debates and the widespread international interest in these events make a strong case for more careful and detailed analysis of the Fast Track Land Reform Programme (FTLRP) without preconceived conclusions. The passage of eleven years since the FTLRP began to be implemented is another trenchant reason to review these processes. The emergence of a range of studies into what has transpired over a lengthy period provides a 'reality check' and an opportunity to extend debates beyond policy prescriptions and their initial implementation to an assessment of what has actually been happening on the ground as a result of the land redistribution that occurred in the early 2000s. The present collection, while rehearsing past events and identifying the social and political forces that drove them, is principally concerned with summarizing the findings of investigations of the range of outcomes. Some material from these case studies has seen publication in one form or another, others are in the works, but nowhere has such a wide range of results been brought together between the same covers. It is hoped that this collection will aid in a much-needed process of rooting the vigorous debate on land reform in Zimbabwe in the available and emerging empirical evidence.

Crucial evidence is now becoming available on some of the economic and social **outcomes** of the FTLRP. The contributions to this collection will be focused more on these broader questions of what diverse pictures are emerging about local livelihoods rather than looking back to explore again the big political issues about the origins of the FTLRP, although this background will be reprised in this overview. Instead the following questions will be addressed: who have been the beneficiaries of the programme and what other groups have been affected (farm workers, former farm-owners, etc.) and how have their livelihoods fared? What new patterns of production have emerged and how successful are they in terms of levels of output, productivity and impacts on poverty? What new 'communities' and social structures have emerged?

Findings on these topics are valuable contributions to the debate, given that diverse views about the likely outcomes of a radical land redistribution emerged at the outset as *a priori* predictions – of agricultural disaster or widespread abundance – matching the polarised stances taken on basically political grounds. Those who condemned the illegality and brutality of the operations were prone to see them as also (and inevitably) ill-advised economically with disastrous effects predicted for agriculture and for confidence, investment, exports and the overall macro-economy. Equally predictably, defenders of radical redistribution and the justice of widely spreading rights to land as 'progressive' (e.g. Moyo and Yeros 2005) tended to expect a long-term expansion of land-based livelihoods and more intensive land use, similar to what happened after the 'Old' Resettlement of the 1980s (see Dekker and Kinsey here). Other logically possible combinations of views on process and outcomes – for instance, that however repressive the means, increased food security and a wider spread of livelihoods might result, or alternatively, that however justified the ending of a racial basis for land rights might be, it would not deliver at the level of production – hardly seem to figure in debates.[1]

This particular polarisation of views about outcomes was not simply a translation of assumptions into predictions, but embodied a deep difference in the way the problem was posed. As Mudege (2008, 12) remarks about debates in the late 1980s that led to the slowing of Old Resettlement, ' …the political problem of lack of access to land was turned … into the technical problem of poor access to relevant knowledge' (by the beneficiaries). Hopefully this careful scrutiny of actual outcomes can move discussion away from predetermined views of what they would be.

There are broader, comparative reasons as to why the Zimbabwe experience deserves careful scrutiny by anyone concerned with land and agrarian reform in Africa, the future of small-scale producers (or 'peasantries') and alternative

---

[1]Honourable exceptions who sought a more nuanced middle ground did exist, notably in the attempt by Hammar, Raftopoulos and Jensen (2003, 17) 'to expose and interrupt such rigid polarities and the essentialisms that underpin them', and in Campbell (2003) and Alexander (2006).

agricultural and rural futures in that continent. First, the sheer scale of the land redistribution in Zimbabwe has surpassed that of almost all other African countries, and thus like all 'extreme' cases it is potentially instructive. This of course challenges Bernstein's conclusion (2003 and 2005) about comparative experience, which sees Zimbabwe as an 'exception' – although he also saw South Africa as an exception (Bernstein 1996). Moreover, this scale has qualitative implications: there has not only been a significant shift in agrarian structure and thus in the scale and character of different rural classes but a major transformation of the agrarian sector as a whole. From a structure where the better half of the agricultural land was in the exclusive hands of large-scale commercial producers, there has been what van der Ploeg (2008) termed, controversially and in another context, a 'repeasantisation' where small-scale, household-based agricultural production is now the predominant form.

Second, however, the Zimbabwe case is not characteristic of the majority of African countries, where systems of land tenure and ownership rooted in pre-colonial societies persisted albeit within an evolutionary transition to capitalism. Its direct relevance is only to that minority of countries where 'settler-colonialism' (the commonly used term in Zimbabwe and elsewhere in southern Africa) was the dominant pattern of agrarian system and of state. This category (see Amin 1976, Mandaza 1986) included Kenya, Algeria, Namibia, Swaziland, Mozambique, Angola and, of course, South Africa. Thus Zimbabwe's experience can be assessed as an instance of that form of transition where there has been acquisition of some part of the large, white-owned holdings and some degree of redistribution, as in all these countries mentioned, usually with some sub-division of those large farm holdings, and the special challenges that such reform poses (Bernstein 2003, 2005, Bush and Cliffe 1984, Cousins 2003, Lahiff and Scoones 2000, Moyo 1995, 2005b, Moyo and Yeros 2005). Zimbabwe can be seen as one variant of this type of transformation; and its similarities and differences make for interesting and potentially explanatory comparison. Indeed, Zimbabwe may hold out lessons for these other former settler colonies in that it is the only one which after a similar, partial land redistribution in the 1980s has undertaken a second round of reform – an option that may provide an exemplar to be followed, or avoided.

Third, of course, there are insights that can be gleaned from Zimbabwe's experience for categories of African countries other than the former settler colonies, and even other developing countries, from comparing the differences in the starting point and the context, in relation to reform of 'traditional', 'customary' or 'communal' land tenure systems, which are central features of rural society in most African contexts, and sometimes found in some version or other on redistributed land. In the post-independence period Zimbabwe pursued changes in the colonial version of 'traditional' land tenure imposed in the areas 'reserved' for African farming and residence, the labour reserves as they have been conceptualised since Arrighi's (1970 and 1973) seminal work. In the 1980s land administration roles were transferred from 'traditional' authorities to local government committees, and then under a 1998 Act back to a version of the traditional authority regime of the post-1964 era of Rhodesian rule. What further changes might be made and how consistent they could be with the tenure rights of new beneficiaries of the second stage of land reform are issues still under debate, and some recommendations from findings from the studies of the FTLRP are discussed below. One opinion is that there should be a single form of land tenure extending across old and new redistributed land, remaining large commercial farms and ranches, and Communal Areas (CAs), but aside from comparing that view with other rationales not based on a singular system, the specific processes and problems of tenure in the communal areas will not be centrally addressed in the collection.

The political and economic context of land reform in Zimbabwe

A first stage of land reform got under way in Zimbabwe immediately after Independence in 1980 and the end of the liberation war that had preceded it (among the many accounts of the war and of how central an issue was land in it see Ranger 1985, Mandaza 1986, Kriger 1992, Bhebe and Ranger 1995) . That process and its legacy in the Old Resettlement Areas (ORAs), as they are labelled here, are placed alongside the recent Fast Track experience in the contribution by Dekker and Kinsey to this collection. Even this early phase was on a significant scale in that a higher proportion of the formerly white-owned farm land in Zimbabwe (23 percent) was transferred to African use by 1996 (Bowyer-Bell and Stoneman 2000; Kinsey 1999, 2003) than in Kenya (15 percent) (Leys 1975, Wasserman 1976, Kanyinga 2000), and considerably more than in the 17 years of land reform in South Africa. Latest available figures from South Africa's new Department of Rural Development and Land Reform (quoted in Umhlaba Wethu 2011) are 6.9 percent from the Land Redistribution programme, similar to Zimbabwe's Old Resettlement, and the Restitution programme together, against a target of 30 percent (for overviews of progress, see Hall 2009, especially p.2; Kleinbooi 2010, 43).

The context of this Old Resettlement also has to be appreciated, which requires the briefest survey of the country's geography. The squashed diamond shape of Zimbabwe consists of a mainly well-watered and fertile core of 'Highveld' at altitudes of over 1000 metres, which is slightly off-centre to the north and east, and corresponds roughly to the three Mashonaland Provinces, with drier terrain descending in steps as one goes down to the country's peripheries in the five other provinces. The highveld was characterised by large-scale farms, mainly owner-managed, growing food (maize, wheat, oilseeds, fruit) and export crops (tobacco, cotton, flowers, vegetables), some under irrigation, plus even larger ranches and irrigated sugar estates and, in recent years increasingly, 'conservancies' for wild-life based tourism in the low-lying areas. Most of the 'reserves', wherein African populations had been dumped at periods from the late nineteenth century until as late as the 1950s (Palmer 1977), now termed 'Communal Areas' (CAs),[2] are in the drier areas, though some are in Mashonaland and fertile parts of Manicaland.

That simplified geography is not just the basis of the agrarian structure which has been transformed but also gave shape to the liberation war. It was fought in the Communal Areas, some of which came under *de facto* control (at least at night) by guerrilla fighters backed by mobilised rural dwellers, from where encroachments and intimidation of farmers near the CAs picked away at the borders between the two types of agricultural areas (Ranger 1985, Sadomba 2009, especially Appendix 4). Much of the land that was bought up and redistributed by government in the 1980s was semi-occupied or made vacant by these forays during the war, which had established 'facts on the ground', which were then regularised by the official Resettlement programme (Moyo 1995, Alexander 2006). It should be remembered

---

[2] For introduction to the considerable literature on reserves, see Palmer (1984).

then that the programme was a consequence of a violent contestation in part over land. But the circumstances of that struggle delivered a great deal of land in the semi-arid areas, classified in Zimbabwe as Natural Regions (NR) IV and V, plus NR III designated as of marginal arable use, rather than much in the high potential NRs I and II (Vincent and Thomas 1961). But, as the Dekker and Kinsey paper shows, summarising studies such as Cusworth and Walker (1988) and Kinsey (1999), the outcomes of the Old Resettlement (OR) process were 'successful' in several respects until the 1990s, notably in making productive use of the land and increasing the number and range of livelihoods, largely as a result of the availability of farming inputs, advice and at times management, infrastructure and services. This account of earlier land redistribution experience has value not merely as a temporal comparison with FTLRP, but it can be used to rebut any crude argument that reductions in output under Fast Track were an inevitable consequence of the replacement of large by small units of production and of 'skilled', white farmers by black smallholders, for their predecessors had shown themselves capable of a different result when conditions were favourable. Of course that argument does not address more subtle concerns of critics about the undeniably real threat to smallholders taking over highly capitalised and technical production in fields like dairying and horticulture, and the contributions will explore those kinds of outcomes. A similar conclusion about the potential of Zimbabwe smallholder production is offered by the entry of communal area farmers in some areas, together with OR farmers, into significant smallholder production of marketable surpluses in crops like cotton, burley tobacco, and sunflowers (Cliffe 1988, Muir-Leresche 2006). The Dekker and Kinsey paper also documents how that 'success' in production and livelihoods levels did not continue into the later 1990s and 2000s, as a result of the macro-economic failure and its consequent denial of inputs like fertiliser and seed and of credit and extension to this sector. Again, this points to the weight that needs to be given to that factor in explanations of poor performance in post-Fast Track output, if even established smallholders were hit so hard.

This Land Reform and Resettlement Programme (LRRP) – referred to here as OR – had run out of steam by the late 1980s. Almost all the land acquired was already in public hands by 1984, although there was a trickle of new people resettled in the 1990s, and the UK budget agreed upon immediately after Independence to finance half of approved schemes under LRRP was never completely used up. But discussion about a further programme of renewed large-scale land redistribution moved from mere rhetoric to actual planning in the late 1990s.[3] Government actually produced a detailed designed and costed LRRP-2 and launched a two-year Inception Phase for it, presenting it at a Donors Conference in June 1998, which endorsed the need for such a programme and agreed upon aims for it. The agreements entered into were followed up by a joint UK/EU mission to examine their possible support

---

[3]Tendi (2010) offers a new explanation for GoZ delay in readdressing land reform in the early 1990s, in addition to the lack of political will or of finances suggested in other analyses: that it was a forswearing by GoZ as a result of a covert understanding, brokered by Commonwealth Secretary-General Anyaoku, to rein in land reform to ease the climate of opinion during South Africa's transition from apartheid. His interviews with Anyaoku and ZANU-PF leaders seem to confirm there was some such agreement, but that does not rule out other explanatory factors, nor is it easily compatible with accounts that suggest GoZ was reluctantly pushed into the FTLRP initiative by pre-emptive action by the liberation war veterans, as argued by analysts such as Sadomba (2009).

for the Inception programme in May 1999, and by a National Stakeholder Workshop in June, brokered by UNDP, which considered a National Land Policy Framework Paper (the Shivji 1998 report). The Inception Phase did receive promises of funding for technical assistance from the US, Norway, Netherlands and Sweden and a loan from the World Bank. Some of the elements that were later to form components of FTLRP were outlined in these documents: the conception of two main groups of beneficiaries, the A1 (smallholder) and the A2 (medium-scale); and a three-tier scheme for extending grazing areas of CAs through incorporation of ranches, comparable to the one Model D scheme implemented under OR (see Alexander 1991, 2006, 130–31, Robins 1994), to be discussed below.

However, funding for this LRRP-2 did not materialise, especially from the UK, the donor country with which the most detailed technical discussions had been taking place. The talks were at an advanced stage and had outlined a plan for transferring some 5 million ha of the remaining farms (about half of the total area eventually obtained in the FTLRP from 2000 to present), acquired by due legal process with compensation, and for which donor contributions would be forthcoming (GoZ 1999). It is still not clear how and why a jointly worked out blueprint was not consummated, and this issue will not be specifically addressed in the present collection. However, the intensity of the rhetoric about who should be blamed became a central dimension of debates over the next decade – Mugabe picturing this as a denial of the UK's historical responsibility and commitments at the Lancaster House Conference that arranged Zimbabwe's Independence; the UK Government seemingly reneging on support of a plan whose targets were not exclusively the 'poor'.[4] Whatever the actual roots, ZANU-PF's main justification of the land occupations from 2000 and of FTLRP when it was formalised in 2002 became and remains an alternative to a donor-supported LRRP-2, and a gesture of defiance to the old colonial power.

Land occupations and the genesis of Fast Track

The emerging pattern of FTLRP differed from the prescriptions of LRRP-2 in the mode of acquisition of land; it was not through the due legal processes as previously specified, involving designation of farms, with the possibility of appeal, and the payment of full value compensation. A replacement Land Acquisition Act 2000, plus regulations proclaimed after enabling changes in the Constitution, allowed for compulsory acquisition without compensation for the value of the land itself, only for the value of 'improvements'. The priorities for the type of land that ought to be acquired also differed, as did the criteria for selecting beneficiaries, although the two main official 'models' on which schemes for allocating holdings did more or less correspond to the A1 and A2 formulae put forward in LRRP-2. The former was closely based on the similarly-labelled Model A used in OR, providing an individual smallholding for residence, homestead garden and field crop cultivation, together

---

[4]The notorious November 1997 letter from Clare Short, UK Secretary for International Development in the new Labour government, discussed in Tendi (2010, 87–93), stated, 'we do not accept that Britain has a special responsibility to meet the costs of land purchase in Zimbabwe'. After further attempts to resolve the differences, the then Minister for Africa, Peter Hain, told the House of Commons Foreign Affairs Committee in May 2000 that release of funding was suspended because the Government of Zimbabwe was running 'a corrupt, inefficient land reform programme'.

with a large area set aside for common grazing to which all members of the new community settled on the former farm would have access (regardless of whether devoting such high proportions of fertile land in Natural Regions I and II was appropriate). This became referred to as the 'villagised' sub-version of A1 as dwellings were clustered around a contiguous settlement area, as opposed to another variant of A1, to be known as 'self-contained', where the grazing was also sub-divided and combined with the individual arable plots to make up small farms. It may be that in the long sweep of history the mode of acquisition is not the main determining factor of outcomes, but the resulting short term circumstances of implementation – hurried, with no clear lines of responsibility, partly through occupation, and with planners trying to play catch-up with little finance for post-transfer support – meant that the mode of acquisition has been a critical factor in the first decade. It should be recognised that some of these factors were present in the OR programme (see Dekker and Kinsey in this collection), if not to the same extent.

Whatever the plans and their roots, it is generally accepted that this process of new land acquisition and redistribution was sparked by incidences of 'occupations', 'incursions' or 'invasions' (the terms used are themselves partisan) of farms inhabited by whites, and in some cases Africans who had obtained farms in the years after 1980. The first such instances occurred in 1998; these were often reined in by the ZANU-PF Government, but began again in earnest in 2000 (Moyo 2001, Moyo and Yeros 2005, Sachikonye 2003a). Varying from political protest to noisy disruption to violent intimidation, beatings and killings, they often dislodged owners and managers, and sometimes their farm workers, and were often named *jambanja* (Chaumba *et al.* 2003). Such incidents and their consequences on the ground are explored in some of the case studies presented here, but they illustrate that occupations were not the only means of acquiring land; in other cases it was a more administrative process. What is more a matter of dispute is the extent to which the farm occupations were a consequence of a coherent strategy articulated and orchestrated by the leadership of ZANU-PF and the Government of Zimbabwe (GoZ), or whether these two institutions had to be forced or manipulated into supporting them by a separate initiative taken by 'veterans', as argued by Moyo and Yeros (2005) and Sadomba (2008). Some who took the former position argued in a more nuanced way that the Fast Track formula inevitably involved a decisive restructuring of state institutions, from the top down to the local level, to implement it, and this marked a shift from a technicist approach to rural issues to a form of state that was partisan and violent (Alexander 2006, Tendi 2010).

The cases discussed in this collection do provide some local evidence of differing processes of occupation and the extent to which these were the key mechanism for land acquisition, without claiming to provide a definitive answer to the larger historical question. Irrespective of the origins of the impetus of the *de facto* occupations of farms, the ruling ZANU-PF party did shift to a policy of encouraging invasions when it was faced for the first time by an emerging, organised political opposition, the Movement for Democratic Change (MDC). The ruling party had to compete with a serious opposition in a 2000 referendum on a new constitution, which included a provision for land acquisition without compensation and which was voted down, in the subsequent parliamentary election in mid-2000 and in the presidential elections of 2002. This No vote in the 2000 referendum may have been an important early trigger for the decision to embark on the FTLRP. Gaidzanwa, as a participant in the 1999 constitutional reform process, recalls that the GoZ-appointed Constitution Commission's original contribution had accepted the principle of compensation for improvements to land. Indeed, the South African legislation on land was actually modelled on the 1990s legislation from Zimbabwe. The FTLR was turned into *jambanja* after the President changed the Clause 57 on land to disclaim any responsibility for the Zimbabwe state to compensate farmers whose land was taken for redistribution. The No vote hardened that resolve as spelled out in the revised Clause 57 in the rejected constitution. Giving the go-ahead to farm occupations and to the use of violence in acquiring land in 2000 was probably seen by ZANU-PF as a means of mobilising rural support and punishing its political foes, the white farmers, 'their' workers and other supporters of MDC (as argued by Alexander 2006). The ZANU-PF government then introduced a process of after-the-fact regularisation from 2002 with the formal introduction of FTLRP.

These differing mechanisms of *de facto* dispossession gave way to a new Land Acquisition Act in 2002 that legalised compulsory purchase without payment for the value of the land. (A legal commitment to pay for the value of improvements made to the land was written in, but in practice this has occurred only on a very limited scale.) And an official FTLR Programme got under way in July 2002, specifying what land was targeted for acquisition. Whatever had been the role of veterans or even spontaneous movements onto farms, once state control of past occupations and future designations was established, power over the process of redistribution of the land acquired was spread to include a melange of other actors: party and government leaders (local and national), traditional authorities as well as veterans (sometimes operating through district land committees, sometimes self-appointed). Who was then to get land was not simply a matter of satisfying popular need, and exercising patronage among rural people. Certain targets for percentages of beneficiaries from designated categories of people, such as veterans, women and former farm workers, had been mooted in earlier discussions like the 1998 Donors Conference on LRRP-2, but only the first category were specified in formal guidelines like GOZ 1998. But within any such designations actual allocations were seen as a means of satisfying the interests of higher, and also middle, levels of politicians, officials and local notables. How the differing local patterns of selecting beneficiaries played out, with what mixture of class interests satisfied, is documented in some of the cases here, but the overall outcomes were certainly varied.

The amount of farm land targeted for acquisition was over 10m. hectares (Utete 2003), about twice what had been specified in LLRRP-2 (GoZ 1998). Targets were also specified in the Programme for the proportion of A1 and A2 holdings. Most of this land had been acquired and sub-divided by the later part of 2003 and the first of a number of land audits was then made by a GoZ-appointed Committee (Utete 2003). This reported that at that date 4,324 farms had been redistributed covering 6.4m. ha., and 34 percent of this area had been allocated to 7,260 beneficiary households who received A2 (medium-scale) holdings and 66 percent to 127,192 holders of A1 smallholdings. Gaidzanwa, who was a member of the Committee, recalls the impression of considerable unevenness of the extent of the process and reliability of the reports across provinces. Those contributions to this collection that present more up to date national figures (Moyo and Scoones *et al.*) suggest slight increases in A1 beneficiaries by 2009 (145,000 is the lowest estimate) and perhaps a doubling of A2. Some case studies obtained numbers from provincial and district offices of AREX (Agricultural Extension Service, Ministry of Agriculture, formerly and now once again in 2010 AGRITEX), or surveyed samples, and their findings show regional variations around these national proportions – although these investigative efforts tend to confirm the observation that, 'Data are hard to get hold of and often inconsistent, and the situation has been very fast-moving with official statistics not covering contested areas and new invasions' (Scoones *et al.* 2010, 33). Inconsistencies are compounded by the fact that some A1 occupants still have no offer letters.

Since 2003, the land reform is changing in shape and emphasis as the political struggles emerge and change. Some politicians use their muscle to dispossess recipients of land, especially A1; but some 'land grabbers' have in turn been dispossessed of land by war vets and others so the redistribution continues! Again, some of the small farmers and elite land grabbers are renting land back to the white farmers so the formal offer letters may not always reflect the actual use of land on the ground.

A further limitation of the local findings from micro-studies lies in the fact that often each pre-existing farm was targeted for sub-division on the basis of a single model, A1 or A2, so reliable figures for the overall balance between types of holdings are only possible at the district or provincial level not from a study of one or a few 'schemes'. Still, compared with the national ratios, where A2 beneficiaries made up less than 10 percent of the total, and received about 25 percent of the land, findings reported here range from the semi-arid Masvingo Province where only about 3 percent of beneficiaries were under the A2 model, but received 23.7 percent of the acquired land, reflecting that the larger areas needed to make a medium commercial enterprise from ranching (2009 figures from Scoones *et al.* 2010, 33); to the high value commercial farming heartland with its developed infrastructure and proximity to the capital, where in Mazowe District, the A2 farms formed a higher percentage, and they received 211 of the former farms compared with only 113 allocated as A1 (Matondi forthcoming). Overall 70 percent of all A2 farms allocated in the country were in the most productive and accessible three Mashonaland Provinces, according to aggregates referred to by Moyo in this collection.

The broad proportions between these two main categories have to be interpreted with care as the distinctions between, for instance, the larger A1 self-contained and smaller A2 were in practice blurred (see Scoones *et al.* 2010 for examples). The differences were further qualified and made more complicated by other additional categories of 'beneficiaries'. There was, for instance, a 'Three-Tier Scheme' seen as suitable for a few areas in the drier south and west of the country, whereby the acquired land was to be used for livestock ranching, and was now made available as a common grazing area by rotation to groups and communities in neighbouring CAs, essentially a variant of the Model D scheme in one location under the OR Programme. According to evidence from case studies done by Moyo (this collection), AIAS (2007) and Matondi (forthcoming) Three-tier was only pursued in two districts in Matabeleland South Province: Mangwe where 13 farms were acquired for this purpose, and Matobo (14 farms). But, as with the original Model D (Robins 1994, Alexander 2006), when it came to implementation there was controversy about the precise aim, potential beneficiaries and patterns of land use. Official plans sought originally to restrict access to beneficiaries with more livestock which they would be expected to build up into a commercial beef herd – restrictions and aims which have been contested by communities, by encroachment as well as verbally, to the point where it became more common to modify the model to make grazing land available to whole communities. But then there seemed to be little communal management and something closer to open access.

In some arable areas various residual provisions (as spelled out by Chambati in this collection) were made for those former farm workers who remained on the land: through participation in occupations or via applications, or simply staying put, as well as *de facto* setting aside of very small plots, essentially 'allotments'. Fox *et al.* (2007) report similar practices in Kadoma of farm workers being allocated plots half the 6 ha. of those of A1 beneficiaries or less. Zamchiya also cites instances of space being made for farm workers after initial allocations, and other cases of farm workers being subsequently displaced.

Another significant set of exceptions to all these formulae was those who received the whole of a pre-existing farm, or even more than one. This was sometimes within and at the time of the original allocation process, sometimes through subsequent, extra-legal displacement of beneficiaries or of unallocated acquisitions of other state land. This process of enforced land grabbing still goes on and is perpetrated by those who might accurately be described as 'cronies'; indeed political muscle is a requirement for such acquisition. Some of the processes and the politically well-connected who are involved are documented in some of our case studies. Estimates of the scale of this pattern are made for Masvingo Province as a whole, where Scoones *et al*. (this collection) report 58 farms (111,330 ha.) as having been allocated as whole units under FTLRP, plus subsequent land grabs. Moyo (2011b, 261, 2011d) has recently estimated those who he designated 'land grabbers', which in his definition comprised those who have acquired farms of between 500-1,000 ha. (roughly a third of the average size of former 'large-scale commercial' holdings – although that does aggregate across the very wide range of what was 'average' in different Natural Regions). Accordingly, he estimates that 80 percent of the 3,000 holdings of this size or larger are now black owned. He suggests that his (rather arbitrary) category includes 'urban and rural based professionals, public and private sector executives, other petty bourgeois elements and black capitalists'. Marongwe (this collection) offers detailed evidence that the reallocation of smaller holdings originally acquired, by occupation or allocation, as A2, and displacement of beneficiaries in favour of more politically connected people was 'quite widespread' after 2002.

For sure, 'cronies' did get some farms. But that is a far cry from assertions that all or most of the land went to them, and much depends on the definition of this political category. A significant number of farms have been acquired by prominent elite figures either by getting whole or large parts of pre-existing farms through FTLRP or by informal and often strong-arm methods, and this is an on-going process. Short of the promised Land Audit (see below) it is hard to estimate the scale. The references in the above paragraph give some indicators. But any statement that beneficiaries of FTLRP are confined solely to 'cronies' in the sense of those in the immediate power circles at the core of the ZANU-PF regime, who received farms on a scale of the previous owners, is indeed a 'myth' on both counts, as Scoones *et al.* assert in their study (this collection); and there is a consensus among contributors that smallholder farms now predominate in number and overall area, and thus a major agrarian restructuring has occurred, even if there is a difference in estimates between authors and regions.

Elite 'land grabbing' has been a dominant trend in another former settler-colony, Kenya. If its two decades longer resettlement experience is any clue to the future of Zimbabwe, there is likely to be a continued and powerful thrust by such interests, but it is also likely to be contested by more democratic forces, especially at the local level. So the eventual outcome may not be the same as in Kenya, not only in the amount of elite grabbing, but also as regards the resulting agrarian structures (Kanyinga 2009, Rutten and Owuor 2009). One tendency in that country has been for many large landowners to settle for informal renting out of land to smallholder tenants rather than persist in being capitalist farmers. The studies presented here do provide a few anecdotal references to the borrowing (Mwenezi, reported in Scoones *et al.* 2010), leasing or selling off (of A1 grazing land in Kadoma reported by Chigumira 2009) of redistributed land, even though such things are proscribed by the laws governing FTLRP. Such tendencies might be prompted by what seems to be a pattern that large farms and even medium A2 holdings are underperforming compared with A1 smallholders, examined here. It may take perhaps a decade or so before it is clear how far any such forms of landlordism become general.

Another form of political patronage beside that of regime cronies receiving large farms was detected in some areas. Marongwe's contribution (this collection) stresses the party political bias favouring those applicants who sought medium-sized A2 holdings in the effectively peri-urban setting of Goromonzi: civil servants, members of the security forces, ZANU-PF members and those who were the clients of party notables were privileged. However, villagised A1 beneficiaries formed the majority of the larger sample used in his study (Marongwe 2009), as they did by and large nationally, although party patronage even operated among them. Moyo (this collection) and Murisa (this collection) among other contributors to this collection note similar bias in the actual processes of allocation of A2 to government and party notables elsewhere, but the beneficiaries in those areas more distant from Harare were drawn more from middle and/or local level institutions. Zamchiya (this collection) takes his analysis of allocation processes in Chipinge further by looking specifically at bias in selection of A1 beneficiaries on self-contained and villagised schemes. He argues that this large segment were drawn principally from local ZANU-PF members and supporters. Though such land recipients may be 'poor' there is a bias toward those with 'political capital'. Mkodzongi (2011) observed a similar process in Mhondoro in Midlands Province but interprets the implications differently, pointing out that anyone might claim party membership or to be a supporter for opportunistic reasons including jumping the queue for land, as Zimbabweans have done at least since the pre-Independence elections in 1979 and 1980 when voters were urged 'to eat Muzorewa but vote Mugabe' (personal observation, Cliffe 1980). Scoones *et al.* (2010, 54) and Mujere (this collection) similarly point to tendencies for people to assume several identities which they deploy to suit circumstances. One essential difference between such accounts lies in their interpretation of how deep party identity goes, though of course the depth of that identity might vary between areas, not all of which are marked by the same historical dominance of ZANU-PF as in a district such as Goromonzi. Of course even if patronage on party lines has not been as permanent and exclusive as ZANU-PF elites intended, such a process entrenches the partisan and patronage-based institutional character of the state and limits the free expression of political views.

The lack of clarity about who the beneficiaries are and what are the continuing dynamics of land acquisition and contestation was recognised in the political process from 2008 leading to a power-sharing government that included ZANU-PF and the two factions of MDC. The Global Political Agreement (GPA) made two pronouncements on land issues. It specifically called for a Land Audit 'for the purpose of establishing accountability and eliminating multiple farm ownership' (GOZ 2008 Article 5.9a), although this has yet to be initiated. But the GPA also set out the broader context within which future policy on land reform should be framed. It asserted that FTLR was irreversible, stating that:

"while differing on the methodology of acquisition and redistribution, the parties acknowledge that compulsory acquisition and redistribution has taken place" (Article 5.4) but in fact differences over acquisition are now basically matters for the past. They did agree that "the primary obligation of compensating former land owners for land acquisition rests on the former colonial power" (GOZ 2008 Article 5.6).

This formula, to which both partners have signed up, presents a particular challenge to prospective donors to land and agricultural programmes and to the international community generally in considering whether to resume normal economic relations with Zimbabwe. There is no doubt that the economic prospects have changed with the dollarization that halted hyper-inflation, and the recent expansion of mineral exports, making extra revenues available and reducing reliance on western largesse. The political circumstances have also changed with the inauguration of coalition government, but the partnership remains an unequal one and policy differences, not always along the party divide, prevent action on key issues. There has, for instance, been protracted discussion about a revision of land tenure, in all sectors but especially with regard to fast track areas, but still no draft legislation. It is one of the aims of this collection to present findings that might illuminate such debates.

The scale and type of land redistribution through FTLR, and the numbers and categories of people involved, begin to indicate the possible longer run outcomes of processes initiated eleven years ago. But what differences FTLRP has made to land production systems, the changed livelihoods pursued, the resulting overall agrarian and class structures will depend on what has taken place on the ground, in a range of different agro-ecological and political settings, in the past decade. The contributions to this collection provide as comprehensive a set of findings as have yet been assembled in one place on these outcomes, covering beneficiary origins, the fate of agricultural production and pursuit of other livelihoods, and the dynamics of the emergence of new communities and agrarian structures. An attempt will be made here to summarise, compare and contrast findings on these topics.

The case studies

Three major, longitudinal studies have been carried out in Zimbabwe covering the whole period of FTLRP, each involving surveys at different dates carried out in eleven of the country's 53 districts by three institutions, the African Institute for Agrarian Studies (AIAS), Harare; the Ruzivo Trust (RT), originally under the Centre for Rural Development, Harare; and a group from the University of Zimbabwe, AGRITEX and the Institute of Development Studies (IDS), University of Sussex, Brighton, UK, coordinated by the Institute for Poverty, Land and Agrarian Studies (PLAAS) at the University of Western Cape in South Africa. We have been fortunate to get three summary studies drawing on this work from Moyo and Chambati (AIAS) and Scoones et al. (IDS). In addition several other researchers have provided work based on individual studies all of which have involved fieldwork in one or more locations.[5]

---

[5]Two other sets of case studies are available under the Livelihoods after Land Reform programme: http://www.lalr.org.za/zimbabwe/zimbabwe-working-papers-1

Chronologically, mention can first be made of Dekker and Kinsey who have provided a summary of the Old Resettlement (OR) beginning in 1980 and they report on findings from the same sample of beneficiary households who have been surveyed over the whole of that period. Their contribution offers not only invaluable comparative material on that earlier process and evidence of broadly 'successful' production and livelihoods outcomes in the first decade, thereby refuting views that simply dismiss the potential of smallholder production and capabilities of Africans on *a priori* grounds, which were voiced then and are still repeated. They also give evidence of how this performance was not maintained in the second decade or so, pointing to possible long term issues such as new generations that might affect sustainability, but also emphasising the impact of economic stresses on inputs and markets from structural adjustment in the 1990s and the economic meltdown of the 2000s on these established farmers.

Moyo has drawn together material from six district studies of outcomes from all but two of the regions (outside the capital, Harare) conducted by the African Institute for Agrarian Studies (AIAS). Most of these data are presented in aggregate or average form, but many tables provide figures for each district as well. He combines that with a decade-long observation of a wide range of policy documents and processes nation-wide. He goes on to look at national developments in production, investment and marketing over the last couple of years and claims to discern, finally, signs of recovery in the contribution of the reformed sector.

Scoones *et al.* present a summary and up-date of their book length study (2010)[6] reporting on developments in differing local environments in the southern province of Masvingo whose previously large holdings in drier conditions were given over more to ranching, wildlife and (in some areas) irrigated agriculture.

Another set of contributions, while examining local case studies and providing evidence of production, livelihood and social outcomes, take as a main focus the consequence to a particular social group: women in one study; farm workers in another. Other studies also provide snippets of information on these two categories. Mutopo sets her localised study in the historical and national context of how far **women** were beneficiaries of old and new land redistribution, referring to their limited land access rights in pre-colonial and colonial society which was not dramatically changed despite women being involved in the liberation struggle. That review of overall evidence in comparing the target of 20 percent suggested by the 1998 Donors Conference, for instance, with actual proportions of women who received land in their own name, is further substantiated by Moyo and Scoones *et al.* in their contributions, and in their other studies (Moyo 2009, Scoones *et al.* 2010), and in official reports (Utete 2003). Other, gender-focused studies like WLZ (2007) and Goebel (2005) and local case studies, like Mutopo's, confirm the narrow openings for women to be granted land officially in their own right but also show how women farmers who manage to get use of land have often had to acquire it indirectly and by wheedling it out of spouses or wider family or political influentials; and having got land have difficulty in utilising it fully, experiencing the usual problems of scarcity of inputs, draught power, water, labour and secure tenure even more severely than men. Mutopo goes on to use an even more micro-focus to explore options faced by women on a former ranch in the south-eastern district of Mwenezi. She brings out the enterprising way some of them have been able to build on the security offered by

---

[6]Access to this and related material: www.zimbabweland.net

their having land, in their own right or via husbands, and on the narrow opportunity of growing a surplus of beans, nuts and vegetables in a semi-arid area to deliver their produce to local markets in towns and over the South African border. She shows how realising these new economic prospects in turn empowered them in terms of controlling their lives more through involvement in a wider social universe and of changing gender relations at home.

Chambati utilises the AIAS survey sources and national data to undertake an exploration of the outcomes of the FTLRP for **farm workers**. He offers a broader analysis, not just of what has happened to them, but of the differing kinds of labour relations that have emerged to replace the paternalistic farmer/farmworker relationship that typically characterised previous patterns of 'belonging' on the former large farms (Rutherford 2003, 2008). He acknowledges that many early reports such as Sachikonye (2003) and FCT (2003) noted that disappointingly low proportions of beneficiaries were former farm workers, seldom above 10 percent, but points out that were no specified targets in the FTLRP policy document (GOZ 2001). He suggests that on the basis of a different ratio, the proportion of sampled farmworkers who took part in invasions and those who got land, as a group they fared better than might be thought. He also shows how some received tiny plots or stayed in their houses on allocated holdings, and how they can now be seen as making up a new kind of agricultural labour for new land holders, but often not tied to one farm, sometimes working in itinerant organised groups and on occasions able to bargain for reasonable wages with labour-starved A2 farmers. A complex new regime of labour relations has been emerging.

Three studies build on further local cases covering a range of agro-ecological and political circumstances to look at beneficiary selection. All cover localities within districts covered by one of the three main studies, so cross checks and comparisons of findings are available to the discerning reader, although this Introduction will only draw attention to some particularly interesting and contested issues. These cases use follow-up surveys to provide more depth to the findings from official land audits, like Utete (2003), and from the three institutions' studies on the **selection of beneficiaries**. They spell out what actually happened in allocating land rather than just the laid-down procedures, and reassess who the beneficiaries were. As mentioned above, Marongwe shows that in one A2 scheme in Goromonzi, located conveniently close to Harare, actual distribution did not accord with the formal process of application nor the procedures for selection according to technical competence and by a land committee. Rather they were made on an *ad hoc* basis, depending on local political circumstances, and were politically partisan with the result that the great majority of actual recipients did not meet the criteria laid down and were not on the original lists. He also argues that this privileging of political and governmental elites resulted in new farmers not having technical capacity, which added to their problems of financing inputs and labour required by the scale and type of production envisaged on A2 farms, and the underutilisation of land at least up to 2003. Those consequences are projected not so much on the basis of production outcomes as on notions that underpin continuing debate, on the assumed potential of 'real farmers', a category which is not clearly defined (Cousins and Scoones 2009). Marongwe's focus is primarily on A2 farmers, and smaller A1 farms are discussed only in cases where their land is being grabbed by the A2 elites. Zamchiya also portrays a pattern of politicisation of the selection and allocation processes in a very different district in the eastern border areas, Chipinge. As indicated above this picture differs in that he detects a party political bias even among beneficiaries in self-contained and villagised versions of A1. Mkodzongi (2011) enters this particular debate in arguing that there is little that is fixed about the categories of ordinary party 'members' or 'supporters' at the local level. But Zamchiya partly anticipates such counter evidence by suggesting that demonstrations of loyalty to ZANU-PF are persistent conditions for maintaining rights to land in his district.

Mujere offers a different perspective on the selection of beneficiaries and its consequences. Moving away from considerations of patronage and political bias in the process of allocation, he focuses on the subjective distinctions between 'strangers' and 'locals' among new farmers in A1 village schemes in the lower potential, but heavily populated Masvingo district of Gutu (also covered in Scoones *et al.*). In the process of exploring the interaction between these different categories and those in other studies, he brings out the issue that, although FTLRP did not involve an explicit commitment or a sub-component, like the Restitution Programme in South Africa, many would-be beneficiaries were motivated by claims of restoring access to specific land from which their kin or their ancestors had been dispossessed. To that extent relations between different land recipients were, and remain, matters of their 'belonging'. Moreover the identities that came into play involved complex sets of claims about who had rights to land. As well as the division between those with some historical claim as opposed to 'strangers', mostly from more distant localities, there were those between restitution claimants. These could be based on ethnicity, such as the predominant group in the district, the Kalanga, or those based on autochthony. And the latter could base claims to 'their' land on the basis of their community having been displaced within living memory, often involving the loss of chiefdoms – the right to return – or the spiritual claim that their ancestors were buried there – the right conferred by 'bones'. Conflicting claims at the start of FTLRP as to who should receive an allocation have evolved into disputes as to who should control the schemes now in place. These latter involve traditional leaders and not just the new farmers themselves, but also interact with the committees of seven put in place to run schemes, and the District Land Committees to whom they are supposedly answerable under government directives but not under existing legislation, and other government and political bodies. Sketching in this dimension of complexity about the new realities of land relations deepens understanding of the outcomes. But the study does not identify what mechanisms might resolve these persistent and potentially debilitating disputes. This perspective does call attention to the need to clarify not just legal land tenure rights but the provision of land administration systems, including dispute resolution procedures.

Murisa's particular focus in two Mashonaland districts goes beyond individual households and how they are making out to consider what patterns of **social organisation** are emerging, especially those that provide cooperating networks for farmers. The very act of resettlement under FTLRP, as with the earlier OR, brought together in one place a set of people from diverse localities and backgrounds (see also Barr 2004). With much local variation across his case studies, Murisa finds that people from nearby and distant communal areas, from towns and former farms sharing little if any past contact or other ties, now find themselves neighbours and facing the challenge of developing local institutions for specific functions (to run social facilities and farm support among other things). How far they have started to build some networks in place of former 'farm communities' is explored in passing in several case studies, but local farmer groups are potentially a vital contribution to building new 'social capital', given the unfamiliar challenges to small and medium size farmers of coping with larger scale technology, like irrigation, tobacco curing and other crop processing, in a context where mechanised but also simple inputs are scarce, and where credit to pay for them and access to new markets is difficult. Murisa documents the presence in his areas of farmers' groups pooling efforts to access inputs, reach markets, obtain tilling services, and also engage in cooperative production activities in irrigation and larger-scale operations such as wheat harvesting. They also join hands in resisting encroachments from A2 and politically connected farmers and in seeking greater tenure security together. His findings reveal that such initiatives do not always survive or prove beneficial. They are sometimes subject to corruption and nepotism (as with cooperatives in many rural areas), with an inverse correlation to size of group.

Overall, the studies in this collection provide first-hand information and observation from 14 different districts in seven of the eight provinces of the country outside Harare. The exception not covered is Matabeleland North. A neighbouring district, Mangwe, in Matabeleland South, is only covered in the form of returns to a sample survey there reported along with those from other sites in Moyo's contribution. The underplaying of these areas is to be regretted, especially for political reasons as they are generally seen along with Manicaland as the rural core for the opposition MDC, but will hopefully be covered by future research.

The eastern province of Manicaland, which contains the whole gamut of agro-ecological conditions found in Zimbabwe, none of which can be regarded as 'typical', is the subject of only one study (Zamchiya) with only limited data on post-settlement economic and social outcomes. The high potential heartland of former settler agriculture in Mashonaland's three provinces is the target of seven investigations in four districts. And four districts in each of the other (more marginal for arable) provinces, Masvingo and Midlands, are also included.

Of course no one case study, whether of a former farm, scheme, sample surveys from a locality or wider district – in any country – can give a 'representative' assessment of the impact of a widespread upheaval such as Zimbabwe's FTLRP. The significance of the findings in the book length account of Scoones *et al.* (2010) has already been dismissed by some reviewers (e.g. Andoh 2010) on the grounds that Masvingo is so distinctly different, even though four differing 'clusters' are surveyed, an argument which does not challenge any of its findings; it just serves as a caution before drawing any wider generalisation of conclusions or lessons, as the book is careful to do, and as Andoh's review explicitly calls for. One aim of this collection is to enhance the value of every one of the case studies in placing them alongside each other in order to have a better, though not foolproof, basis for such generalisations – as will be attempted in comparing findings and considering future implications in the rest of this Introduction.

Findings

New agrarian structure

One basic conclusion with which all broadly agree, but with differing assessments of details and actual numbers, is that the 'dualist' pattern of land use inherited from the settler colonial past, with a large, capitalist white sector and African smallholder reserves, modified to a degree in the 1980s and 1990s, has now undergone fundamental structural change. The overwhelming form of agriculture is small-scale, relies mainly but not exclusively on household labour, and produces for domestic consumption as well as the market. In fact that racially based dualism of land use organisation was always over-simplified since the establishment from the 1930s of a middle stratum of 'small-scale commercial or African Purchase Area farms', owned by individual African farmers outside the communal area system, with a hired labour force (Cheater 1984). The A2 farms created under FTLRP were estimated (Moyo in this collection, quoting GoZ figures) in 2010 to constitute 22,373 farms. However, Matondi (forthcoming) quotes a lower figure again derived from official figures of only 16,386 – a significant difference pointing up the need to complete the Land Audit agreed in the GPA. Whatever the exact number, these can be considered to be the other part of the middle band of what is now in Moyo's phrase a 'tri-modal' structure. The top stratum in 1980 had consisted of some 6,000 white-owned large commercial farms, and even larger estates and ranches, occupying some 15m. ha., the better half of the country's agricultural land. This was reduced to 12.5m. ha. by the 1990s, and through FTLRP is now down to perhaps 3m. ha. or less. There has been a corresponding increase in the area devoted to smallholder production which included only the CAs in 1980 and which ran to 16 m. ha., but now represents 62 percent of the farming area (CAs plus OR plus A1).

The processes of FTLRP and the new structure emerging have also to be understood as involving class reformation. Many of the very largest of the former commercial farms, especially the ranches and agro-estates, most under corporate ownership, remain in private hands (Moyo 2011d). Together with the few remaining farms owned still by white individuals, many of them now operating on reduced areas – estimated by Moyo (this collection) at 200–300 nationally; 4 percent of the land area in Masvingo Province (Scoones *et al.* 2010) – these represent a continuing but much reduced class of large capitalist farmers directly employing a labour force. Formerly there were about 6000 such farmers and the evidence in keeping track of them comes from studies like Selby (2006) and the still existent Commercial Farmers Union (CFU) who conducted a survey of former members. An interview with two CFU officials (Cliffe, personal communication, 2010) revealed that over 60 percent of their survey had no desire to resume farming in Zimbabwe, even if it were feasible; most cited age as the decisive factor. It is known that some have moved to farm in countries as diverse as Malawi, Zambia, Mozambique, South Africa and as far afield as Nigeria (Hammar 2010).

The findings of these studies agree that there has also been a reduction of those who were farm workers, but offer a more complex and differential pattern to that of the wholesale dispossession of this class pictured in some accounts of the early 2000s, and even a different overview to those few studies that were done at that time (Sachikonye 2003b, FCT 2001). A significant minority remain on the estates and others live on remaining or newly acquired large farms; another proportion has retained or found a place as permanent workers on new farms, especially A2. Moyo and Scoones *et al.* report that a majority of the latter hire permanent labour, although Chambati suggests that a new form of wage labour where workers are no longer tied to one farm but may be itinerant, working in groups sometimes, are more characteristic of new labour relations. A large part of the hired labour is in fact casual. Former farm workers were absorbed in FTLRP schemes: in Masvingo they constituted 11 percent of A1 beneficiaries; in Mazowe a smaller proportion; in one scheme surveyed by Chigumira (2009) the proportion was 30 percent, while in her other two areas of study they were negligible. In other schemes some farm workers received a holding that was but a small proportion of the standard A1 plot (Chambati). Elsewhere some have hung on to their residences, in the midst of new villages or as 'squatters' on new holdings.

Who benefitted?

Reference has been made above to varied findings about the selection of beneficiaries and allocation of holdings, and to the different conclusions about how far the outcome skewed the list of those who got land politically, either in the narrow sense of highly placed cronies, or a more widespread preference for ruling party supporters. One crucial dimension is that of **gender.** At the donors' conference of September 1998 (discussed earlier), a quota of 20 percent for women was adopted. However, this quota did not become formal policy and was not included in the Inception Phase Framework Plan 1999–2000 and has not been put into statute (Jeanette Manjengwa, personal communication). There is a consistency in the reports about what proportion of actual beneficiaries (i.e. named on the offer letter in their own right) were women: Utete 2003 offered national figures of 18 percent of A1 and 12 percent of A2. In Mazowe these were 13 percent and 11 percent, respectively (Matondi forthcoming); very similar figures from a 2007 Provincial Land Audit are quoted for Masvingo by Scoones *et al.* (this collection). These findings compare with what seem to have been even smaller proportions in ORA allocations in the 1980s (Dekker and Kinsey, this collection). But those like Mutopo (this collection) and Scoones *et al.* who explore the gender dimension in more detail indicate that women received a slightly higher proportion in subsequent reallocations, and also use a variety of stratagems to obtain the use of land even if they were not the recorded beneficiary. These studies suggest that the disproportionality compared even with the ratio of women-headed households in rural communities generally, often as high as 30 percent, is a result of the patriarchal practices at work in civil society (Gaidzanwa 1988, Pankhurst 1991), although there seems to have been little pro-active attempt to reverse such patterns in the formal allocation process (Jacobs 2009).

Beyond these findings of who among certain social groups got land, the evidence that studies in Masvingo, Mazowe and in Moyo's contribution refer to, of national or provincial or district figures, use categories that merge the place of origin and the occupation of beneficiaries, adding up those who are from urban or communal areas, say, alongside civil servants or war veterans. Some of the difficulties of interpreting these figures and even of compiling them are underlined by Matondi (forthcoming) who in attempting to make a classification of 'backgrounds' from full lists from Mazowe District made available by local officials found it was impossible from the records to place 68 percent of households unambiguously in an appropriate box. Of the remainder two thirds were dubbed 'ordinary', clearly used as a residual category not in formal occupation, thus including farmers but others too. Figures covering Masvingo Province reported by Scoones *et al.* classify 50 percent as 'ordinary' (defined as those without a formal occupation pre-land reform) from rural areas and 18 percent from urban areas. Chigumira (2009) in a comparable set of case studies found over 30 percent on one A1/villagised scheme were from CAs, but only tiny numbers on an A1/self-contained and an A2 scheme, while some 30 percent of A1 were from urban areas, compared with 70 percent of the A2. In some of these cases, and also in Mazowe (Matondi forthcoming), there was a smaller proportion of veterans than the 15 percent that was laid down as a target. All the cases citing such evidence of 'backgrounds' categorised significant though varied minorities of beneficiaries as civil servants and from the security services, with higher percentages among A2. All of these reports use such figures of 'ordinary' (or some similar category) as evidence that the great majority of land recipients were not members of privileged strata of society and many were 'poor' (as of course were a very high proportion of Zimbabweans by the end of a decade of economic collapse). Zamchiya's figures for Chipinge, the only case study from Manicaland, are an exception: 50 percent listed as civil servants (including security branches), 22 percent traditional authorities (including headmen, village heads, messengers and family members); veterans were 17 percent and only 11 percent were 'other ordinary'. Different studies define categories of beneficiaries in inconsistent ways, making direct comparison difficult. Different figures thus do not necessarily reflect regional variations, although these are of course inevitable.

Outcomes for production and livelihoods

To make any assessment of the impact of FTLRP on production or livelihoods invites the making of some kind of comparison, typically of a before-and-after kind. Various comparative indicators are used in the studies. One used by Moyo and by Scoones *et al.* traces the change in total output of various crops since 2000 using national figures. These show a general downward trend but with exceptions such as cotton, small grains, round nuts and beans. Moyo updates this picture by pointing to what might at last signify a reversal of some of the declines in the seasons after 2008, coinciding with the end of the worst period of macro-economic crisis. What might be a significant component of this shift is tobacco which not only recorded growth of production after a sharp decline in 2000–04 but a **surge in producers** from some 3,000 registered in 2000 to some 50,000 mainly small growers by 2010. These examples might also point to the role of contract farming when normal input supplies are limited, and in particular the Chinese corporate enterprises engaged in it, especially in cotton and tobacco (see Moyo in this collection).

Other contributors offer some account of output at local levels (Mazowe, Goromonzi and Zvimba) or even on what were individual farms and now consist of many smaller ones (Chigumira 2009 and Mutopo). Inevitably outcomes are varied between sites and within samples, although evidence of the central priority given to maize comes out and is consistent with the findings from the larger surveys of Moyo and Scoones *et al.* as are the significant proportions achieving the measure normally used as the threshold to ensure grain self-sufficiency, of 1+ tonne per household. Collecting data on before and after situations has proved very difficult given the chaos and confusion of FTLR, the intense politicisation of land, etc. However, the literature contains two individual small cases where new settlement farmers have been compared with nearby CA farmers and they offer contrasting findings: Zikhali (2008b) found that a sample of FTLRP used 'significantly more fertiliser, and draught power per hectare' than a sample in Chiweshe CA and realised three times higher yields (although their land was of higher potential and they got better access to inputs and extension). On the other hand, Chamunorwa (2010) found that among samples in three Mashonaland East districts fertiliser use on A1 farms was slightly below those of CA farmers, and that the former had poorer yields for commercial crops, but better for food.

But there are limitations inherent in any such comparisons: the figures indicate correlations between trends and the process of fast track; they do not prove causal relations. They do not compare impact of other simultaneous events or processes, especially the massive macro-economic decline and a range of accompanying factors such as declining public services, including those to agriculture, and worsening access to markets, credit, and inputs, which will have affected all agriculture including what had been the growing sectors of old resettlement areas and communal areas. In addition such comparisons do not compare like with like. Production figures may relate to the same physical areas but now measure output under different agrarian systems. For instance, former ranching areas are now producing grain and vegetable crops (Mutopo); and the livestock numbers and herd composition and the objectives of husbandry change from the simple commercial ranching logic of maximising profits by optimising off-take and/or quality to a more varied pattern when livestock fulfil a more complex range of functions including self-subsistence, draught power, bride-wealth, sources of finance for key household expenditure and a means of saving, especially when money is worthless as it was during the hyperinflation up to 2009 (Mavedzenge *et al.* 2009).

A further difficulty in using output data as a basis for identifying 'outcomes' of fast track land reform is that such a formulation can easily descend into implying a causal relationship. But the very finding that the output record varies greatly between crops, regions and strata of beneficiaries in the same area suggests the need for exploring what intervening variables might have explanatory power. In particular there is a strong consensus among contributions that most beneficiaries report facing major constraints to realising the potential of their land and other resources because of lack of timely access to seed, fertilisers and other inputs into production, to credit, reliable markets with guaranteed fair prices, to labour, and to extension services, as well as the limited opportunities for non-farming livelihood sources given the massive downturn in the economy. The fact that such factors have affected other sectors of agriculture like the OR areas, reported by Dekker and Kinsey as we have seen, reinforces the possibility that such contextual factors may be responsible for poor performance rather than land redistribution *per se*. Scrutinising evidence that can illuminate that issue is of course important in clarifying future options for policy and practice.

A different approach is to look at **livelihoods** and not just farm output. Before and after numbers of livelihoods can be instructive. Several of the studies of areas and individual farms and schemes show that in more marginal areas more self-provisioning households of new settlers and casual labour can be provided for than the number of owner and worker families under the old farm system. But Scoones *et al.* and some of the individual case studies provide valuable evidence of the varied types of livelihoods and in turn the class dynamics that have emerged; the former summing up the outcome in terms of 10 percent 'dropping out', a third 'hanging in', 20 percent 'stepping out', while just over a third are 'stepping up'. Similar categories which combine a range of measures and outcomes are used in other studies, e.g. by Matondi (forthcoming), to show complex patterns of 'success' and 'failure' in securing livelihoods which flesh out the simple counting of those households producing more than a ton of maize.

What problem of land tenure security?

An issue that is hotly debated in the process of formulating a new Land Tenure Act in Zimbabwe is how big a priority is securing land rights in the new (and Old) resettled areas. On this question there is a major polarisation between the findings of some contributors. Several of them explore the actual processes of selecting beneficiaries and allocating land, as we have seen. Matondi (forthcoming) goes beyond documenting what happened and who benefitted to probe what those procedures mean for the presence or absence of tenure security. Surveys from Mazowe, and other districts in Mashonaland investigated by his team, ask respondents their views about the degree of insecurity and its causes. He tends to see insecurity of tenure as inherent in the redistributed holdings: A1 beneficiaries who, at best, have offer letters or some even less formal occupancy, and A2 who were promised leases that have not always been issued. The rights conferred by these forms of title are rendered less secure by the realities where 'regulations' passed by local power-holders interfere, and by several people sometimes being allocated the same plot. These research findings are used to support the conclusion that insecurity is a root problem and directly affects investment in land and productivity, although the survey data presented on people's attitudes seem to focus mainly on what they state they are doing to fend off or prepare for insecurity, partly begging the question, and does not explore, for instance, the link between tenure rights and credit. At the other end of the spectrum, Moyo, reporting on AIAS (2009) research, seems to downplay the problem of insecurity, offering as evidence responses to a different kind of attitudinal survey that asks people whether it is a problem that hinders them in utilising their land or in investing in it. But it also presents data about whether people are indeed investing in their land, and finds that they are not seriously deterred by only having a 'permit' or no document at all. These answers suggest that the problem is overstated, which in turn lends support to a policy stance that resists individualisation of tenure, and the need to seek alternative forms of securing credit not based on property. Another source of potential insecurity, not based on the actual documents related to the piece of land, is raised by Mujere's contribution. He records conflicts over land arising from the competing entities claiming power to allocate new holdings and mediate disputes over rights; on the one hand traditional authorities claiming jurisdiction on the basis of their customary role and the 1998 Traditional Leaders Act, which gave them some duties in Old Resettlement, and the new Land Committees set up under FTLRP regulations but so far without legislated authority. He argues that formal legislation to clarify which bodies have final juridical authority and a single channel for resolving disputes is needed if continuing disputes over access to land are to be avoided.

Existing legislation allows for a wide range of tenure forms, ranging from individual, freehold title to regulated leases to permits to 'communal tenure' (Moyo 2008). Several contributions point to the formal lack of clarity that exists over land rights on the FTLR schemes: A2 beneficiaries were supposed to be allocated land after successful applications, and then receive tenancies on a 25 or 99 year lease; many received land other than through that process and few have yet got leases. A1 allocates were only entitled to 'permits' to use land, comparable to OR land, which were not transferable and could be withdrawn.

Conclusions

This attempt to bring together some of the concrete investigations that have been conducted in Zimbabwe in the last decade, and to bring out in the last section above some of the key findings, will be rounded off by posing two issues about the further implications of this work. The first is to ask what lacunae are revealed by laying out this cross section of work. What further investigation is needed to fill crucial gaps or to resolve apparent differences in findings? The second matter is to review how far these materials illuminate the long term prospects for Zimbabwe after FTLR, matters that relate to current policies but go far beyond them.

Future research priorities

There is a clear need for widening of empirical knowledge, both geographically and thematically. The most obvious spatial gap in knowledge is with regard to Matabeleland: there are no case studies in the Matabeleland North Province, and only some limited surveys by AIAS and Ruzivo Trust in the Mangwe District of the Matabeleland South Province. The logic for more case studies goes beyond simply having a more representative area sampling. These provinces are distinct environmentally; most of them lie in the drier Natural Regions IV and V. Moreover, the large scale commercial enterprises that have been absorbed within FTLRP were very largely devoted to livestock ranching, which in turn implies a different kind of transformation in agrarian structure. How far has that outcome been the spawning of small or middle size commercial ranches, solely devoted to livestock and based on a logic of optimising off-take of beef for internal or export markets? Or is a more mixed farming pattern emerging based on a herding strategy of meeting local milk and meat demand, and supplying draught oxen and other four-legged inputs and complementarities to appropriate cropping? But two dimensions of the political context are potentially determinant of outcomes, likely to be different from those in other former ranching areas like those in Masvingo. First, the OR Programme had been a contentious issue. The fighting between ex-fighters of ZAPU and the new Zimbabwe army and the eventual brutal repression of civilians generally in this region in the 1980s interrupted implementation until 1987. Much land purchased by GoZ remained undistributed for several years; in particular the Model D scheme which was to make ranch land available as an extension of grazing to existing CAs was not implemented for several years and was resisted by local communities, as was Model A resettlement (Alexander 2006). The contemporary variant of Model D always seemed likely to be a target for resistance. Three-tier schemes have been limited to Matabeleland, and as far as our studies report, to two districts in Matabeleland South, Mangwe and Matobo. But it seems they did not benefit small commercial ranchers, as intended. Local pressures forced a shift to what amounted to open access to those with livestock in neighbouring CAs. But more investigations are needed into what mechanisms have been set up to manage these new grazing areas, and whether they work. Assessment of the Three-tier variant is in turn just one of the questions that need to be answered as part of the second difference in context: what are the consequences of introducing FTLRP in a region that has a long history of opposition to the former ruling party?

A more general imperative for widening the territorial reach of case studies would be that of carrying out new ones outside the central core of Mashonaland.

Researchers have tended to do field work in areas situated conveniently close to Harare – for the same reasons as the national politically influential see the desirability of being within daily commuting range of the city. This proximity is itself likely to mean such samples are biased toward certain kinds of beneficiaries, as well as being typical only of areas with good infrastructure, reliable climate, more advanced agriculture (with all that means in terms of challenges and potential). A comparison even of the studies presented here does suggest that more peripheral areas – like Masvingo Province (by Scoones *et al.*), the Mangwe surveys in Matabeleland South by AIAS and Ruzivo (only reported on in comparative tables in this collection by Moyo, but see AIAS 2009 and Matondi forthcoming), Chipinge in Manicaland – have a smaller proportion of A2 farms and beneficiaries with lower education and that are more likely to be from CAs. There is also a need to study production and livelihood outcomes in these marginal areas but also in middling areas of NR III which are also underrepresented. Of course national surveys, such as the proposed Land Audit, but also ones that cover land use and economic outcomes, will also be invaluable in giving a representative picture in years to come.

The set of studies presented here do cover a wide range of themes: mechanisms of land allocation, origins of beneficiaries, their political and occupational status, the fate of special categories – women (see also WLZ 2007), farm workers – in FTLRP, the patterns of land use, of livelihood generation, the economic constraints and potential, the evolution of new social relations and conflicts. A range of possible broadenings of those agendas for research can be mentioned. Although many studies provide evidence of the extent to which women have been among official and unofficial recipients of land and a few, like Mutopo, have explored how some women have pursued livelihood options, other issues of gender can be added. For instance, in reports on access to land for women, that category is seldom differentiated. More exploration of sub-categories such as women heads of households would be valuable, as well as the question of who is responsible for managing new farms. Matondi (forthcoming) quotes the bare statistics that women do so in 46 percent of A1 plots, but more such data, including probing decisions about land use, investment, etc., are required. More broadly there is need to investigate whether the division of labour and gender relations are in any way different from those in CAs, and to link that analysis with changing patterns of migration and availability of other livelihoods. Such matters are also vital in exploring policy options for (re-)developing the fast tracked land and the rural and national economies.

Another dimension of 'divided families' that demands investigation are relations between the FTLR areas and the CAs: how widespread is the maintenance of two homesteads and is this changing? What economic and social linkages are there? For instance, the AIAS Baseline Survey (2007) reports that in their Mangwe sample 31 percent of new farmers keep livestock belonging to others. Is this common elsewhere and on what relational or contractual basis (mutual aid, kin obligation, rental)? What patterns of management of common land in A1 villagised schemes, perhaps in conjunction with patterns in nearby CAs, have emerged; in particular how do the special cases of Three-tier schemes manage their affairs? More generally there is need for studies within the CAs, of the impact of FTLRP to assess such issues as to whether FTLR has realised any 'decongestion'.

A central question with policy implications is to chart changing patterns of production and marketing. Only the three longitudinal studies of the institutions (Moyo and AIAS, Scoones *et al.* and Matondi and Ruzivo Trust) have offered detailed quantitative evidence on such matters. A replication of such studies would aid future planning and would also thoroughly test Moyo's hypothesis that 'an anticipated corner has been turned' and production levels and number of benefitting livelihoods are increasing.

Future scenarios

Recovery of economy and livelihoods?

A number of recommendations for action are made in the contributions, some at odds with each other. It is to be hoped that this collection can prompt some potential applicable ideas for policy but also further debate around options – especially about ways forward beyond what most contributors see as a policy 'impasse'. There are several suggestions for improving performance, livelihoods and daily life in fast track areas: infrastructure on and off farms, timely supply of inputs and credit, extension advice, technical improvements appropriate to new types of smallholder and middle-scale farms, farmers' organisations and social institutions. Most contributors, implicitly at least, seem to believe that past experience in Zimbabwe and favourable land resources indicate that another 'agricultural revolution' is a possibility (c.f. the challenging title of Rukuni *et al.* 2006). There is also agreement that the realisation of any such potential depends on a **political** process as well as technical prescriptions. Here a major divide then opens up: some contributors, like other writers, feel that a major shift in the holders of political power and in reform of existing institutions is a **prerequisite** to any prospect of implementing alternative approaches (see Habib 2011). Similarly, donors seem to believe it would be premature to commit resources to the present government, even after coalition, and to the FTLR sector.

Even bilateral and international agencies are providing significant amounts of farming and livelihood support in Zimbabwe but apart from UN agencies they still do not engage with GoZ to provide input support as opposed to emergency aid. Moreover, their support programmes to small farmers through NGOs do not touch those smallholders in the FTLR areas as these lands are still legally contested. The continued application of such conditionality is not a legal or logical formula, and flies in the teeth of the principles accepted by both sides of the GPA. Farm support through NGOs cannot provide some of the essentials for the sustained period needed for recovery – like research, extension, and widespread credit (Tupy 2007).

On the other hand Moyo and Scoones *et al.* among the contributors propose that political reform with emphasis on rights and redress as well as redistribution is essential – but as an accompanying dimension of technical and economic recovery and improvement not as a preliminary stage. Moyo's recommendations include the view that the nature of politics is taking new forms going beyond a one-dimensional inter-party struggle to a more complex expression of class struggles involving chiefs, lineage leaders, farmers old and new, bureaucracies and old and new international economic actors. Both he and Scoones *et al.* call for strong cooperating and representative organisations of producers, as does Murisa.

Future land tenures?

As indicated above the tenure forms on both A1 and A2 schemes have a temporary nature and will need eventual regularisation: on the former allocatees simply have permits or even only letters of allocation; on the latter leases are promised and have yet to be issued. But alternative measures for FTLR land are being considered alongside a process of enacting reform of tenure governing other sectors, remaining large commercial farms and estates, communal land and OR. In current debates, some analysts advocate forms of freehold title not only for fast track land holders but for all landholders, mainly through conversion of leaseholds and permits to Deeds of Grant, with an option of maintaining the 99 year leases for A2 farmers (Rukuni *et al.* 2009). For others, leasehold and permit systems for the moment offer working security for A2 and A1 farmers respectively (Scoones *et al.* 2010, Moyo 2008). Reform of communal tenure, which was last enacted as recently as the 1998 Act which involved traditional leaders in the allocative role of local government, is also being debated. A hybrid approach that combines communal tenure with legally binding arrangements as recommended by the Rukuni Commission of 1994 (Rukuni 1994) is still favoured by some, though no longer by Rukuni (2009) himself. The present ambiguity of whether traditional authorities also have some powers in Old Resettlement Areas, as stipulated in the Traditional Leaders Act 1998, or even by extension to FTLRP farm areas, needs clarification and is further complicated by proclamations of new regulations under the 1998 Act. Moyo (2008) and Scoones *et al.* (2010) also advance proposals under which communal areas and those newly made available through state action like FTLRP should be subject to some form of community control and not completely individualised and alienated for all time – but stipulate that such control should be through some democratic form rather than through 'traditional' leaders. (For a survey of earlier patterns and proposals about land in communal areas see Ranger 1993; for options being considered in contemporary debates see MLLR 2009). The Brooks Poverty Institute report of 2009 called for the creation of a unified land tenure system, but held back from advocating any particular tenure form (Chimhowu 2009). Of course all land tenure systems exist in a wider political-social-economic context, and the Zimbabwean experience clearly demonstrates that tenure security depends crucially on the wider political setting, land administration capacity and the ability to assure the rule of law. Without these, no tenure regime can assure security.

Similar debates over having a single land tenure system have also been central features of debates in Namibia and South Africa, although both have resisted arguments for complete individualisation across the board. How some form of community voice to temper individual rights should be enshrined in some parts of the system of law and administration over land resources remains unresolved, however, and in South Africa there is at present a vacuum as their Communal Land Rights Act to follow a traditional leaders formula for this has been declared unconstitutional.

There is no clear-cut evidence available in these or other studies that could definitively settle these debates as to the most appropriate policy, even for the land reform areas, nor is it likely that survey data of the here and now could ever be a foolproof guide to future policy (Andersson 2007). However, the material does help illuminate the realities on the ground as opposed to the political or legal formulae. Comparative studies that have sought to compare outcomes from different tenure systems can be potentially instructive although the Africa material is equally inconclusive (Bruce and Migot-Adholla 1994, World Bank 2003) – although they do tend to cast doubt on the idea that there is one ideal solution. An informed debate in Zimbabwe could well look particularly at one country's experience. Kenya was also a settler colony but is the only country in sub-Saharan Africa that has implemented individual freehold titling, both in the communal areas and resettlement land holdings, and thus where there is historical evidence over 50 years of what does happen when that happens (Shipton 2009, Mckenzie 1993). Wanjala (2000, 98) characterises the situation that developed in Kenya when the lending, which individualised titling of land was supposed to promote, led to massive defaults:

> As the banks have moved in to realize their securities by exercising their statutory right of sale, the social consequences have forced the government (the very advocate for the system) to come up with ambiguous directives through the office of the Chief Justice directing that no sales of agricultural land should be conducted unless the Provincial Administration has given its consent.

Thus individual tenure has not facilitated agricultural credit nor provided security of tenure through the market, and does not even operate in practice (see also Berry 1993, Platteau 2000, Mackenzie 1993). A cautionary tale for Zimbabweans.

Future dynamics of reformed agrarian structures

These contributions concur that there has been major transformation of the agrarian structure in Zimbabwe since 2000, and that the central characteristic of that outcome is that smallholder production now predominates. But these structures are not yet set in stone. Other possible long term scenarios might emerge from new forms of class struggle, as Moyo anticipates, by efforts of the beneficiaries themselves and by the unresolved political struggles for reform.

Scoones *et al.* point to one emerging dynamic in on-going agrarian restructuring by pointing to the range of economic units and associated classes which have emerged from FTLRP. There are specialist entrepreneurs providing services such as mechanised operations and technical advice to down-sized commercial farms and A2 farms, as well as intact corporate estates and ranches; a wide range of types and strata of smallholders, and of permanently and casually employed types of workers reproducing themselves in a variety of ways and located on existing large farms, in the interstices of new communities. Anecdotal evidence indicates that there have been many unpredicted social and economic interactions between these elements: former commercial farms do contract ploughing for people who occupied their land; different types of farmers periodically swap use of land rather than set aside 'fallow' for tobacco; former farmers and managers act as 'mentors'; new marketing mechanisms have evolved like the side-stepping of the classic tobacco auctions by state capitalist dealers from East Asia. Further blossoming of such interchange will no doubt emerge spontaneously but the multiplier effects could be increased with informed planning and creative institutional flexibility.

Several contributions report another dynamic: continued struggle to confirm de facto or assert new rights to land, especially efforts by large A2 owners, often politically connected, to oust smaller A2 and A1 farmers. Moyo stresses how this is likely to persist and thereby poses a question of whether encroachment by elites will reach a scale that represents a further structural transformation. One future scenario would be that such accumulation by a new class could signal a return to a basic dualism: black capitalist farmers replacing large scale white capitalist farmers. But some of our studies suggest a more complex future: evidence of A2 farms leasing out their unutilised land to get an immediate return. Such a trend might be reinforced as

A1 farmers seek to defend their terrain, and their off-spring and land seekers from the CAs use squatting as a tactic for acquisition. It is possible that actual long term outcomes might be a complex combination of forms of landlordism together with partially successful resistance by existing and would-be smallholders. In general terms this has been the path that has emerged in Kenya but what actual mosaic may arise in Zimbabwe will depend on different overt political struggles.

References

AIAS Baseline Survey 2007. *Inter-district household and whole farm survey database*. Harare: African Institute for Agrarian Studies.

Alexander, J. 1991. The unsettled land: the politics of land redistribution in Matabeleland, 1980–90. *Journal of Southern African Studies*. 17.4: 581–610.

Alexander, J. 2006. *The unsettled land: state-making and the politics of land in Zimbabwe, 1893–2003*. Oxford: James Currey.

Amin, S. 1976. *Unequal Development*. Brighton: Harvester.

Andoh, S. 2010. Review of Scoones *et al*. *Africa Today*. 57.4: 125–29.

Andersson, J.A. 2007. How much did property rights matter? Understanding food insecurity in Zimbabwe: a critique of Richardson. *African Affairs*, 106(425), 681–690.

Arrighi, G. 1970. Labour supplies in historical perspective: a study of the proletarianisation of the African peasantry in Rhodesia. *Journal of Development Studies*, 6(3): 197–234.

Arrighi, G. 1973. The Political Economy of Rhodesia. *In*: G. Arrighi and J. Saul, eds. *Essays in the Political Economy of Africa*. New York: Monthly Review Press.

Barr, A. 2004. Forging effective new communities: the evolution of civil society in Zimbabwean resettlement villages. *World Development*, 32.10: 1753–66.

Bernstein, H. 2003. Land reform in southern Africa in world historical perspective. *Review of African Political Economy*. 96: 203–26.

Bernstein, H. 2005. Rural land and land conflicts in sub-Saharan Africa. In: Moyo and Yeros, eds.

Berry, S. 1993. No condition is permanent. The social dynamics of agrarian change in sub-Saharan Africa. Madison: University of Wisconsin Press. Bhebe, N. and Ranger, T. eds.1995. *Society in Zimbabwe's liberation war*. London: James Currey.

Bruce, W. and S. Migot-Adholla, eds. 1994. *Searching for land tenure security in Africa*. Washington: World Bank.

Bush, R. and L. Cliffe. 1984. Agrarian policy in migrant labour societies: reform or transformation in Zimbabwe? *Review of African Political Economy*. 29: 77–94.

Campbell, H. (2003). *Reclaiming Zimbabwe: the exhaustion of a patriarchal model of liberation*. Claremont: Philips.

Chambati, W. 2011. From land dispossession to land repossession: restructuring of agrarian labour relations in Zimbabwe. *Journal of Peasant Studies*. Forthcoming.

Chamunorwa, A. 2010, Comparative Analysis of Agricultural Productivity between Newly Resettled Farmers and Communal Farmers in Mashonaland East Province, Livelihoods after Land Reform Working Paper 8, Harare. www.lalr.org.za.

Chaumba J., I. Scoones and W. Woolmer 2003. New politics, new livelihoods: agrarian change in Zimbabwe. *Review of African Political Economy*. 98: 585–608.

Cheater A. 1984. *Idioms of accumulation: rural development and class formation among freeholders in Zimbabwe*. Gweru: Mambo Press.

Chigumira, E. 2009. My land, my resources: assessment of the impact of FTLRP on natural environment, Kadoma District, Zimbabwe, Working Paper 14, *Livelihoods after Land Reform Programme*, PLAAS, University of Western Cape. Available www.lalr.org.za.

Cliffe, L. 1988. Zimbabwe's agricultural success and food security. *Review of African Political Economy* 43: 4–25.

Cliffe, L. 2000. The Politics of Land Reform in Zimbabwe. In: T.A.S. Bowyer-Bower and C. Stoneman, eds. *Land reform in Zimbabwe: constraints and prospects*. Ashgate: Aldershot, UK.

Cousins, B. and I. Scoones. 2010. Contested paradigms of 'viability' in redistributive land reform: perspectives from southern Africa. *Journal of Peasant Studies* 37 (1): 31–66. http://www.tandfonline.com/doi/abs/10.1080/03066150903498739

Cusworth, J. and Walker, J. 1988. Land Resettlement in Zimbabwe: A Preliminary Evaluation. London; ODA Evaluation report EV434, Overseas Development Administration.

Davies, R. (2004). Memories of underdevelopment: a personal interpretation of Zimbabwe's economic decline. In: B. Raftopoulos and T. Savage, eds. Zimbabwe: injustice and political reconciliation. Weaver Press: Harare.

Fox, R., Chigumira, E and Rowntree, K. 2007. On the Fast Track to Land Degradation? A case of the impact of the Fast Track Land Reform Programme in Kadoma District Zimbabwe. Geography 92.3: 212–24.

FTZ 2001. The impact of land reform on commercial farm workers' livelihoods. Harare: Farm Community Trust of Zimbabwe.

Gaidzanwa, R. 1988. Women's land rights in Zimbabwe. Harare: Rural and Urban Planning Department, University of Zimbabwe. Goebel A. 2005. Gender and land reform: the Zimbabwe case. Montreal and Kingston: Magill and Queens University Presses.

Government of Zimbabwe 1990. ESAP Policy Document. Harare: Government Printers.

Government of Zimbabwe 1998. The inception phase framework plan of the second phase of the land reform and resettlement programme. Harare: Ministry of Land and Agriculture.

Government of Zimbabwe 2003. National Economic Revival Programme: Measures to address the current challenges, Ministry of Finance and Economic Development, February 2003.

Government of Zimbabwe (GoZ) 2006. Preliminary national A2 land audit report. Ministry of Lands, Land Reform and Resettlement/SIRDC, Government of Zimbabwe, Harare.

Government of Zimbabwe 2011. Indigenisation and Economic Empowerment Act [Act 14/ 2007]. The Indigenisation and Economic Empowerment (General) Regulations, 2011 were published in General Notice 114 of 2011. Government of Zimbabwe.

GPA (Global Political Agreement) 2008. Agreement between the Zimbabwe African National Union-Patriotic Front (ZANU-PF) and the two Movement For Democratic Change (MDC) formations, on resolving the challenges facing Zimbabwe. Harare: September.

Habib, A. 2011. Foreword: Reflections on the Prerequisites for a Sustainable Reconstruction in Zimbabwe. In: Hany Besida (ed), Zimbabwe: picking up the pieces. Palgrave Macmillan, New York.

Hall, R., ed. 2009. Another countryside?:policy options for land and agrarian reform in South Africa. Cape Town: PLAAS, University of Western Cape.

Hammar, A. 2008. In the name of sovereignty: displacement and state making in post-independence Zimbabwe. Journal of Contemporary African Studies, 26.4: 417–34, October.

Hammar, A. 2010. Ambivalent mobilities: Zimbabwean commercial farmers in Mozambique. Journal of Southern African Studies, 36: 2, 395–416.

Hammar, A., McGregor, J. and Landau, L. 2010. Introduction. Displacing Zimbabwe: crisis and construction in southern Africa. Journal of Southern African Studies, 36: 2, 263–83.

Hammar, A., B. Raftopoulos and S. Jensen, eds. 2003. Zimbabwe's unfinished business: rethinking land, state and nation in the context of crisis. Harare: Weaver Press.

Herbst, J. 1990. State politics in Zimbabwe. Harare: University of Zimbabwe Publications.

Jacobs, S. 2009. Gender and Agrarian Reforms. Routledge: Abingdon and New York.

JAG/RAU 2008. Land, retribution and elections: post-election violence on Zimbabwe's remaining farms 2008. A report prepared by the Justice for Agriculture (JAG) Trust and the Research and Advocacy Unit. May.

Kanyinga, K. 2000. Redistribution from above: the politics of land rights and squatting in coastal Kenya. Research reort no. 119. Uppsala: Nordiska Afrikainstutet.

Kanyinga, K. 2009. The legacy of the white highlands: Land rights, ethnicity and the post-2007 election violence in Kenya. Journal of Contemporary African Studies, 27:3: 325–44.

Kinsey, B. 1999. Land reform, growth and equity: emerging evidence from Zimbabwe. Journal of Southern African Studies, 25.2: 173–96.

Kinsey, B. 2003. Comparative economic performance of Zimbabwe's resettlements models. In: Roth and Gonese. Eds.

Kinsey, B. 2005. Fractionalising local leadership: created authority and management of state land in Zimbabwe. In: S. Evers, M. Spierenburg and H. Wels, eds. Competing jurisdictions: settling land claims in Africa. Leiden: Brill.

Kleinbooi, K. 2010, *Review of land reforms in southern Africa 2010*. Cape Town: PLAAS, University of Western Cape.

Kriger, N. 1992. *Zimbabwe's guerrilla war: peasant voices*. Cambridge: Cambridge University Press.

Leo, C. 1984. *Land and Class in Kenya*. Toronto: University of Toronto Press.

Leys, C. 1975. *Underdevelopment in Kenya: the political economy of neo-colonialism*. Berkley: University of California Press.

Mackenzie, F. 1993. 'A Piece of Land Never Shrinks: Reconceptualising Land Tenure in a Smallholding District, Kenya', in T.J. Basset and D. Crummey (eds), *Land in African Agrarian Systems*. Madison: University of Wisconsin Press.

Magaramombe, G. 2003. *An overview of vulnerability within the newly resettled former commercial farming areas*. Harare: Drfat report prepared for UN Humanitarian Coordinator.

Magaramombe, G. 2010. Displaced in place: agrarian displacements, replacements and resettlement among farm workers in Mazowe District. *Journal of Southern African Studies*, 36(2): 361–75.

Mamdani, M. 2008. Lessons of Zimbabwe. *London Review of Books*, 30.23: 17–21.

Mandaza, I. ed. 1986. *Zimbabwe: the Political Economy of Transition: 1980–1986*. Dakar: CODESRIA.

Matondi, P., ed. (forthcoming). *Inside the political economy of redistributive land and agrarian reforms in Mazoe, Shamva and Mangwe districts in Zimbabwe*. London: Zed Press.

Mavedzenge *et al.* 2009. *Development and Change*. http://www.ingentaconnect.com/content/bpl/dech/2008/00000039/00000004/art00005;jsessionid = 3rqt9hljijqdh.alexandra.

Mhone, G.C.Z. 1996. The socio-economic crisis in Southern Africa (Botswana, Malawi, Zambia, Zimbabwe. *Africa Development* 21: 267–278.

Migot-Adholla, S., Place and Kosura. 1994. Security of tenure and land productivity in Kenya. *In*: W. Bruce and S. Migoy-Adholla, eds. *Searching for land tenure security in Africa*. Washington: World Bank.

Ministry of Finance Budget Speech 2010. *National land policy framework paper*. Ministry of Finance of Zimbabwe. Available from: http://www.zimtreasury.org/downloads/834.pdf. [Accessed 22 December 2010].

Mkodzongi, G. 2011. Land grabbers or climate experts? farm occupations and the quest for livelihoods in Zimbabwe.Paper European Conference of African Studies 4, Uppsala.

MLA 1998. *National land policy framework paper*. (Shivji report) Harare: Ministry of Lands and Agriculture.

MLLR 2009. Memorandum To Cabinet By The Minister Of Lands And Rural Resettlement Hon. H. M. Murerwa (M.P) On The Update On Land Reform And Resettlement Programme, Ministry Of Lands And Rural Resettlement, September 2009.

Moore, D. 2005. *Suffering for territory: race, place and power in Zimbabwe*. Harare: Weaver Press.

Moyo, S. 1995. *The land question in Zimbabwe*. Harare: SAPES Books.

Moyo, S. 2005a. Land policy, poverty reduction and public action in Zimbabwe. *In*: A. Haroon Akram-Lodhi, S.M. Borras Jr and C. Kay, eds. *Land, Poverty and Livelihoods in an Era of Globalisation: Perspective from developing and transition countries*. Routledge ISS Studies in Rural Livelihoods, Routledge. London and New York, pp. 344–82.

Moyo, S. 2005b. The Politics of Land Distribution and Race Relations in Southern Africa. *In*: Bangura, Yusuf; Stavenhagen, Rodolfo (Eds.): "Racism and Public Policy" New York: Palgrave Press.

Moyo, S. 2011a. Three Decades of Agrarian Reform in Zimbabwe: Changing Agrarian Relations. *Journal of Peasant Studies*. Forthcoming

Moyo, S. 2011b. Land concentration and accumulation by dispossession: redistributive reform in post-settler Zimbabwe. Forthcoming Journal of Agrarian Change.

Moyo, S., 2011c. Agrarian reform and prospects for recovery, in Hansy Baseda (ed) *Zimbabwe: Picking Up The Pieces*. Palgrave: MacMillan.

Moyo, S. 2011d. Land concentration and accumulation after redistributive reform in Zimbabwe. *In; Review of African Political Economy*, 38. 128. June: 257–76. Moyo, S., Rutherford. B and Amanor-Wilks, D. 2000. "*Land Reform and Changing Social Relations for Farm Workers in Zimbabwe*", Review of African Political Economy 84(18), pp. 181–202, ROAPE Publications Ltd.

Moyo, S. and Yeros, P., 2005. The Resurgence of Rural Movements under Neoliberalism. *In*: Sam Moyo and Paris Yeros eds. *Reclaiming the Land: The Resurgence of Rural Movements in Africa, Asia and Latin America*. London: Zed Books.

Moyo, S. and Yeros, P., 2007. The Radicalised State: Zimbabwe's Interrupted Revolution. *Review of African Political Economy*, 111, 103–21.

Moyo, S., Chambati, W., Murisa, T., Siziba, D., Dangwa, C., Mujeyi, K. and Nyoni, N. 2009. *Fast Track Land Reform Baseline Survey in Zimbabwe: Trends and Tendencies, 2005/06*. African Institute for Agrarian Studies (AIAS). Harare.

Mudege, N. 2008. *An ethnography of knowledge: the production of knowledge in Mupfurudzi resettlement scheme, Zimbabwe*. Leiden: Brill.

Muir-Leresche, K. 2006. Agriculture in Zimbabwe. *In*: M. Rukuni *et al.* Zimbabwe's agricultural revolution revisited. Harare: University of Zimbabwe Publications, pp. 99–108.

Murisa, T. 2010. Emerging forms of social organisation and agency in the newly resettled areas of Zimbabwe: the cases of Goromonzi and Zvimba districts. Thesis (PhD). Rhodes University.

Murisa, T. 2011. Emerging Forms of Social Organisation in Newly Resettled Areas of Goromonzi and Zvimba Districts. *Journal of Peasant Studies*. Forthcoming.

Mustapha, R. 2011. *Review of African Political Economy*.

Mutopo, P. 2011. Women's struggles to access and control land and livelihoods after Fast Track Land Reform in Mwenezi District, Zimbabwe. *Journal of Peasant Studies*. Forthcoming.

Palmer, R. 1977. *Land and racial domination in Rhodesia*. Heinemann, London.

Palmer, R. 1990. Land reform in Zimbabwe, 1980–1990. *African Affairs* 89.335: 163–81.

Pankhurst, D. 1991. Constraints and incentives in 'successful' Zimbabwean peasant agriculture: the interaction between gender and class. *Journal of Southern African Studies*. 17.4: 611–632.

Platteau, J.P., 2000. 'Does Africa Need Land Reform?'. *In*: C. Toulmin and J. Quan eds. *Evolving land rights, policy and tenure in Africa*. London: International Institute for Environment and Development and Natural Resources Institute.

Raftopoulos B. and A. Mlambo, Eds 2009. *Becoming Zimbabwe: a history from the pre-colonial period to 2008*. Weaver Press, Harare.

Ranger, T. 1985. *Peasant consciousness and guerrilla war in Zimbabwe*. James Currey: London.

Ranger, T. 1993. The communal areas of Zimbabwe. *In*: T. Bassett and D. Crummey, eds. *African agrarian systems*. Madison: University of Wisconsin Press.

Robertson, J., 2011. A Macroeconomic Policy Framework for Economic Stabilisation in Zimbabwe. *In*: Hany Besida (ed), *Zimbabwe: Picking Up The Pieces*. Palgrave Macmillan, New York.

Robins, S. 1994. Contesting social geography of state power: a case study of land use planning in Matabeleland, Zimbabwe. *Social Dynamics* 20.2: 119–37.

Roth, M. and F. Gonese, eds. 2003. *Delivering Land and Securing Rural Livelihoods: Post-Independence Land Reform and Resettlement in Zimbabwe*. Harare: Centre for Applied Social Sciences, University of Zimbabwe.

Rukuni, M. (1994). *Report on the Inquiry into Appropriate Agricultural Land Tenure Systems.(Rukuni Report). Executive Summary for His Excellency the President of The Republic of Zimbabwe, Volume 1: Main Report; Volume 2: Technical Reports*. Harare: Government Printers.

Rukuni, M., P. Tawonenzi and E.K. Eicher with M. Munyuki-Hungwe and P. Matondi eds. 2006. *Zimbabwe's agricultural revolution revisited*. Harare: University of Zimbabwe Publications.

Rukuni, M., J. Nyoni and P. Matondi. 2009. *Policy Options for Optimisation of the Use of Land for Agricultural Productivity and Production in Zimbabwe*. Harare: World Bank.

Rutherford, B. 2003. Belonging to the Farm(er): Farm Workers, Farmers, and the shifting politics of citizenship. *In*: A. Hammar, B. Raftopoulos and S. Jensen, eds, *Zimbabwe's Unfinished Business: Rethinking Land, State and Nation in the Context of Crisis*. Harare: Weaver Press.

Rutherford, B. 2008. Conditional belonging: farm workers and the cultural politics of recognition in Zimbabwe. *Development and Change* 39 (1): 73–99.

Rutten, M. and S. Owuor. 2009. 'Weapons of mass destruction: Land, ethnicity and the 2007 elections in Kenya'. *Journal of Contemporary African Studies* 27:3: 305–24.

Sachikonye, L. 2003a. From 'growth with equity' to 'fast track' reform: Zimbabwe's land question. *Review of African Political Economy* 30.96: 227–40.

Sachikonye, L. 2003b. *The Situation of Commercial Farm Workers after Land Reform in Zimbabwe,* report prepared for the Farm Community Trust of Zimbabwe.

Sadomba, W. 2008. *War veterans in Zimbabwe's land occupations: complexities of a liberation movement in an African post-colonial settler society,* Ph. D. Wageningen University, June.

Sadomba, W. 2011. War veterans-led occupations (1998–2008): politics and land in Zimbabwe. *CODESRIA NWG Book Chapter.* Forthcoming.

Sadomba, W. 2011. War veterans-led occupations (1998 to 2008): Zimbabwe land occupations and politics. *In*: Sam Moyo and Walter Chambati (eds) *Review of Zimbabwe Land and Agrarian Reform 2000–2005.* National Working Group (NWG) book forthcoming.

Scoones, I., Marongwe, N., Mavedzenge, B., Murimbarimba, F., Mahenehene, J. and Sukume, C., 2010. *Zimbabwe's Land Reform: Myths and Realities.* Harare and London: Weaver Press, Jacana Media and James Currey.

Selby, A. 2006. *Commercial farmers and the state: interest group politics and land reform in Zimbabwe.* University of Cambridge, Ph.D.

Shipton P. 2009. *Mortgaging the ancestors: ideologies of attachment in Africa.* New Haven: Yale University Press.

Shivji, I., L. Gunby, S. Moyo and W. Ncube. 1998. *National Land Policy Framework.* Harare: Government of Zimbabwe.

Spierenburg, M. 2004. *Strangers, spirits and land reforms: conflicts about land in Dande, northern Zimbabwe.* Leiden: Brill.

Sukume, C. and S. Moyo. 2003. *Farm sizes, decongestion and land use: implications of the Fast Track Land Redistribution Programme in Zimbabwe.* Harare: AIAS Mimeo.

Tendi, B-M. (2010). *Making history in Mugabe's Zimbabwe: politics, intellectuals and the media.* Oxford: Peter Lang.

Tupy, M.L. 2007. A Four-step recovery Plan for Zimbabwe. CATO Institute. Available from: http://www.cato.org/pub_display.php?pub_id=8191 (Date accessed: 9 October 2008).

Umhlaba Wethu 11 2011. Land Barometer. Cape Town: PLAAS, University of Western Cape.

UNDP 2008. Comprehensive Economic Recovery in Zimbabwe: A Discussion Document. Harare: UNDP.

USAID 2010. Zimbabwe Agricultural Sector Market Study. Weidemann Associates, Inc., Washington.

Utete 2003. Report of the Presidential Land Review Committee, under the chairmanship of Dr Charles M.B. Utete, Vol 1 and 2: Main report to his Excellency The President of The Republic of Zimbabwe, Presidential Land Review Committee (PLRC)., Harare.

van der Ploeg 2008. *The new peasantries: struggles for autonomy and sustainability in an era of empire and globalization.* London: Earthscan.

Vincent, V. and Thomas, R. 1961 *Agricultural Survey of Southern Africa.* Harare: Government Printers.

Wanjala, S. 2000. Problems of land registration and titling in Kenya: Administrative and political pitfalls and their possible solution. *In*: S. Wanjala, ed. *Essays on land law: the reform debate in Kenya.* Nairobi: Faculty of Law, University of Nairobi.

Wasserman, G. 1976. *Politics of decolonization: Kenya Europeans and the lans issue 1960–1965.* Cambridge: Cambridge University Press.

WLZ 2007. A rapid appraisal in select provinces to determine the extent of women's allocation to land during the land reform and resettlement programme, 2002–2005. Unpublished paper, Women and Land in Zimbabwe (WLZ), Harare.

WFLA 2009. The socio-economic and political impact of the Fast Track Land Reform programme on Zimbabwean Women. A Research extract.

World Bank 2003. *Land policies for growth and poverty reduction.* Oxford and Washington: Oxford University Press and World Bank.

World Bank 2006. *Agricultural Growth and Land Reform in Zimbabwe: Assessment and Recovery Options.* Report No. 31699-ZW, Harare, World Bank.

Zikhali, P. 2008. Fast Track Land Reform and agricultural productivity in Zimbabwe. *In: Land reform, trust and natural resources management in Africa.* Economic Studies 170, University of Gothenberg.

Zimbabwe Human Rights NGO Forum 2007. Adding insult to injury: A Preliminary Report on Human Rights Violations on Commercial Farms, 2000 to 2005. A report prepared by the Zimbabwe Human Rights NGO Forum and the Justice for Agriculture Trust [JAG] in Zimbabwe. June 2007.

Lionel Cliffe has been involved in land reform debates on Zimbabwe since the Geneva Conference in 1976, and authored an FAO report on *Options for Agrarian Reform in Zimbabwe* (1986). His recent involvement, which facilitated his involvement with this Collection, benefitted from an Emeritus Research Fellowship from The Leverhulme Trust for work on 'Comparisons of Land Reform in former settler-colonies: Kenya, Zimbabwe and South Africa', support for which is gratefully acknowledged. He co-authored the basic text *Zimbabwe: Politics, Economics and Society* (1989) with Colin Stoneman. He has worked on and in East and southern Africa for 50 years, including Tanzania (as the first Director of Development Studies, University of Dar es Salaam), Kenya, Zambia, Namibia, Eritrea. He was a founding editor of the *Review of African Political Economy*. He was elected a Distinguished Africanist by the African Studies Association UK in 2002. He is Emeritus Professor of Politics of the University of Leeds.

Jocelyn Alexander is Professor of Commonwealth Studies at Queen Elizabeth House, Dept of International Development, University of Oxford. She has been carrying out research on southern African history and politics for over 20 years and is the author of *The Unsettled Land: State-making and the Politics of Land in Zimbabwe, 1893–2003* and co-author of *Violence and Memory: One Hundred Years in the 'Dark Forests' of Matabeleland*, as well as numerous articles and book chapters.

Ben Cousins holds a DST/NRF Research Chair in Poverty, Land and Agrarian Studies at the University of the Western Cape, and is located at the Institute for Poverty, Land and Agrarian Studies (PLAAS). He worked in agricultural training and extension in Swaziland (1976–1983) and Zimbabwe (1983–1986), and has carried out research on land, agriculture and rural social dynamics in Zimbabwe (1986–1991) and South Africa (1991–2011). His main research interests are the politics of land and agrarian reform, agrarian change and agro-food regimes, land tenure and common property resources. He has edited or co-edited 5 books on these topics, including *At the crossroads: Land and agrarian reform in South Africa into the 21ˢᵗ century* and (with Aninka Claassens, co-editor) *Land, power and custom: Controversies generated by South Africa's Communal Land Rights Act*.

Rudo Gaidzanwa is Professor of Sociology and Dean of the Faculty of Social Studies, University of Zimbabwe. She has served on the Constitution Commission, 1999–2000, and the Utete Committee Land Audit in 2003, and has researched and published on land and gender, economic empowerment and poverty issues since 1981. She has also consulted with UN agencies such as UNICEF, UNIFEM, UNDP and international and national NGOs on land, gender, empowerment and poverty issues.

Changing agrarian relations after redistributive land reform in Zimbabwe

Sam Moyo

Redistributive land reform and agrarian reforms since 2000 progressively changed some of Zimbabwe's agrarian relations, particularly by broadening the producer and consumption base. However they fuelled new inequities in access to land and farm input and output markets. These complex structural changes are explored using a series of surveys, secondary sources and official documents. Findings show that exploitative agrarian labour practices continue despite the diversification of labour towards numerous farms and other enterprises. Agricultural output declined primarily due to reduced inputs and credit supplies, and frequent droughts, but has been rising since 2006. Increasing export production now involves more producers, driven by the diversification of agrarian merchants and contract farming. Agro-industrial capital has gradually increased its domestic operations in the supply of inputs and marketing, especially after re-liberalisation in 2008. Many new farmers accumulate assets although some struggle for social reproduction. Agrarian politics now entail new struggles over agrarian markets, land and labour rights.

Introduction

Much of the literature on both sides of a polarised debate on agrarian change in Zimbabwe has focused narrowly on the Fast Track Land Redistribution process and its immediate consequences for agricultural production. Many only saw a linear decline in agricultural outputs and loss of formal employment (Robertson 2011 is a typical, recent example of one side on this issue). It is commonly claimed that Zimbabwe had been the breadbasket of Southern Africa. Yet it was an irregular food exporter and importer while South Africa met regional food deficits (Moyo 2010). The decline in output is attributed, often implicitly, merely to the replacement of skilled large-scale farmers with subsistence producers and the loss of private property rights (e.g. Richardson 2005). This assumption tends to point to an alternative presently being proposed to 'rationalise' FTLR structures through green revolution-type reforms which integrate small farmers into agribusiness (USAID 2010).

I thank Ndabezinhle Nyoni, Paris Yeros, Lionel Cliffe and four anonymous reviewers for their comments.

These perspectives neglect to examine broader issues of the changing uses of land and labour by varied farming classes with differentiated access to agrarian markets. The displacement of white farmers and farm workers is highlighted (Hammar *et al.* 2010) but changes in the agrarian labour relations are not (see Chambati 2011 in this collection). Moreover, changing agrarian relations in a labour reserve economy are inevitably not only a consequence of what happens in relation to land reform but also of what is happening with urban production and labour markets. There are few efforts to examine changes in the agrarian markets (e.g. as does Scoones *et al.* 2010) and the recent strategies of capital in response to shifts in policy and global markets (Moyo 2011b). In addition to the issues of changing agrarian relations and the impact of agricultural markets and capital, there is a need to study state intervention carefully to elucidate its contribution to agrarian change. Instead media-based reports tend to be restricted to highlighting corruption and patronage in favour of ZANU-PF (Zimbabwe African National Union-Patriotic Front) elites and the exclusion of opponents, missing the substantive class and regional dynamics in which these are embedded.

The Government of Zimbabwe argues that it supported farmers and intermediaries in a non-partisan manner, in the face of droughts, structural constraints on credit and fiscal capacity and political isolation and sanctions from Western nations (RBZ 2007a). Such constraints existed but policy implementation was riddled by incoherence, inconsistencies and class contradictions (Moyo and Yeros 2007, 2009). The underlying issue was how to finance the strategy in favour of peasants while retaining state autonomy.

These questions require empirical attention in the context of a progressive vision of transforming settler-colonial agrarian relations. Land redistribution alone does not automatically yield egalitarian agrarian relations. Progressive agrarian reforms seek increased productivity among small producers to increase food and other supplies to home markets and contribute to industrial diversification and increased employment (see Patnaik 2003). Articulated national development requires trade protection and subsidies (Chang 2009). Implementing such a vision requires producer cooperation (Moyo and Yeros 2005) against dominant capital, which prioritises externally-oriented production and markets while depressing commodity prices.

Articulating such a perspective, this paper provides a macro-picture of Zimbabwe's changing agrarian relations since 2000. After outlining the agrarian history and policy context (detailed in Moyo and Yeros 2005 and Moyo 2011b), it describes the unfolding land and labour relations (also detailed in Moyo 2011a and Chambati 2011 in this collection) to illuminate the agrarian class structure. The orientation of agricultural production is then examined to elucidate its class and regional character. This enables us to explore the trajectory of uneven access to agrarian input and output markets. This highlights the reconfiguration of agrarian markets and the re-insertion of capital through contracts focused on export-oriented production and the class differentiation of farm investment and productivity. Finally, the paper looks forward and projects possible future outcomes by outlining the way farmers are reorganising themselves for state support, and access to markets and land (see also Murisa 2011 in this collection).

This paper relies on research undertaken through the African Institute for Agrarian Studies (AIAS) between 2002 and 2010 on land reform beneficiaries and wider agrarian processes. Various district surveys were undertaken between 2002 and

2004,[1] and a national baseline survey was administered in six districts between 2005 and 2006.[2] This captured data on land tenure, socio-economic characteristics of beneficiaries, production, markets, labour and farmer organisational issues from 2089 land beneficiaries and 761 farm workers. That data is cited as AIAS (2007), while its analysis was reported in Moyo *et al.* (2009). Follow-up field work undertaken between 2007 and 2009 is also cited.

The context of agrarian reform

Equitable agrarian reform in Zimbabwe was compromised during independence negotiations in 1979 in favour of power transfer and liberal democratic reform (Habib 2011). Settler-colonial accumulation by dispossession from 1890 created a labour reserve economy (Amin 1972) dependent on cheap domestic and foreign migrant labour (Arrighi 1973). Peasant farming, rural small-scale industry and commerce were repressed through extra-economic regulations and taxes, but this did not create full-scale proletarianisation (Bush and Cliffe 1984, Yeros 2002). Racial and class inequalities in the agrarian relations were consolidated by discriminatory subsidies to large-scale farmers (Moyo 2002) and narrow import substitution export-led strategies. The consequent rise and fall of the peasantry is well documented (Weiner 1988).

Independent Zimbabwe inherited a racially skewed agrarian structure and discriminatory land tenures dominated by 6000 white farmers and a few foreign and nationally owned agro-industrial estates, alongside 700,000 peasant families and 8000 small-scale black commercial farmers (Table 1). From 1980 Zimbabwe pursued a market-based land reform programme. By 1990 land redistribution had resettled over 70,000 families on about three million hectares (Cliffe 2000). As far as this went it was considered successful in meeting production and livelihood targets (Cusworth and Walker 1988) but the land redistribution was inadequate (Moyo 1995). It left regressive agrarian labour relations on the 75 percent of large farms which were not redistributed (Amanor-Wilks 1995).

The adoption of the Economic Structural Adjustment Programme (ESAP) in 1990 further slowed down land redistribution, encouraged renewed land concentration and foreign land ownership and fuelled export-oriented production, including extensive landholdings for eco-tourism (Moyo 2000). State agrarian subsidies and social welfare transfers to peasants were reduced. This reversed the production gains realised by the top 20 percent of the peasantry, leading to declining maize yields from 1991 (Andersson 2007). ESAP exposed farmers to volatile and monopolistic world markets and reinforced unequal production relations (Moyo 2000).

Wider social dislocations emerged as fiscal capacity dwindled (Bochwey *et al.* 1998). Increased rural landlessness and retrenchment of urban workers extended land hunger (Yeros 2002). Wage repression led to extensive strikes and protests by industrial and agricultural workers between 1994 and 1998 (Sachikonye 2003, Rutherford 2003). Unprecedented political conflicts emerged within and outside the

---

[1] For details on the districts, which have diverse agro-ecological potentials, mixed farming systems and ethno-regional contexts, see Moyo *et al.* (2009, 8–13).

[2] These include Chipinge, Chiredzi, Goromonzi, Kwekwe, Mangwe and Zvimba. The survey was accompanied by key informant interviews, group discussions, field observations and other open interviews. Secondary evidence and information gathered during two land audits (Buka 2002, Utete 2003) complement the survey data.

Table 1. Agrarian structure: estimated landholdings from 1980 to 2010.

Farm categories	Farms/households (000s)						Area held (000 ha)						Average farm size (ha)*		
	1980		2000		2010		1980*		2000*		2010*		1980	2000	2010
	No	%	No	%	No	%	Ha	%	ha	%	ha	%			
Peasantry	700	98	1,125	99	1,321	98	16,400	49	20,067	61	25,826	79	23	18	20
Mid-sized farms	8.5	1	8.5	1	30.9	2	1,400	4	1,400	4	4,400	13	165	165	142
Large farms	5.4	1	4.956	0.4	1.371	0.1	13,000	39	8,691.6	27	1,156.9	4	2,407	1,754	844
Agro-Estates	0.296	0.1	0.296	0.02	0.247	0.02	2,567	8	2,567	8	1,494.6	5	8,672	8,672	6,051
Total	714	100	1,139	100	1,353	100	33,367	100	32,726	100	32,878	100	46.7	28.7	24.3

Source: Adapted from Moyo 2011a (table 1).
Notes: *The average farm size of the peasantry includes common grazing lands.

ruling ZANU-PF (Moyo and Yeros 2007). Authoritarian rule escalated and elections, involving the newly formed MDC, were bitterly contested from 2000, often with violence (Raftopoulos and Mlambo 2009).

These contradictions ignited popular land occupations from 1997 (Moyo 2001), many of which were led by liberation war veterans (Sadomba 2008). Many of these entailed violence and most white farmers were replaced (JAG/RAU 2008). Government condoned the occupations but sought to co-opt them (Moyo 2001). Eventually it gained control over them through state land expropriations and by redistributing land to over 150,000 families in two types of schemes under the Fast Track Land Reform Programme (GoZ 2001).[3]

This departure from neoliberal prescriptions on land reform only partially contested the prevailing wisdom that agricultural growth and viability require large-scale farm sizes (Cousins and Scoones 2010) since the redistribution programme provided for black commercial farmers on relatively large plots, although these are much smaller than those of the erstwhile white farmers. The reliance on cheap labour was not abandoned as A2 farms presumed labour supplies from landless workers. Nonetheless, the process rowed against the current of escalating land alienation in Africa and drew political opposition and Western sanctions.

Facing external pressures and economic decline, the Zimbabwe state radicalised economic policy between 2001 and 2007 while negotiating 'normalisation' with capital by allowing agro-industrial estates and agri-businesses to operate within a heterodox policy framework (Moyo and Yeros 2007, Moyo 2011b). State controls over agricultural commodity markets, trade and financial markets were introduced in 2001 (World Bank 2006). Subsidies targeted distressed industries, including agro-industries to improve the supply of inputs to farmers (RBZ 2006). Agricultural inputs and food prices were regulated. The parastatal Grain Marketing Board (GMB) monopolised grain buying.[4] Genetically modified seeds were actively prohibited and open-pollinated seed was encouraged. Mechanisation subsidies to counter labour shortages and expand cropped areas were escalated from 2006. The state attempted to expand agricultural production on public estates (Moyo 2011b). The Look East Policy was introduced in 2005 to diversify foreign markets. Agrarian labour policy however allowed for low wages (Chambati 2011).

The increasing shortages of foreign currency for imports and of consumption goods – the result of declining production and rising interest rates, which made credit unaffordable – elicited expansionary fiscal interventions. The excessive printing of money and the uncontrolled use of parallel currency markets by the state and businesses generated hyperinflation. This in turn led to aggressive price controls and fuelled underground markets, while sanctions were escalated.

These simultaneously defensive and proactive policies did not extensively oust capital nor fully socialise production relations partly because of the embeddedness of some ruling party leaders and other influential actors in capital and the rootedness of commodity production relations. The plan was opposed ideologically and in practice

---

[3]The A1 model targeted landless and poor families, providing land use permits on small plots for residence cropping and common grazing, while the A2 scheme targeted new 'commercial' farmers, providing larger individual plots on long-term leases to beneficiaries supposedly with farming skills and/or resources (including for hiring managers).

[4]*Statutory Instrument 235A of 16 July 2001.* Grain Marketing (Controlled Products Declaration) (Maize and Wheat) Notice, 2001 in terms of section 29 of the Grain Marketing Act (*Chapter 18:14*).

by big business (see Moyo 2011b). Due to conflicting class and political interests, factions of the ruling party elite and other factions of domestic capital and international capital (represented by black nationals) clashed over the allocation of widely spread and limited public resources and subsidies. These measures affected the interests of opposing classes and politicians, with some seeking to evade or benefit from them. Planning deficiencies were evident. Corruption emerged within and outside ZANU-PF as various classes competed for access to the subsidies and rents. Patronage often included or excluded both political opponents and supporters. These events widened the fractures within ZANU-PF and fuelled the violently contested elections of 2008 (Moyo and Yeros 2009).

Facing dramatic economic collapse and a political stalemate the economy was liberalised in 2008 (RBZ 2009) and a 'power-sharing' government was formed through regional mediation (GPA 2008). Controls on agricultural markets, the capital account and trade, and off-budget subsidies were abandoned, and the economy was 'dollarised' (RBZ 2009). Foreign investors were invited, albeit within the 'Indigenisation' policy requiring domestic control of majority shares (GoZ 2011). These policy shifts, which resulted from the agency of diverse landholders and classes, subsequently elicited diverse responses from them. This complex, contradiction-riddled and confrontational process reshaped various dimensions of the agrarian relations which had been fundamentally restructured by land redistribution.

Changing agrarian structure: land and labour relations

Fast track land redistribution undermined the underlying logic of settler-colonial agrarian relations founded on racial monopoly control over land that deprived peasants of land-based social reproduction and compelled cheap agrarian labour supplies. Redistribution reversed racial patterns of land ownership and broadened access to land across the ethnically diverse provinces, while replacing most private agricultural property rights with land user rights on public property (Moyo 2011a). This represents socially progressive agrarian change. But it also generated new inequities based on uneven land ownership and control of labour.

Zimbabwe's agrarian structure comprises four relatively distinct farm categories, based on differences in land size, forms of land tenure, social status of landholders and capacity to hire labour. These include peasants, also called small producers here; middle-sized farmers; large-scale capitalist farmers; and agro-industrial estates, plantations, and conservancies. Redistribution enlarged the peasantry and expanded the number of mid-sized farms, while downsizing the number, farm size and area of large-scale capitalist farms, as well as of the agro-industrial estates (Table 1). Landlessness remains, especially among some farm labourers. This structure has created a platform for new agrarian class formation processes and struggles.

Agrarian relations among the peasantry continue to be defined mainly by self-employment of family labour towards producing foods for auto-consumption and selling some surpluses, as well as various non-farm work and short-term wage labour. The peasantry have differentiated capacities to hire limited labour, and some provide labour services to others. Most of the families hold customary rights to arable and homestead plots and common grazing areas in Communal Areas, while their A1 beneficiary counterparts hold state permits for similar family and common land rights. About 30 percent of the A1 beneficiaries are made up of urban workers and a few former farm workers (Moyo et al. 2009). Eighteen percent of the land beneficiaries retain homes and plots in Communal Areas to diversify their reproduction and production; using extended family resources and sharing land with extended family members and subletting to others is commonly practised (Moyo *et al.* 2009).

Over 31,000 middle-scale and large-scale capitalist farmers, most of whom are blacks, now exist. Two thirds of them received land as A2 beneficiaries in all provinces on varied land sizes (Table 2), with varied farm assets (Table 5). These farmers rely on relatively larger amounts of hired labour than on family labour (see Chambati 2011). They hold land through tenures amenable to market transactions, mainly through leases, while a few retain freehold title. The majority of them originate from the middle class, including currently or formerly employed professionals, small non-farm capitalists and rural 'elites', including chiefs and some better-off peasants, as well as some working class people (AIAS 2007). Those with larger-scale farms tend to be better educated and linked to employment and business (Moyo 2011a), and are better placed to negotiate political power and mobilise resources. A few hire farm managers, while some rent land claiming their land sizes are too small to be 'viable' (AIAS 2007). Some hold multiple farms (Moyo 2011a).

The agro-industrial estates were reduced to 240 establishments, mostly owned by large-scale capital and covering over one million hectares or three percent of all the farming land (Moyo 2011b). They still hold freehold title to land in vertically integrated enclaves, including tourism conservancies and state estates. They hire large amounts of permanent and seasonal labour (Chambati 2011) and contract an expanding number of outgrowers. The latter comprise small and medium-sized land beneficiaries relying on family and hired labour. The state has retained its plantations and expanded production through partnerships with capital. The indigenisation policy calls for the redistribution of the estates' shareholdings to locals. So far about 39 blacks hold shares in conservancies (*The Herald* 2011a).

The persistence of large-scale landholdings meant the exclusion of potential land reform beneficiaries (Moyo 2011a) and fuels the 'illegal' occupation of lands (MLLR

Table 2. National classification of (A2) farms.

Provinces	Farm sizes						Total
	below 100	101–300	301–500	501–1500	1501–3000	3001–5001+	
Manicaland	1,427	404	100	31	2	1	1,965
Mashonaland Central	1,991	696	184	60	5	3	2,939
Mashonaland East	3,113	1,049	195	95	4	3	4,459
Mashonaland West	6,353	1,421	300	158	16	8	8,256
Masvingo	1,027	86	43	121	49	4	1,330
Mat North	198	221	66	145	35	8	673
Mat South	117	643	321	230	54	1	1,366
Midlands	136	606	428	191	15	9	1,385
Total	**14,362**	**5,126**	**1637**	**1,031**	**180**	**37**	**22,373**

Source: Compiled by author from MLRR (2010) data.

2009). The policy of limiting access to redistributed land for former farm workers was partially motivated by the desire for cheap labour supplies. Landless people and poorer peasants still provide some farm labour services at low wage rates (see Chambati 2011 in this collection). Many landless farm labourers reside precariously on new landholdings, perpetuating exploitation practises via tenancy. However, once the pre-2000 relations of agrarian labour were undermined by land redistribution agrarian labour shortages became more common on the remaining large-scale farms. The number of full-time labourers on capitalist farms has declined and short-term hired labour has expanded on diverse farms and in rural non-farm activities (Chambati 2011, this collection). About 25 percent of the beneficiaries in our surveys retained other jobs, while 27 percent of the land beneficiary household heads are non-residents who combine urban waged labour with farming (Moyo et al. 2009). Policies intended to curb widespread petty natural resource and mineral exploitation by the landholders, landless and others by promoting farm mechanisation failed to address such labour shortages (Utete 2003). These trends confirm that repeasantisation and semi-proletarianisation are simultaneous outcomes of the agrarian reforms (Moyo and Yeros 2005) although we cannot foreclose the trajectory (see O'Laughlin 2002).

Unequal land and labour relations are being consolidated by tendencies towards land concentration through informal land rentals. About 25 percent of the land beneficiaries sublet or share their land (Moyo et al. 2009) without official sanction. Some of these households lack production inputs or face social calamities such as illness or deaths (Mhondoro field interviews 2008). Others sublet land for speculative reasons or seek to maximise incomes from farming partnerships (Mhondoro field interviews 2008). These tendencies suggest the contingency of such land markets. Large-scale re-concentration of agricultural lands is however restricted by state ownership of redistributed lands and natural resources. Moreover A2 farmers have no legal right to evict informally settled farm workers.

Various A2 farmers advocate the commodification of land through freehold tenure, land rentals and sometimes by evicting land beneficiaries but this is actively resisted by small landholders. However, 80 percent of the surveyed beneficiaries perceived their land tenure to be secure. About 16 percent of them had been threatened with eviction and many had successfully resisted this, while seven percent had once been evicted, mostly from A1 and highly prized peri-urban landholdings (Moyo 2011a). However, land conflicts mainly arise from unclear boundary definition, double allocation of some plots by the state and disputes over the use of common natural resources. This affected 21 percent of the beneficiaries in 2006, mostly in higher agro-ecological potential and peri-urban areas, some of whom putatively retained Communal Area homes as insurance against eviction (Moyo 2011a).

Redistributive land reform did not reverse the regressive agrarian relations evoked by patriarchal power relations. Land access biases against women, youth and immigrants and the exploitation of female labour through male control of products are common (see also Makura-Paradza 2010). While more women secured their own land than in previous reforms, men and husbands still dominate agrarian transactions (WLZ 2007, Moyo 2011a). The extension of traditional leadership to newly redistributed areas widens these contradictions (Murisa 2010), although this cannot be over generalised. Legally, however, the new land user rights are derived from the state and not custom. Agrarian relations are still coloured by power relations derived from ethno-regional identity. Land redistribution re-linked people with their original 'homes and ancestral spirits', providing scope to re-mobilise lineage-based ethnic ties and territoriality (Mazoe focus group discussion 2005, Mkodzongi 2011). Often these affinities are used to exclude those defined as not belonging, but this operates unevenly among the provinces and peri-urban areas (Moyo 2011a).

Despite the progressive outcomes of land redistribution, agrarian relations are imbued with salient struggles over access to land and labour. These are amplified in various production processes.

Changing agrarian production relations

Extensive land redistribution is expected to alter the structure and orientation of agricultural production, including generating a transitional decline in output as new classes of farmers mobilise resources to establish themselves in relation to changing markets and state interventions. Indeed, the output of Zimbabwe's main agricultural commodities started declining in 2002 and then selectively began to rise from 2006. The number of farmers involved in producing diverse commodities and the overall cropped area expanded substantially, while yields generally declined. This trend affected the commodities differentially as some had been predominantly grown by small-scale farmers, while others had been grown mainly on large-scale farms and plantations on land which was unevenly redistributed. Moreover some commodities are produced predominantly for export markets and are externally financed. A new uneven class and regional production structure has emerged restructuring some of Zimbabwe's social relations of agrarian production, while consolidating others.

Expansion of food production among the peasantry

The output of maize, Zimbabwe's main staple grain produced mostly by peasants, declined severely in an erratic long run pattern associated with droughts (Figure 1). National maize yields per hectare fell to about 50 percent of the 1990s average, while the average yields of A1 maize producers were half those realised by A2 producers, and land beneficiaries in wetter agro-potential areas realised twice the yields of those in drier regions (AIAS 2007). However, the output of small food grains such as sorghum and millet increased (Figure 2) although their average yields fell 40 percent below the 1990s average (MIMAD 2010a). The output of wheat, which before 2000 was predominantly grown by large-scale farmers, declined dramatically (Figure 2) and the area cropped to wheat fell from an average of 58,000 hectares in the 1990s to 18,200 hectares in 2010.

However, the numbers of food grain producers expanded after the Fast Track Land Redistribution Programme, leading to a major increase in the national cropped area dedicated to food grains from 1,794,527 hectares in 1999 to 2,655,687 hectares by 2011 (MAMID 2010a). Land reform beneficiaries dedicated 78 percent of their cropped land to food grains (Moyo *et al.* 2009). This shifted the orientation of production and use of prime lands away from exports to the staple grains prioritised by peasants.

The outputs and cropped areas of oilseeds such as soyabeans and sunflowers, which target the home market but were mainly produced by large-scale farmers

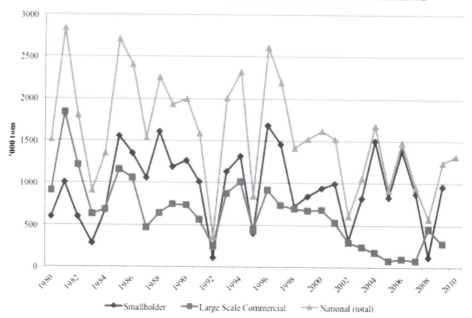

Figure 1. Sub-sectoral maize production trends (1980–2010) in Zimbabwe.
Source: MAMID (2010a, 2010b).

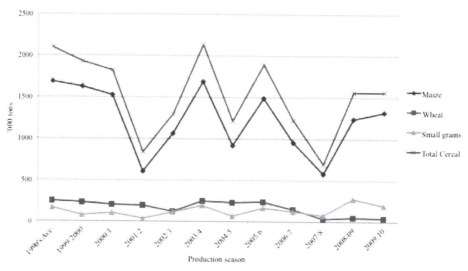

Figure 2. Cereal crops output: 1990s vs. 2000s average.
Source: FAO (2009), MIMAD (2010a, 2010b, 2011).

before 2000, also declined but later they experienced a limited up-turn (Figure 3). The output of groundnuts and edible beans, however, increased, and these crops continued to be grown mainly by peasants. About 21 percent of the land beneficiaries grew groundnuts, while around six percent of them grew beans and soyabeans (Moyo *et al.* 2009). This suggests that the numbers of higher value food producers has expanded and that peasant production is diversifying.

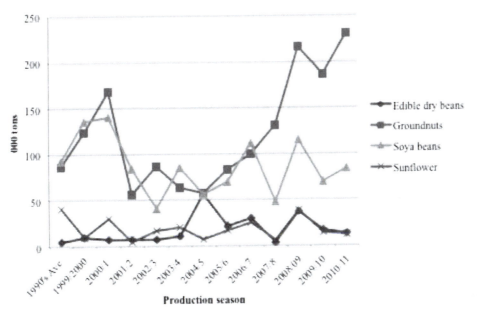

Figure 3.   Oilseed output: 1990s vs. 2000s average.
Source: TIMB Statistical data, Cotton Ginners Association (2010), MAMID (2011).

Resilience of export oriented production?

In terms of financial value, the largest share of agricultural output decline occurred among export commodities which had been predominantly produced on large-scale landholdings. In 2004 tobacco output had fallen by 72 percent from the 1990s average (Figure 4) but by 2011 its output was rising substantially (Figure 4). Its yields and planted area declined by 55 and 20 percent respectively during the period and fertiliser utilisation per hectare was halved by 2004 (TIMB2010,MIMAD2011). But over 50,000 farmers, including 30,000 peasants, are now producing tobacco on smaller cropped areas, compared to about 700 large-scale and 3,500 small producers around 2002 (TIMB 2010, MIMAD 2011). The A2 farmers are realising higher yields as they use more inputs (Moyo *et al.* 2009). Some remaining white farmers continue to produce tobacco.

The output of cotton, which is largely exported and was predominantly produced by small farmers, surpassed the 1990s average by 49 percent by 2011 (Figure 4). This reflects a sizeable expansion of small-scale producers in a commodity whose production was already entrenched, the cotton's tolerance of drought, and the continuity of contract-based inputs supplied by capital. On average four percent of the land beneficiaries grew cotton, while 21 percent grew it in Chiredzi district, despite its limited cotton growing history (Moyo *et al.* 2009). Established agro-industrial production structures and technocratic wisdom were being challenged by farmers in the new milieu (see also Scoones *et al.* 2010).

The outputs of plantation export commodities only began to decelerate around 2004 but these were rising by 2011 (Figure 5). Sugar output fell by 20 percent from the 1990s average levels in 2006 and then by 50 percent during hyperinflationary conditions between 2007 and 2008, only for the rate of decline to decelerate by 2011 (EU 2009, RBZ 2011). The structure of sugar production hardly changed as the area cropped by the estates was hardly reduced while the outgrowers' cropped area

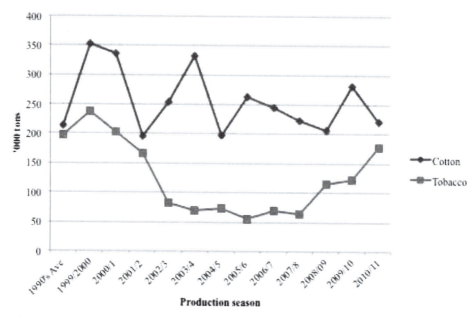

Figure 4.   Key exports' output trends: 1990s average vs. 2000s.
Source: TIMB Statistical data, Cotton Ginners Association (2010), MAMID (2011).

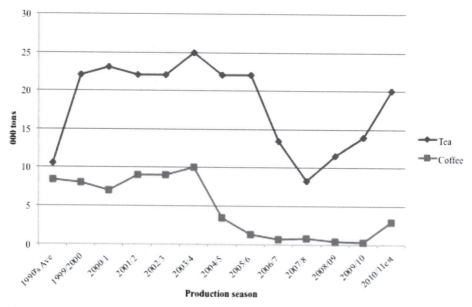

Figure 5.   Tea and coffee output trends in Zimbabwe.
Source: IMF (2005), Zimbabwe Tea Growers Association data, Zimbabwe Coffee Mills data.

declined substantially (EU 2009, RBZ 2011). The number of sugar outgrowers grew from 47 whites with an average of 147 hectares each in 2000 to about 560 with average hectarages of 10 to 30 hectares (Moyo 2011b).

The number and area of large-scale farm and estate producers of coffee and tea had declined by 2006, but tea output patterns were anomalous compared to other crops. By 2010 tea output levels were well above the 1990s averages, but this level was lower than the peak reached prior to 2007 (Figure5). Meanwhile coffee output had mainly declined by over 90 percent in 2010, with its cropped area falling by over 30 percent, among the outgrowers, whose numbers almost doubled (Zimbabwe Coffee Mills).[5] Remaining tea estates are now diversifying production towards macadamia nuts, pineapples and passion fruit, ostensibly due to labour shortages and lower prices, while outgrowers in tea and sugar growing areas are also producing some foods due to input shortages (EU 2009, USAID 2010).

The production of formally marketed dairy and beef was previously dominated by large-scale farms. The national cattle herd fell by 19 percent by 2009 (Table 3). Most cattle are now held by many smaller-scale and middle-scale farmers practising mixed farming, which provides various foods, draught power and manure. Yet 58 percent of the land beneficiaries had no cattle, while 36 percent had between three and above 20 cattle, and the rest had one to two beasts (AIAS 2007). However, 70 percent of the households in drier agro-ecological regions, officially considered ranching areas, had cattle. Milk output also fell by 66 percent from 2001 to 2010 (MAMID 2011).

There were rapid declines in pork production by 2005 but pork output had risen substantially by 2009, while the numbers of goats were stable compared to sheep, which are not a common food (Figure 6). Less than 25 percent of the land beneficiaries had small ruminant livestock, while two percent of them had piggeries by 2006 (MAMID 2011). Formally marketed pork production was previously dominated by over 100 LSCF producers and agro-industrial plants, but by 2010 it involved over 250 smaller producers.[6] Overall the diversification of livestock producers was accompanied by lower quality, breeding stocks and calving rates since investments in breeding, animal health and pen-feeding had declined (FAO 2009).

Diversification of the production base and class bias

In general agricultural production patterns during the 2000s became more differentiated in class and regional terms, while export-oriented output rose faster than food. The production base was restructured by introducing more

Table 3. Cattle numbers by farming sector: 2001 – 2011.

Sector	2001	2003	2005	2009	2011
A2/LSCF	1,291,110	453,418	519,028	442,080	453,385
Communal	4,398,081	3,994,830	3,604,361	3,692,196	3,529,739
Resettlement & A1	505,360	717,969	844,800	919,616	1,020,070
Small scale	23,565	180,648	219,424	167,828	147,559
Total	**6,418,116**	**5,296,865**	**5,187,613**	**5,221,720**	**5,156,753**

Source: MAMID (2010a, 2010b), FAO (2009).

---

[5]Personal communication, 2010.
[6]Interview with Theo Khumalo of Colcom, April 2011, Harare.

producers into all commodities and expanding the overall cropped area substantially, despite the decline of yields for most crops and livestock. Numerous producers earned farming incomes and provided their own food. A process of income re-distribution was underway, although this favoured an expanded range of middle- to larger-scale farmers, mainly in wetter regions. While more farmers are now producing exports, they are however well below 15 percent of all the farmers.

Average land utilisation rates among land beneficiaries were at 40 percent (AIAS 2007) comparing favourably with former large-scale farming areas (World Bank 1991). Around 2006 about 54 percent of the beneficiaries cropped less than three hectares, while only 14 percent cropped more than 10 hectares (Table 4). Thus middle-scale and larger-scale landholders cropped proportionately less land than the peasants. Their pre-2000 counterparts usually cropped below 700,000 hectares, despite employing more formal labourers (World Bank 1991).

While before 2000 agricultural production was predominantly export oriented, the incomes realised were concentrated among a few large farmers, alongside domestic and foreign capital. Peasants cropped much more land and used more labour towards producing various foods largely for auto-consumption. The latter

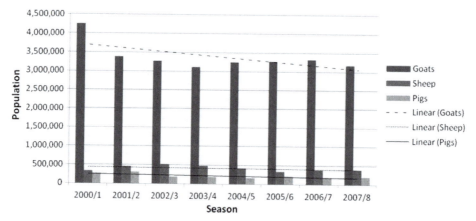

Figure 6. Small stock production.
Source: MIMAD (2010b).

Table 4. Total cropped area by farm size in selected new resettlement areas.

Cropped area (ha)	1-19 No.	1-19 %	20-49 No.	20-49 %	50-99 No.	50-99 %	100-299 No.	100-299 %	300+ No.	300+ %	*Total* No.	*Total* %
0	177	18.8	122	19.8	30	16.3	49	27.5	11	37.9	*389*	*19.9*
0.1-1	90	9.5	44	7.2	13	7.1	27	15.2	–	–	*174*	*8.9*
1.1-3	285	30.2	145	23.6	27	14.7	37	20.8	3	10.3	*497*	*25.5*
3.1-5	222	23.5	105	17.1	24	13.0	23	12.9	3	10.3	*377*	*19.3*
5.1-10	97	10.3	92	15.0	40	21.7	12	6.7	–	–	*241*	*12.4*
10+	73	7.7	107	17.4	50	27.2	30	16.9	12	41.4	*272*	*13.9*
Total	*944*	*100.0*	*615*	*100.0*	*184*	*100.0*	*178*	*100.0*	*29*	*100.0*	*1950*	*100.0*

Source: AIAS Household Baseline Survey (2007).

trajectory has been consolidated, but land and labour productivity continue to be low. Consequently, domestic food production has on average been 35 percent short of requirements, with wheat, soyabeans and dairy faring worst, while the output of pulses has expanded. Zimbabwe continues to have some relatively more secure food enclaves, reflecting uneven production and productivity patterns. Small producers in the southern districts continue to face regular grain deficits, while those in the wetter regions increased their production of pulses and cash crops, particularly tobacco and various vegetables.

Class and regional biases in the capacities of various farmers to produce relatively larger areas of high value crops and export commodities are being reproduced. A broader base of capitalist farmers has emerged, while the plantations are consolidating their vertical integration into world markets using more outgrowers. More peasants and middle-scale producers are now slowly expanding the supply of more diverse foods and raw materials to the home and export markets. This also entails the reinsertion of large foreign capital into Zimbabwe's restructured agrarian markets.

Differentiated access to agricultural inputs and markets

Agricultural productivity generally declined due to reduced and uneven access to inputs and output markets. This particularly affected smaller producers who nonetheless deployed their labour to expand cropped areas. Access to inputs was also constrained by reduced public and private agricultural finance, leading to the diversification of input supply and commodity marketing arrangements by capital.

Uneven access to agricultural inputs

Maize seed was in short supply during 2003 and 2006. Three transnational companies had dominated hybrid maize seed production by 1999 through contracts with about 200 larger-scale growers whose land was redistributed. By 2010 numerous medium- and large-scale farmers were being contracted to produce seed, unravelling the previous oligopoly. Similarly many more middle-scale farmers were producing tobacco seedlings and meeting current demand (TIMB 2010). Shortages of potatoes and vegetable seeds persisted, leading to supplementary imports. By 2011 over 80 percent of all the farmers were using commercial hybrid seeds now grown by more farmers contracted to capital, which retained dominance in the privatised bio-genetic industry.

Fertiliser application in Zimbabwe in the 1990s averaged 30 kg/ha. This was halved by 2004 (FAO 2009) and applied to larger cropped areas. Around 2006, 50 percent of the land beneficiaries were utilising inorganic fertilisers, mostly for maize, tobacco and cotton production, with relatively more A2 beneficiaries using fertilisers, and 20 percent of A2 beneficiaries using pesticides (Moyo et al. 2009). The wetter agro-ecological regions used more fertilisers. Nationally less fertiliser was applied by small farmers than by the larger farms (Figure 7). Export crops used the largest share (Tripathy et al. 2007). Most A1 farmers were using animal manure rather than fertilisers (Moyo et al. 2009), bearing in mind the rise of inputs prices globally (Moyo 2010). Less than 10 percent of farmers also adapted to the rising price of inputs and access problems by adopting conservation farming to optimise absorption of water and fertiliser (FAO 2011).

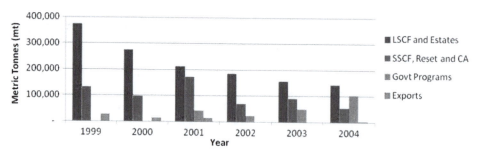

Figure 7. Sub-sectoral utilisation of locally produced fertiliser.
Source: Triparthy *et al* (2007).

The use of productivity enhancing inputs on livestock was mainly found in the drier southern provinces. Twelve and 21 percent of the land beneficiaries used stock feeds and veterinary chemicals respectively, although 34 percent of the pork producers used them (Moyo *et al.* 2009). Only 35 percent of the cattle producers used public dipping and veterinary services, with slightly more A1 beneficiaries using public dips (Moyo *et al.* 2009). This limited livestock productivity. The re-insertion of large capital in the livestock sector was relatively limited because of the continued and dispersed control of breeding stock by former large-scale farmers and the pervasiveness of uneven regional investment.

Agricultural production continues to depend mainly on rain-fed farming. Only five percent of the national cropped lands are irrigated and plantations control 57 percent of this, while small producers control 30 percent (World Bank 2006). About 17 percent of the land beneficiaries had one form of irrigation facilities, while 28 percent of the A2 farmers had irrigated crops compared to 14 percent of A1 farmers, and only 10 percent of both groups had invested in irrigation (Moyo *et al.* 2009). Irrigation was slightly more common in Chipinge and Chiredzi (Moyo *et al.* 2009) where more irrigation facilities were pre-existing on plantations. Some irrigation facilities were disabled by departing landowners and land occupiers and numerous dams remain underutilised (World Bank 2006). This uneven class and regional distribution of irrigation facilities is also associated with export production.

Agricultural productivity is also constrained by low and uneven access to farm machinery. Most peasants still depend on labour-intensive ox-drawn traction and hand weeding. Only 49 percent of the land beneficiaries had access to animal-driven ploughs, while less than 20 percent had access to power-driven equipment, with only six percent of the A1 beneficiaries having access to tractors, compared to 36 percent of the A2 farmers (Moyo *et al.* 2009). Over 70 percent of the A2 farmers who used tractors owned them (ibid). Small landholders own less than 22 percent of the national tractor fleet, farm equipment and machinery (MAEMI 2009). Public or private draught power hire services are limited, and the government's mechanisation programme added only 3217 tractors to the national stock (MoF 2010). During the 2000s limited fuel subsidies were provided, but these mainly benefited A2 farmers.

About 30 percent of the land beneficiaries had on-farm and off-farm infrastructure, including some left on the redistributed farms (Table 5). Some A1 farmers shared farm houses and stores for social services, while A2 farmers gained these individually. Given the limited availability of credit, these investments are significant. While access to subsidised inputs during the 2000s reached diverse farmers, the outcome favoured larger-scale export farmers with better access to markets (AIAS 2007). Thus, only some resource rich farmers had access to inputs and this limited the capacity of most to hire labour (see Chambati 2011) and invest. The capital intensity of farming is uneven and influences wider agrarian relations. Those few larger-scale farmers using motorised traction include both high and low intensity labour hirers, while the majority hire little labour and are poorly capitalised (see Chambati 2011). Class biases in the control of land, labour and access markets have no doubt also been shaped by unequal political connections and social status, as well as the re-configuration of such markets.

The reconfiguration of agrarian markets

The production of farm inputs such as fertilisers by domestic industry was also falling by 2003, as were imports (Figure 8). Many agro-industries did not adapt to

Table 5. Productive investment in newly resettled areas.

Type of investment	A1 model		A2 model		Total	
	No	%	No	%	No	%
Homestead	1089	66.0	206	47.0	1295	62.0
Irrigation equipment	168	10.2	48	11.0	216	10.3
Farm equipment & machinery	111	6.7	39	8.9	150	7.2
Storage Facilities	123	7.5	30	6.8	153	7.3
Livestock	200	12.1	79	18.0	279	13.4
Tobacco barns	22	1.3	6	1.4	28	1.3
Electricity	5	0.3	2	0.5	7	0.3
Worker housing	123	7.3	62	14.2	185	8.9
Plantations & orchards	12	0.7	2	0.5	14	0.7
Environmental works	18	1.1	5	1.1	23	1.1

Source: AIAS District Household Baseline Survey (2005/06); N=2089.

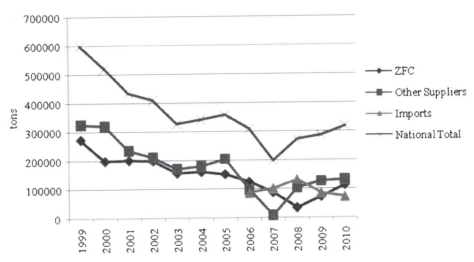

Figure 8. National fertiliser supply (2001 – 2010).
Source: Compiled by AIAS from Zimbabwe Fertilizer Company (ZFC) 2010 data.

the changing demand structure. Agrarian merchants and fertiliser producers reduced local operations, and increased operations in neighbouring countries. As interest rates rose, agri-business increasingly depended on subsidised foreign currency and credit in exchange for reduced prices and turned to manufacturing inputs on pre-paid contracts for large-scale exporters.

More fertiliser and tractors were imported from China and Iran through concessional loans and from South Africa and elsewhere in cash. New indigenous importers were contracted to supply government input schemes. Similarly, domestic producers and suppliers of machinery and implements increased supplies when contracted by the government (see RBZ 2007a). This led to the recovery of some agro-industrial capacity (CZI 2010). However, often these schemes were alleged to have fiddled with prices, quality and distribution.

Around 2006, 30 percent of the A1 land beneficiaries relied on subsidised seed compared to 20 percent of A2 beneficiaries, while in high potential districts such as Chipinge 50 percent relied on subsidised seeds, compared to nine percent in remote Chiredzi (Moyo *et al.* 2009). Between 68 and 87 percent of the land beneficiaries, depending on the district, purchased their inputs in markets (Moyo *et al.* 2009). A marginally larger proportion of A2 compared to A1 farmers benefited from government inputs. Over 96 percent of the few land beneficiaries who used agro-chemicals for cotton and tobacco production bought them in markets. Similarly, less than one percent of the land beneficiaries received government livestock inputs, aside from dipping services. International donors played a limited role in subsidising inputs until 2009. However, while such state subsidies were limited in scale the level of support substantially surpassed that provided during the 1990s.

Formal cattle markets, whose exportation had substantially declined by 2004 due to reduced slaughters and livestock disease, became spatially and socially disaggregated. Meat trade increasingly involves direct sales between producers and new abattoirs, retailers and consumers (Scoones *et al.* 2010). Livestock production on input sub-contracts and leased grazing created new tied markets (Zvimba field observation 2008).

Government support for grain marketing also increased compared to the ESAP era, but its monopoly became unsustainable. Grain procurement relied on limited subsidies and expensive credit while the grain sold to millers was highly subsidised. Small millers mushroomed in tandem, competing and colluding with established large agribusiness to hike prices (News24 2003). Meanwhile farmers received low grain prices and often delayed payments (FAO 2009). Many producers, consumers, millers and traders circumvented the controlled market leading to high prices in parallel markets and unstable supplies. Few A2 producers 'obeyed' the marketing regulations, evoking patriotism while expecting future subsidies (Mhondoro District interview 2009). This limited the potential welfare transfers to deficit areas, despite the substantial fiscal outlays which were supplemented by humanitarian aid.

Around 55 percent of land beneficiaries sold their maize to the GMB, while the rest used local markets. Over 22 percent of all the beneficiaries sold their edible beans to the GMB, with slightly more A1 beneficiaries doing so (Moyo *et al.* 2009). Over 35 and 57 percent of the soyabeans produced by the A1 and A2 land beneficiaries respectively were sold to the GMB, which used soyabean trading to generate income, while around 31 percent of this crop was retained for own use (Moyo *et al.* 2009). The rest was sold to agro-processing firms, contractors and local agro-dealers, including some which had secured subsidies. Distance from markets, limited state capacity, and manipulative trading practises by the growing and variegated merchant classes in markets generated various contradictions around state interventions.

Most of the tobacco and cotton was being sold by land beneficiaries to contractors and private buyers in 2006, while over 65 and 55 percent sold their sugar and tea in Chiredzi and Chipinge respectively to agro-processing plantations. Between 22 and 40 percent of the A2 producers of cotton and tobacco had kept their output that year because prices were poor, while 12 and 24 percent of the A1 and A2 farmers could not secure 'independent' markets for their sugar, allegedly because buyers 'condemned' it (Moyo *et al.* 2009).

In 2006, the RBZ partially relaxed the policy of withholding large proportions of foreign currency earned by exporters (RBZ 2006). Contract farming benefited from this and became central to the marketing of most commodities, excluding grains, building upon previous experiences with cotton and barley. A few black tobacco merchants piloted such contracting and later sold the businesses to transnationals. The number of tobacco contractors grew to 12 in 2010 from fewer than three in 2003, including four black firms, four new multi-racial contractors and four foreign contractors from China and elsewhere (TIMB 2010). Some of these were subcontractors of Western transnationals such as British American Tobacco. Cotton contractors and buyers also increased and included Indians and Chinese, with the latter up scaling tobacco contracting.

Understanding the role of contract farming in class formation processes is complicated by its unclear association with land sizes and assets owned. Some contract financiers prefer peasants and middle-scale producers because they are less able to resist lower price margins compared to larger-scale producers, who generally have higher social standing and fare better in procuring inputs using their own income, credit and subsidies. Conflicts arose between farmers and contractors over the depressed prices on offer and many 'resisted' this by side marketing their contracted commodities to other merchants who had not provided contract inputs to them. But they bought such outputs at marginally higher prices. New sugar outgrowers fought with plantation managers over transportation charges and the pricing of inputs and outputs (*The Herald* 2011b).

After market liberalisation and dollarisation the limited contract production of foods such as soyabeans faced competition from subsidised imports from the West and genetically modified based grain and oilseeds imports from South Africa. This drove farmgate prices down, reducing local producer incomes. To some extent this propelled many small producers needing cash to reinvest and meet consumption needs into tobacco contracts. This explains the second temporal decline in the soyabean output levels around 2009.

The government revived the Agricultural Marketing Authority (AMA) in 2004 to better regulate markets and introduced new contract farming regulations in 2009 (USAID 2010). By 2011 however the large-scale Zimbabwe Commercial Farmers Union and the CBZ bank, backed by the Ministry of Trade and Commerce and funded by some donors, were establishing Commodity Exchange in Zimbabwe (COMEZ). The AMA feared this would drive speculation on food and increase foreign influence (*The Herald* 2011c). Class-based struggles over these changing agrarian markets remain potent and open with finance being critical.

Access to finance for farming

Throughout the 2000s the volume of agricultural finance from domestic sources and external concessional loans and aid from Western donors fell sharply compared to provisions during the 1990s (RBZ 2007b). Private agricultural credit declined from over US$315 million in 1998 to about US$6 million in 2008 (MAEMI 2009). European trade credit and the agricultural commodity bonds market had virtually disappeared by 2005. Government credit through Agribank had averaged around US$25 million per annum between 2000 and 2007 and had peaked at US$104 million in 2004, but it declined to below US$3 million in 2007 (MAEMI 2009). Declining revenues limited budgetary allocations and led to larger-scale money printing. This credit constraint fuelled the diversification of the forms and sources of agriculture finance from the mid-2000s.

Subsidised credit, inputs and foreign currency supplies increased (RBZ 2005). Foreign currency was being secured on parallel markets by citizens and government, which fuelled speculative pricing, shortages of goods, and hyperinflation. Around 2004 the government attempted to plug the wheat deficit through subcontracted production and other partnerships with domestic agro-industrial capital on the ARDA estates, using subsidised funds (Moyo 2011b). But this foundered over pricing and profit-sharing disagreements.

Hyperinflation made agricultural credit less 'competitive' than short-term trading (Matshe 2004). The World Bank (2006) attributed decreased financing to the undermining of profitability and investment incentives by market controls. Others argued that the land user rights policy and land disputes created uncertainties for investors and limited credit supplies (Richardson 2005, Rukuni et al. 2009).

During 2006, over 78 percent of the A1 land beneficiaries used their savings, remittances from home and abroad and non-farm incomes to finance farming operations, compared to 83 percent of the A2 farmers. Only 10 percent of them received external financing for production and two percent had access to credit (Moyo et al. 2009). But rising food and input prices after the 2002 drought and global price hikes in 2005 undermined the real incomes of peasants (Wiggins 2005).

Not surprisingly contract farming became central to the financing of smaller and middle-scale farmers (see The Standard 2009) who joined export production to gain access to inputs and increase their earnings. This shifted pre-2000 agrarian relations from the dominance of private credit relationships between large-scale farmers and banks towards bonding more farmers with contracting intermediaries. Before 1986 government had been the major lender (Moyo 1995). When foreign currency and agricultural markets were re-liberalised, agricultural subcontractors escalated such pre-financing arrangements. Private bank credit to agriculture increased to over $300 million in 2010 (MoF 2011), but over 60 percent of this went to contractors (USAID 2010).

China played a leading role in financing agriculture through loans for imported fertiliser, agro-chemicals and tractors and in contracting tobacco and cotton from 2006 (Edinger and Burke 2008). By 2009 the state had lured 'foreign investors' in partnership with domestic capital to produce and process sugarcane (for ethanol) and increase beef exports, using 20-year build-operate-transfer and land-lease arrangements on parastatal lands (Moyo 2011b). Large agribusiness was regaining dominance in agricultural input and output markets, and new agrarian capital from the East and South was seeking to invest in agricultural exports and to supply inputs. Zimbabwe remained a net exporter of capital to the West during the decade (UNCTAD 2008).

These reconfigurations of the agrarian markets and state interventions fuelled contradictions within the state apparatus over its autonomy and the uneven benefits realised by various farming classes. The re-financialisation of Zimbabwe's economy was increasingly orienting agricultural production to exports but diversifying its global integration. The class dynamics of the emerging agrarian relations reinforced the policy shift that enabled the dominance of large foreign and domestic capital, which is now only peripherally engaged directly in production.

Changing farmer organisation

These shifts in the agrarian relations have altered the political landscape and debates on future agrarian reforms. New forms of farmer organisation and protest are emerging at the local, district and national levels, including among women farmers and through commodity associations (Moyo et al. 2009). Agrarian politics are increasingly shaped by international capital in alliance with new large-scale farmers and the established white and emerging black bourgeoisie around agrarian markets within an increasingly neoliberal regime. Renewed social differentiation among the peasantry and the emergence of a middle class of farmers is being shaped by new forms of accumulation and consumption, which are altering the rural political constituency. Continued sanctions and competition for access to the promised fruits of the indigenisation of mining and other businesses foreshadows new intra-class conflicts, especially among various sections of the elite.

Existing farmers and new settlers mobilise common histories and experiences despite cultural and ethno-regional difference in struggles for land and production resources (Mkodzongi 2011). Church, women's and farmers' clubs and local development associations are being adapted by landholders and others to the new farming areas (Scoones et al. 2010). Local liberation war veterans' and collaborators' associations that led land occupations are recreating their leadership roles (Sadomba 2008). Farm labour activism is being restructured to negotiate labour rights, given their limited protection and the ineffectiveness of the national unions (Chambati and Magaramombe 2008). Former white farmers' unions have been reconstituted. The Commercial Farmers Union (CFU) retains a few members and the new Justice for Agriculture (JAG) mobilises to influence policy and international support over 'lost' land and compensation.

Mainstream discourses on the land audit proposed by the Inclusive Government create tensions since multiple landholders and 'unproductive land users' could be evicted, however they raise the hopes of some for access to land. Elitist perspectives imply that evictions will affect resource-constrained land-holders who cultivate smaller areas (WFLA 2009). Some scholars point to ZANU-PF loyalty and patronage as having dominated land allocations, to the extent that mainly chiefs, civil servants and war veterans benefited (see Zamchiya 2011 in this collection) in parts of some districts, as observed in Chipinge. Yet other studies find that a more broadly based class and social process defined land allocations – and in the nature of beneficiaries – despite the ruling party's overall dominance. (Moyo 2011a, Scoones et al. 2010). This suggests continuing differences over aspects of the land reform but it emphasises the importance of thorough empirical research.

Nonetheless, class struggles over access to land persist since exclusion, tenure insecurities and competing claims over some land and natural resources persist (see Moyo 2011a). The more politically influential and wealthier use administrative fiat, ethno-regional sentiments and at times force to expand their landholdings. Poorer landholders compete mostly through lower intensity confrontations among themselves. Some who did not benefit occupy land and poach resources 'illegally'. Many local associations seek to defend their land allocations using kinship ties and other sources of solidarity (see Masuko 2008). Local state authorities, including the bureaucracy and traditional leadership, and ZANU-PF structures are central to mediating the social legitimacy of the land redistribution and adjudicating land conflicts.

The state faces pressure to expand agrarian support and market protection from various middle class and capitalist lobbies in farming, agro-industry and trade, including some formed specifically to access subsidies. Over 70 percent of the local farmers' associations aggregate their resources to tap public extension and inputs support and to negotiate markets (Moyo *et al.* 2009, see Murisa 2010), replicating pre-2000 strategies (see Moyo 2002). But middle-scale and larger-scale farmers within the bureaucracy had greater policy influence and received more subsidies (see Moyo and Yeros 2005). Inter-class grievances emerged even among ZANU-PF supporters, as experienced over farm machinery subsidies. Recognising the input shortages facing peasants, the Inclusive Government slightly increased subsidies to them (MoF 2011), while a ZANU-PF presidential 'well-wishers' fund also supplied substantial inputs in 2010.

Contract farming relations also fomented the establishment of local farming associations, reinforcing the orchestration of farmer organisation by capital, non-governmental organisations (NGOs), and the bureaucracy seeking access to scattered peasants, as occurred during the 1980s. NGOs that had extensively mobilised small farmer cooperation for rural development retreated during the 2000s (Helliker 2008). Their discourses shifted towards governance and rule of law issues, focusing on the displacement of white farmers and farm workers. Some NGOs however defended land beneficiaries facing eviction (e.g. Zimbabwe Lawyers for Human Rights). Few NGOs advocate the redistribution of remaining large landholdings (Moyo 2011b). But donors and NGOs are consolidating their market-oriented aid schemes for input support for only Communal Areas in collaboration with older farmers unions (see FAO 2011).

New agricultural commodity associations, mostly representing middle-scale farmers, have emerged (Moyo 2011b). For instance, the Sugar Cane Farmers (outgrowers) Association in Chiredzi estates clashed with former white landowner outgrowers and estate managers over services and prices. They protested against the European Union-funded National Sugar Adaptation Strategy, which only finances holders of 'uncontested' (i.e. not redistributed) land.

Agrarian politics are also being re-shaped by the changing local administrative and political power relations that resulted from replacing white farmers' control over land, territory and labour. Local influence is now more broadly diffused but the landless are the most vulnerable. Territorial reconfiguration has enabled freer flows of people, goods and services leading to labour mobility, petty trading, gold panning and natural resource extraction. Their regulation is beyond the reach of under-resourced local administrative structures.[7] Local power struggles mainly involve lineage-clan leaders, traditional leaders, farmers and social associations, and local bureaucrats, as the powers that large landowners and war veteran leaders of the land occupations had have been replaced. Sparse local government authorities are ill-equipped to regulate the expanded land administration regime and ubiquitous natural resource and mineral extraction. Hereditary chiefs demand more powers to fill these regulation gaps.[8]

Despite this reconfiguration of agrarian politics some analysts fall back on past discourses shaped by political party polarisation, casting a shadow over progressive politics on egalitarian agrarian reforms. Fortunately most scholars and activists have been coming together to examine the future requirements of agrarian reform, including how to further democratise the land administration system, finance the supply and equitable distribution of inputs and create markets favourable to small producers.

Conclusion

Land reform has restructured Zimbabwe's agrarian relations by reconstituting the agrarian structure, mainly through expanding the numbers of small and middle-scale agricultural producers and reconfiguring the underlying labour relations. The racial and private character of land ownership was largely reversed. While access to land was broadened, large-scale farm holdings and plantations persist with disproportionately more land than is warranted, and this sustains continued landlessness. Agroindustrial estates were marginally restructured by introducing more small-scale outgrowers and through the expansion of public estate production, increasingly in partnership with foreign capital. Pressures to commodify land persist as various elite classes demand the reinstitution of freehold tenure and pursue informal land rentals. While agrarian labour relations are now more dominated by self-employment in diverse farming and non-farm activities, and part-time wage-labour is more common, prevailing wages and incomes remain repressed by low productivity, exploitative commodity markets and slow recovery of production in other economic sectors.

Heterodox interventionist policies introduced from 2002 initially shaped changes in the agrarian markets, including uneven access to subsidised inputs, shortages of goods and hyperinflation. Overall agricultural output had declined, but this began to rise slowly and selectively from 2006. A broader range of producers invested in cultivating much more land mainly for food, although land productivity remained lower than is potentially achievable due to limited access to farming inputs in markets, despite the subsidies. Many more producers are now involved in the production of diverse exports than was the case before 2000, suggesting a degree of income redistribution. However export commodity producers have the greatest access to farming inputs and commodity markets, and such access generally favours larger landholders who can also hire more labour and accumulate more assets.

---

[7]The state has tried but failed to stop popular exploitation of such resources using environmental and mining regulations.

[8]See statement by Chief Charumbira at a COPAC meeting held in Harare, on 4 January 2010.

The supply of farming inputs and credit and the buying of commodities by established agro-industrial conglomerates, merchants, and banks, which had been substantially reduced by 2007, was generally propped up by state subsidies. Input supplies were revamped by the re-liberalisation of markets in 2009. Agricultural production is increasingly being organised through contract farming, focusing on export commodities. Chinese contractors and loans negotiated with the state fuelled the revival and reconfiguration of agrarian markets and lured back capital from the West.

This restructuring of Zimbabwe's agrarian relations has the potential to deepen the autonomy of the peasantry and intensify productivity towards increasing supplies of more nutritious foods and raw materials for the home market. But these gains are limited by the increased export orientation and ascendance of agribusiness over popular markets, limited fiscal capacity and the statist defences against sanctions. While class struggles over land, agrarian markets and labour now entail new and more broadly based forms of smaller producer organisations and protests, these are pitted against an emerging alliance of larger-scale farm producers and foreign monopoly capital linked to expanding multi-racial domestic capital, whose membership traverses the political party lines.

Progressive agrarian change should consolidate the trajectory of smallholder development by securing peasants' control of land and protecting their markets to enhance popular food sovereignty, productivity and cash-earnings from self-employment in farming and non-farm industry. This requires stronger producer and consumer cooperation against agri-business and contractors (from the West, East and South) that link input and credit supply to crop purchases at depressed prices and against trade policies which enhance the dumping of subsidised foods. Expanding the democratic space and regaining policy autonomy towards articulated national development will be critical to the agrarian politics of advancing the peasantry and industrial diversification. Redirecting the expanding rents from mining to support this agenda will be crucial. As elsewhere, agrarian change under contemporary imperialism is neither linear nor even.

References

AIAS 2007. Inter-district household and whole farm survey database. AIAS baseline survey. Harare: African Institute for Agrarian Studies (AIAS).

Amanor-Wilks, D.1995. *In search of hope for Zimbabwe's farm workers. Dateline Southern Africa/Panos Institute.* Harare: AT&T Systemedia.

Amin, S. 1972. *Neocolonialism in West Africa.* Harmondsworth, UK: Penguin.

Andersson, J.A. 2007. How much did property rights matter? Understanding food insecurity in Zimbabwe: a critique of Richardson. *African Affairs,* 106(425), 681–690.

Arrighi, G. 1973. The political economy of Rhodesia. *In*: G. Arrighi and J.S. Saul, eds. *Essays in the Political Economy of Africa.* New York and London: Monthly Review Press, pp. 336–377.

Bochwey, K., P. Collier., J.W. Gunning and K. Hamada. 1998. External evaluation of ESAF: Report by a group of independent experts. Part IV: country profile: Zimbabwe. International Monetary Fund. Washington DC.

Buka 2002. A preliminary audit report of Land Reform Programme.Unpublished report, Government of Zimbabwe.

Bush, R. and L. Cliffe. 1984. Agrarian policy in labour migrant societies: reform or transformation in Zimbabwe? *Review of African Political Economy,* 29, 77–94.

Chambati, W. 2011. Restructuring of agrarian labour relations after Fast Track Land Reform in Zimbabwe. *Journal of Peasant Studies, this collection.*

Chambati, W. and G. Magaramombe. 2008. The abandoned question: farm workers. *In*: S. Moyo, K. Helliker and T. Murisa, eds. *Contested terrain: land reform and civil society in contemporary Zimbabwe*. Pietermaritzburg: S&S publishers, pp. 207–238.

Chang, H. 2009. How to 'do' a developmental state: political, organisational and human resource requirements for the developmental state. *In*: O. Edigheji, ed. *Constructing a democratic developmental state in South Africa: potential and challenges*. Cape Town: HSRC Press, pp. 82–96.

Cliffe, L. 2000. The politics of Land Reform in Zimbabwe. *In*: T.A.S. Bowyer-Bower and C. Stoneman, eds. *Land reform in Zimbabwe: constraints and prospects*. Aldershot, UK: Ashgate, pp. 3–46.

Cotton Ginners Association 2010. Cotton pricing 2010. May 2010.

Cousins, B. and I. Scoones. 2010. Contested paradigms of 'viability' in redistributive land reform: perspectives from southern Africa. *Journal of Peasant Studies*, 37(1), 31–66.

Cusworth, J. and J. Walker. 1988. Land resettlement in Zimbabwe: a preliminary evaluation. London; ODA Evaluation report EV434, Overseas Development Administration.

CZI 2010. Back to basics: post-dollarisation manufacturing recovery. Confederation of Zimbabwe Industries (CZI) manufacturing sector survey, 2010.

Edinger, H. and C. Burke. 2008. AERC scoping studies on China-Africa Relations: a research report on Zimbabwe. Centre for Chinese Studies, University of Stellenbosch, South Africa. March.

EU 2009. Annual action plan 2009 Zimbabwe. National sugar compact. Draft final report. EC sugar facility. March 2009.

FAO 2009. Special report: FAO/WFP crop and food security assessment mission to Zimbabwe, 22 June 2009. Rome: Food and Agriculture Organization of the United Nations; Rome: World Food Programme.

FAO 2011. Conservation Agriculture (CA). *Agriculture Coordination Working Group Journal*, 61, 1–6.

GAIN Report 2010. Zimbabwe: sugar annual. Global Agricultural Information Network (GAIN). USDA Foreign Agriculture Service.

Government of Zimbabwe (GoZ) 2001. Millennium Economic Recovery Programme (MERP).

Government of Zimbabwe (GoZ) 2003. National Economic Revival Programme: Measures to address the current challenges, Ministry of Finance and Economic Development, February 2003.

Government of Zimbabwe (GoZ) 2011. Indigenisation and economic empowerment act [Act 14/2007].

Global Political Agreement (GPA) 2008. Agreement between the Zimbabwe African National Union-Patriotic Front (ZANU-PF) and the two Movement for Democratic Change (MDC) formations, on resolving the challenges facing Zimbabwe. September 2008.

Habib, A. 2011. Foreword: reflections on the prerequisites for a sustainable reconstruction in Zimbabwe. *In*: H. Besida, ed. *Zimbabwe: Picking Up The Pieces*. New York: Palgrave Macmillan, pp. xi–xv.

Hammar, A., J. McGregor and L. Landau. 2010. Introduction. displacing Zimbabwe: crisis and construction in Southern Africa. *Journal of Southern African Studies*, 36(2), 263–283.

Helliker, K. 2008. Dancing on the same spot: NGOs. *In*: S. Moyo, K. Helliker and T. Murisa, eds. *Contested terrain: land reform and civil society in contemporary Zimbabwe*. Pietermaritzburg: S&S publishers, pp. 239–274.

IMF 2005. Zimbabwe: selected issues and statistical appendix. IMF country report No. 05/359. October 2005. International Monetary Fund. Washington, DC.

JAG/RAU 2008. Land, retribution and elections: post-election violence on Zimbabwe's remaining farms 2008. A report prepared by the Justice for Agriculture (JAG) Trust and the Research and Advocacy Unit. May 2008.

Makura-Paradza, G.G. 2010. Single women, land and livelihood vulnerability in a communal area in Zimbabwe. AWLAE Series no. 9. Wageningen: Wageningen Academic.

Masuko, L. 2008. War veterans and the re-emergence of housing cooperatives. *In*: S. Moyo, K. Helliker and T. Murisa, eds. *Contested terrain: land reform and civil society in contemporary Zimbabwe*. Pietermaritzburg: S&S publishers, pp. 181–206.

Matshe, I. 2004. The overall macroeconomic environment and agrarian reforms. African Institute for Agrarian Studies. Mimeo.

Ministry of Agricultural Engineering, Mechanization and Irrigation (MAEMI) 2009. Agricultural engineering mechanization and irrigation strategy framework: 2008–2058. Draft.

Ministry of Agriculture, Mechanisation and Irrigation Development (MAMID) 2010a. Second round crop and livestock assessment report. 9 April 2010. Government of Zimbabwe. Harare: Government Printers.

Ministry of Agriculture, Mechanisation and Irrigation Development (MAMID) 2010b. Agricultural statistical bulletin, 2010. Government of Zimbabwe. Harare: Government Printers.

Ministry of Agriculture, Mechanisation and Irrigation Development (MAMID) 2011. Second round crop and livestock assessment report. 14 April 2011. Government of Zimbabwe. Harare: Government Printers.

Mkandawire, T. 2010. On Tax Efforts and Colonial Heritage in Africa. *Journal of Development Studies* 46, 1647–1669.

Mkodzongi, G. 2011. Land occupations and the quest for livelihoods in Mhondoro Ngezi Area, Zimbabwe. Mimeo.

MLLR 2009. Memorandum to cabinet by the Minister of Lands and Rural Resettlement Hon. H. M. Murerwa (MP) on the update on land reform and resettlement programme, Ministry Of Lands And Rural Resettlement, September 2009.

MoF 2010. Ministry of Finance budget speech 2010. Ministry of Finance of Zimbabwe. Available from: http://www.zimtreasury.org/downloads/834.pdf. [Accessed 22 December 2010].

MoF 2011. The 2011 mid-year fiscal policy review: 'Riding the storm: economics in the time of challenges'. Presented to parliament by Hon T Biti, MP, Minister of Finance. 26 July 2011.

Moyo, S. 1995. The land question in Zimbabwe. Harare: SAPES Books.

Moyo, S. 2000. Land reform under structural adjustment in Zimbabwe; land use change in Mashonaland Provinces. Uppsala: Nordiska Afrika Institutet.

Moyo, S. 2001. The land occupation movement and democratization in Zimbabwe: contradictions of neo-liberalism. *Millennium Journal of International Studies*, 30(2), 311–330.

Moyo, S. 2002. Peasant organisations and rural civil society in Africa: an introduction. *In*: Moyo S. and B. Romdhane, eds. *Peasant organisations and democratisation in Africa*. Dakar: Codesria Book Series, pp. 1–26.

Moyo, S. 2005a. Land policy, poverty reduction and public Action in Zimbabwe. *In*: A. Haroon Akram-Lodhi, S.M. Borras Jr and C. Kay, eds. *Land, poverty and livelihoods in an era of globalisation: perspective from developing and transition countries. Routledge ISS studies in rural livelihoods*. London and New York: Routledge, pp. 344–382.

Moyo, S. 2010. The agrarian question and the developmental state in Southern Africa. *In*: O. Edigheji, ed. *Constructing a democratic developmental state in South Africa: potential and challenges*. Cape Town: HSRC Press, pp. 285–314.

Moyo, S. 2011a. Three decades of agrarian reform in Zimbabwe. *Journal of Peasant Studies*, 38(3), 493–531.

Moyo, S. 2011b. Land concentration and accumulation after redistributive reform in post-settler Zimbabwe. *Review of African Political Economy*, 38(128), 257–276.

Moyo, S. 2011c. Agrarian reform and prospects for recovery. In H. Baseda, ed. *Zimbabwe: picking Up the pieces*. Palgrave MacMillan, pp. 129–155.

Moyo, S. and P. Yeros. 2005. Land occupations and land reform in Zimbabwe: towards the National Democratic Revolution. *In*: Sam Moyo and Paris Yeros, eds. *Reclaiming the land: the resurgence of rural movements in Africa, Asia and Latin America*. London: Zed Books, pp. 1–64.

Moyo, S. and P. Yeros. 2007. The radicalised state: Zimbabwe's interrupted revolution. *Review of African Political Economy* 111: 103–121.

Moyo, S., W. Chambati., T. Murisa., D. Siziba., C. Dangwa., K. Mujeyi and N. Nyoni. 2009. *Fast Track Land Reform baseline survey in Zimbabwe: trends and tendencies, 2005/06*. African Institute for Agrarian Studies (AIAS). Harare.

Murisa, T. 2010. Emerging forms of social organisation and agency in the newly resettled areas of Zimbabwe: the cases of Goromonzi and Zvimba districts. Thesis (PhD). Rhodes University.

News24, 2003. Zimbabwe food price showdown. 18 July. New24 (South Africa). http://www.news24.com/xArchive/Archive/Zim-food-price-showdown-20030718 [Accessed 20 December 2010].

O'Laughlin, B. 2002. Proletarianisaion, agency and changing rural livelihoods: forced labour and resistance in colonial Mozambique. *Journal of Southern African Studies*, 28(3), 511–530.

Patnaik, U. 2003. Global capitalism, deflation and agrarian crisis in developing countries. UNRISD Social Policy Paper Series 2003.

Raftopoulos B. and A. Mlambo, eds. 2009. *A history from the pre-colonial period to 2008: becoming Zimbabwe*. Weaver Press: Harare, pp. 201–232.

RBZ 2005. Monetary policy statement. Issued in terms of the RBZ Act Chapter 22:15 Section 46. by Dr G. Gono. Governor of the Reserve Bank of Zimbabwe. 20 July.

RBZ 2006. Sunrise of currency reform. Monetary policy review statement for first half of 2006. Reserve Bank of Zimbabwe. Monday 31 July 2006.

RBZ 2007a. The launch of phase 1 of agricultural equipment acquired under the RBZ farm mechanisation programme, 11 June 2007. Reserve Bank of Zimbabwe.

RBZ 2007b. The impact of sanctions against Zimbabwe. A supplement to the mid-term monetary policy review statement. October 2007. Reserve Bank of Zimbabwe.

RBZ 2009. Exchange Control Directive Rk:39 issued in terms of Section 35 (1) of The Exchange Control Regulations Statutory Instrument 109 of 1996. 23 February 2009.

RBZ 2011. Monetary policy statement, January 2011. Reserve Bank of Zimbabwe.

Richardson, C. 2005. The loss of property rights and the collapse of Zimbabwe. *CATO Journal 25, no (Fall 2005)*. Available online at HYPERLINK "http://www.cato.org/pubs/journal/cj25n3/cj25n3-12.pdf" www.cato.org/pubs/journal/cj25n3/cj25n3-12.pdf Accessed September 2010.

Robertson, J. 2011. A macroeconomic policy framework for economic stabilisation in Zimbabwe. *In*: Hany Besida, ed. *Zimbabwe: picking up the pieces*. New York: Palgrave Macmillan, pp. 83–105.

Rukuni, M., J. Nyoni and P. Matondi. 2009. *Policy options for optimisation of the use of land for agricultural productivity in Zimbabwe*. Harare: World Bank.

Rutherford, B. 2003. Belonging to the farm(er): farm workers, farmers, and the shifting politics of citizenship. *In*: A. Hammar, B. Raftopoulos and S. Jensen, eds. *Zimbabwe's unfinished business: rethinking land, state and nation in the context of crisis*. Harare: Weaver Press, pp. 191–241.

Sachikonye, L. 2003. *The situation of commercial farmers after land reform in Zimbabwe, report prepared for the Farm Community Trust of Zimbabwe*.

Sadomba, W. 2008. War veterans in Zimbabwe: complexities of a liberation movement in an African post-colonial settler society. PhD thesis. Wageningin University, Netherlands.

Scoones, I., N. Marongwe., B. Mavedzenge., F. Murimbarimba., J. Mahenehene and C. Sukume. 2010. *Zimbabwe's land reform: myths and realities*. Harare: Weaver Press.

Sukume, C. and S. Moyo. 2003. *Farm sizes, decongestion and land use: implications of the Fast Track Land Redistribution Programme in Zimbabwe*. Harare: AIAS Mimeo.

The Herald 2011a. Zimbabwe announces new wildlife based land reform programme. 9 March 2011. Zimpapers.

The Herald 2011b. Zimbabwe: sugarcane farmers, Tongaat Hulett clash. 23 March 2011. Zimpapers.

The Herald 2011c. Zimbabwe: Agriculture Stakeholders Hit Out at Commodity Exchange. 2 February 2011. Zimpapers.

The Standard 2009. Government bungling stifles contract farming, 23 September.

TIMB 2010. End of 2010 tobacco marketing season final report. Tobacco Industry Marketing Board, October 2010.

Tripathy, D., C. Sukume, J. Nyoni and T. Dube. 2007. *Grain pricing and Marketing in Zimbabwe: strategic options*. Report initiated and submitted to the Ministry of Agriculture.

UNCTAD 2008. *UNCTAD Handbook of statistics 2008*. New York and Geneva: United Nations.

USAID 2010. *Zimbabwe agricultural sector market study*. Weidemann Associates, Inc.

Utete 2003. Report of the presidential land review committee, under the chairmanship of Dr Charles M.B. Utete, Vol 1 and 2: Main report to his Excellency The President of The Republic of Zimbabwe, Presidential Land Review Committee (PLRC). Harare.

Weiner, D. 1988. Land and agricultural development. *In*: Stoneman, ed. *Zimbabwe's prospects: issues of race class and capital in Southern Africa*. London: Mcmillan, pp 63–89.

WFLA 2009. The socio-economic and political impact of the Fast Track Land Reform programme on Zimbabwean women. A research extract.

Wiggins, S. 2005. Southern Africa's food and humanitarian crisis of 2001–04: Causes and lessons. A discussion paper for Agricultural Economic Society Annual Conference, Nottingham, UK, 4–6 April.

WLZ 2007. A rapid appraisal in select provinces to determine the extent of women's allocation to land during the land reform and resettlement programme, 2002–2005. Unpublished paper, Women and Land in Zimbabwe (WLZ), Harare.

World Bank 1991. *Zimbabwe: agriculture sector memorandum: Vol 1 and Vol 2. No. 9429-Zim*. Washington, DC.

World Bank 2006. *Agricultural growth and lLand reform in Zimbabwe: assessment and recovery options, Report No. 31699-ZW*. Harare, World Bank.

Yeros, P. 2002. The political economy of civilisation: peasant-workers in Zimbabwe and the neo-colonial world. Thesis (PhD). University of London.

Sam Moyo is Executive Director of the African Institute for Agrarian Studies (AIAS) in Harare, Zimbabwe, and is the current President of the Council for the Development of Social Research in Africa (CODESRIA). His publications include: *African land questions, agrarian transitions and the state: contradictions of neoliberal land reforms* (2008); *Land reform under structural adjustment in Zimbabwe: land use change in the Mashonaland Provinces* (2000); *The land question in Zimbabwe* (1995); [co-edited with Paris Yeros]: *Reclaiming the land: the resurgence of rural movements in Africa, Asia and Latin America* (2005); [co-edited with Kojo Sebastian Amanor]: *Land and sustainable development in Africa* (2008); [co-edited with Paris Yeros]: *Reclaiming the nation: the return of the national question in Africa, Asia and Africa* (2011).

Zimbabwe's land reform: challenging the myths

Ian Scoones, Nelson Marongwe, Blasio Mavedzenge, Felix Murimbarimba, Jacob Mahenehene and Chrispen Sukume

Most commentary on Zimbabwe's land reform insists that agricultural production has almost totally collapsed, that food insecurity is rife, that rural economies are in precipitous decline, that political 'cronies' have taken over the land and that farm labour has all been displaced. This paper however argues that the story is not simply one of collapse and catastrophe; it is much more nuanced and complex, with successes as well as failures. The paper provides a summary of some of the key findings from a ten-year study in Masvingo province and the book *Zimbabwe's Land Reform: Myths and Realities*. The paper documents the nature of the radical transformation of agrarian structure that has occurred both nationally and within the province, and the implications for agricultural production and livelihoods. A discussion of who got the land shows the diversity of new settlers, many of whom have invested substantially in their new farms. An emergent group 'middle farmers' is identified, who are producing, investing and accumulating. This has important implications – both economically and politically – for the future, as the final section on policy challenges discusses.

Introduction

Zimbabwe's land reform has had a bad press. Images of chaos, destruction and violence have dominated the coverage. While these have been part of the reality, there is also another side of the story. There have been important successes which must be taken into account if a more complete picture is to be offered. This paper argues that the story is not simply one of collapse and catastrophe; it is much more nuanced and complex, with successes as well as failures.

As Zimbabwe moves forward with a new agrarian structure, a more balanced appraisal is needed. This requires solid, on-the-ground research aimed at finding out

The work on which this paper is based was carried out as part of the Livelihoods after Land Reform in southern Africa (www.lalr.org.za) which was coordinated by Professor Ben Cousins at the Institute for Poverty, Land and Agrarian Studies at the University of the Western Cape and funded by the UK Economic and Social Research Council and Department for International Development ESRC/DFID Joint Scheme for Research on International Development (Poverty Alleviation) (Grant No. RES-167-25-0037). We would like to thank the many people across Masvingo province, particularly farmers in our study sites, who have helped with this research over the years. We would also like to thank three anonymous reviewers, as well as the Special Issue editors, for their insightful comments on an earlier version. For further information see www.zimbabweland.net

what happened to whom and where and with what consequences. This was the aim of work carried out in Masvingo province since 2000 and reported in the book, *Zimbabwe's Land Reform: Myths and Realities* (Scoones *et al.* 2010). This paper offers an overview of the main findings.[1] The question posed for the research was simple: what happened to people's livelihoods once they got land through land reform from 2000? Yet, despite the simplicity of the question, the answers are extremely complex.

The research involved in-depth field research in 16 land reform sites; in some sites over a decade from 2000. The research sites were located in four research 'clusters' across the province, involving a sample population of 400 households. Masvingo is a relatively dry province, with average annual rainfall ranging from around 1000mm to under 300mm. Former land uses of the research sites included livestock ranches (sometimes with limited irrigation plots), with low capitalisation and limited labour forces, as well as, for several A2 sites, irrigated farms, including those linked to sugar estates in the lowveld. Different types of survey were conducted in these sites between 2000 and 2010, including a full census (N = 400 in 2007–08), a stratified sample survey (N = 277 in 2008, stratified according to cattle ownership, a key indicator of wealth) and individual household biographies (N = 110, across all 'success groups'), as well as more qualitative observations in all sites. Records of crop production, sales and inputs were carried out each season, and livestock and other asset holdings were monitored, in all cases through recall interviews. A detailed study of investment took place in 2007–08, involving assessments for all households. The research team included a group of field researchers resident in Masvingo province – in Masvingo, Hippo Valley and Chikombedzi. The study area stretched from the relatively higher potential areas near Gutu to the sugar estate of Hippo Valley to the dry south in the lowveld, offering a picture of diverse agro-ecological conditions (Figure 1). What we found was not what we expected. It contradicted the overwhelmingly negative images of land reform presented in the media, and indeed in much academic and policy commentary besides. In sum, the realities on the ground did not match the myths so often perpetuated in wider debate.

Most commentary on Zimbabwe's land reform insists that agricultural production has almost totally collapsed, that food insecurity is rife, that rural economies are in precipitous decline, that political 'cronies' have taken over the land and that farm labour has all been displaced. The reality however is much more complex. In our research we needed to ask far more sophisticated questions: Which aspects of agricultural production have suffered? Who is food insecure? How are rural economies restructuring to the new agrarian setting? And who exactly are the new farmers and farm labourers? By countering one myth of disaster and catastrophe, we must of course be wary of setting up an opposite myth of rose-tinted optimism. Instead, we should offer an empirically-informed view based on detailed analysis and solid evidence. This has been the aim of our research, and we offer a taste of the findings below.

Of course a focus on Masvingo province gives only a partial insight into the broader national picture. With most land being previously extensive ranch land, with pockets of irrigated agriculture outside the sugar estates, it is clearly different to the

---

[1]The paper draws from the book, as well as a series of feature articles prepared for the Zimbabwean newspaper (see http://www.ids.ac.uk/go/news/zimbabwe-s-land-reform-ten-years-on-new-study-dispels-the-myths). See www.zimbabweland.net

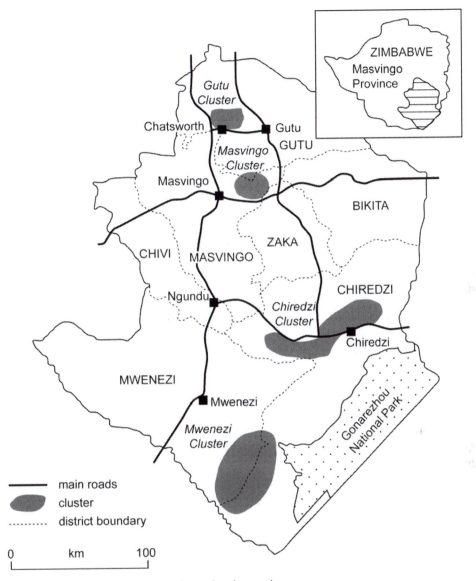

Figure 1.   Map of Masvingo province, showing study areas.

Highveld around Harare, where highly capitalised agriculture reliant on export markets did indeed collapse and where labour was displaced in large numbers (Sachikonye 2003). But the picture in the new farms of Masvingo is not unrepresentative of broad swathes of the rest of the country, as research across multiple districts, from Mazoe to Mutoko to Mangwe is now showing (Moyo *et al.* 2009, Matondi 2010; plus contributions to this collection from across the country, see Cliffe *et al.* for an overview).[2] Smallholder farmers have dominated the

---

[2]The papers produced as part of the Livelihoods after Land Reform small grants competition show an extraordinary range, but again some important common themes (http:// www.lalr.org.za/zimbabwe/zimbabwe-working-papers-1).

allocations under the 'Fast Track Land Reform Programme' (FTLRP), and many are doing well; surprisingly so given the parlous economic conditions over much of the past decade.

But any good analysis must take a differentiated view, unpacking diverse trajectories; successes as well as failures. Contributions to this collection demonstrate this diversity, much of which is encompassed in the Masvingo study. For example Zamchiya (this collection) shows how some high-value sites were captured by elites in Chipinge, while in peri-urban Goromonzi, Marongwe (2009, this collection) shows how allocation of A2 farms was manipulated by political connections. These were all patterns observed in Masvingo province, as shown below. But individual cases should not detract from a balanced overall assessment, and only a broader overview can provide a firm basis for thinking about future policy. Subsequent sections of this paper offer such an overview of the key findings from the Masvingo study, starting with a summary of the broader national changes before focusing on the results from Masvingo province.

A radical change in agrarian structure

Across the country, the formal land re-allocation since 2000 has resulted in the transfer of land to nearly 170,000 households by 2010 (Moyo 2011a, 496). If the 'informal' settlements outside the official 'fast-track' programme are added, the totals are even larger.

Events since 2000 have thus resulted in a radical change in the nation's agrarian structure (Table 1). At Independence in 1980, over 15 million hectares were devoted to large-scale commercial farming, comprising around 6000 farmers, nearly all of them white. This fell to around 12 million hectares by 1999, in part through a modest, but in many ways successful, land reform and resettlement programme, largely funded by the British government under the terms of the Lancaster House agreement (Gunning *et al.* 2000).

The Fast Track Land Reform Programme, begun in 2000, allocated to new farmers over 4500 farms making up 7.6 million hectares, 20 percent of the total land area of the country, according to (admittedly rough) official figures. In 2008–09 this

Table 1.   Changes in the national distribution of land, 1980–2009.

Land category	1980 Area (million ha)	2000 Area (million ha)	2009 Area (million ha)
Communal areas	16.4	16.4	16.4
Old resettlement	0.0	3.5	3.5
New resettlement: A1	0.0	0.0	4.1
New resettlement: A2	0.0	0.0	3.5
Small-scale commercial farms	1.4	1.4	1.4
Large-scale commercial farms	15.5	11.7	3.4*
State farms	0.5	0.7	0.7
Urban land	0.2	0.3	0.3
National parks and forest land	5.1	5.1	5.1
Unallocated land	0.0	0.0	0.7

Source: derived from various government sources and compiled by the African Institute of Agrarian Studies (Scoones *et al.* 2010, 4). *Note: includes all large commercial farms, agro-industrial estate farms, church/trust farms, BIPPA farms and conservancies.

represented over 145,000 farm households in A1 schemes and around 16,500 further households occupying A2 plots (Rukuni *et al.* 2009)[3], equivalent to 11.7 percent of farm households in the country.[4]

Overall, there has been a significant shift to many more, smaller-scale farms focusing on mixed farming, often with low levels of capitalisation. Much of this expansion of agricultural activity is on land which was previously under-utilised (Moyo 1998), and in the Masvingo case often involving a transfer of land which was used for ranching at low stocking rates with limited herding labour to more intensive smallholder crop and livestock production. But this is not to say that large-scale commercial units no longer exist. Especially important in Masvingo province is the estate sector, including for example the major sugar estates in the lowveld. These largely remained intact following land reform, with out-grower areas being transferred to sub-divided A2 plots. Today, there are still around 3.4 million hectares under large-scale farming, some of it in very large holdings, such as the 350,000 hectare Nuanetsi ranch in Masvingo province.[5] There are, however, perhaps only 200 white-owned commercial farmers still operating across 117,000 ha nationally, complemented by 950 black-owned large scale farms on 530,000 ha (Moyo 2011a, 514). Most white-owned farms were taken over, with a substantial number of farm workers displaced.

Two main 'models' have been at the centre of the land reform process since 2000 – one focused on smallholder production (so-called A1 schemes, either as villagised arrangements or small, self-contained farms) and one focused on commercial production at a slightly larger scale (so-called A2 farms). Much larger A2 farms, replicating the large-scale farms of the past, have also been created, many later in the land reform process (Moyo 2011a, 2011b).

In practice, the distinction between these models varies considerably, and there is much overlap, with some self-contained A1 schemes, for example, being very similar to smaller A2 schemes. As Zamchiya's study (this collection) shows, processes of land allocation, rather than their administrative definition, have more importance in understanding who ended up on the land and what happened next. Most A1 schemes, and all 'informal' land reform sites, were allocated following land invasions starting from 2000.[6] These had diverse origins, but were usually (but not always) led by war veterans and involved groups of people from surrounding communal areas and nearby towns (Chaumba *et al.* 2003, Moyo 2001). More formal allocation of

---

[3]Total land allocations have continued to increase. Moyo (2011a, 497) claims that total land allocations amount to nearly 9 million ha by 2010, as more larger A2 farms have been allocated.

[4]Nationally, the agrarian structure now includes smallholder farms (made up of communal area households (82.1 percent), old resettlement households (5.4 percent) and A1 households (10.5 percent) medium-scale farms (made up of old small-scale commercial farms (0.6 percent) and A2 farms (1.1 percent)) and large-scale farms, conservancies and estates (0.4 percent). In other words, today 98 percent of all farms can be classified as smallholdings (Scoones *et al.* 2010, 6).

[5]Moyo (2011a, 514–7; 2011b, 262) divides this between large scale farms, including white and black owned farms and new A2 farms (1.5 million ha, including about 500,000ha allocated to 200 very large scale A2 farms) and estates/conservancies/institutions (1.2 million ha). However, in the absence of a detailed land audit these aggregate figures are prone to error, and remain estimates.

[6]Exceptions exist, such as the case studied by Zamchiya (this collection) which involved A1 self-contained schemes allocated by the state and not through land occupations, showing more similarity with the A2 pattern elsewhere.

plots happened later, with the pegging of fields and settlement sites as part of the 'fast track land reform programme' (FTLRP) and the issuing of 'offer letters'.[7]

Depending on the pressure on the land, the local demands and often the discretion of the planning officers, A1 sites were demarcated as villages (with shared grazing and clustered homesteads) or 'self-contained' plots, with houses, arable fields and grazing within a single area. The 'informal' A1 sites, by contrast, were usually organised in line with local preferences. These sites, with their origins in land invasions, took on a particular social and political character, organised initially by a 'Seven Member Committee', often with a war veteran base commander in the lead. Later these became village committees, and were incorporated into chiefly authorities and local government administrative systems (Scoones *et al.* 2010, 188–212). By contrast, A2 schemes were allocated later (from around 2002) as a result of business plan applications to the Provincial and District Land Committees.

The membership of each of these 'schemes' is thus highly dependent on the politics of the allocation process, with final outcomes highly contingent on local situations. The land invasions, while of course highly politicised, and supported by ZANU-PF (Zimbabwe African National Union – Patriotic Front) and the military, had diverse origins and participation: some were made up largely of poor, local villagers from nearby areas, while others involved a smaller group of organised war veterans (Hammar *et al.* 2003, Marongwe 2003). In our Masvingo cases, each site was different (Scoones *et al.* 2010, 46–51). A2 schemes by contrast required a formal application process, and officially there were strict criteria for acceptance. Many who applied were civil servants, often linked to the agriculture ministry, who had few strong political connections if any; although in some instances they were able to manipulate the administrative procedures in their favour[8]. The most obvious, and often blatant, corrupt practices linked to political patronage were associated with the later allocation of larger A2 farms, especially around the time of the 2008 elections when the struggle for power and the deployment of political patronage by the ZANU-PF elite was at its height. Moyo (2011a, 514) reports 53 large-scale A2 farms in Masvingo province across 110,719 hectares.

In sum, the land reform has resulted in a very different farming sector, with a radically reconfigured agrarian structure. While the old dualism has been disrupted, with many more smallholder farmers on the land, elements of the large-scale farming structure remain. Yet despite these major changes, the new setting, as we show below, is not without considerable entrepreneurial dynamism and productive potential. This major restructuring of course has had knock-on consequences for the agricultural sector as a whole. Any radical reform will of course have a transitional phase, as production systems, markets and trading priorities readjust (Kinsey 2003), but the key questions are how long will the transition take, and what form will the agriculture sector take in the longer term? In the period between 2001 and 2009, national production of wheat, tobacco, coffee and tea all declined, as did the export of beef (Scoones *et al.* 2010, 148). Compared to 1990s averages, wheat production decreased by 27 percent and tobacco production by 43 percent, with more dramatic declines from 2006 (Moyo 2011a), yet there was some recovery, particularly for tobacco, in 2010 and 2011 (Anseeuw *et al.* 2011). Equally, national maize production

---

[7]These are documents providing a permit to occupy the land, but no formal title or lease.
[8]Political manipulation of allocation processes was more common in areas close to towns (Marongwe, this collection) and where high value crops were at stake (Zamchiya, this collection).

has become more variable because the reduction of irrigation facilities and significant droughts have resulted in shortages, with average production over this period down by 31 percent from 1990s levels. However, other crops and markets have weathered the storm and some have boomed. Aggregate production of small grains (sorghum and millet) has exploded, increasing by 163 percent compared to 1990s averages, while cotton production has also increased, up 13 percent on average by 2009. The agricultural sector therefore has certainly been transformed, and there are major problems in certain areas, but it certainly has not collapsed.

Transforming land and livelihoods in Masvingo province

Aggregate figures – with all the necessary caveats about their accuracy – only tell one part of the story. To get a sense of what is happening in the fields and on the farms, we need a more local focus. Only with such insights can we really begin to understand the impacts of Zimbabwe's land reform. In this section, we zero in on Masvingo province in the central south and east of the country.

In Masvingo province about 28 percent of the total land area was transferred as part of the FTLRP, according to 2009 official figures. Much of this land was previously cattle ranches, with limited infrastructure, low levels of employment and only small patches of arable land outside the irrigated lowveld areas. This was taken over by over 32,500 households on A1 sites (making up 1.2 million hectares) and about 1,200 households in A2 areas (making up 371,500 ha), alongside perhaps a further 8,500 households in informal resettlement sites, as yet unrecognised by the government. Although there is much variation, the average size of new A2 farms is 318 hectares, while that of A1 family farms is 37 hectares, including crop and grazing land[9]. At the same time one million hectares (18.3 percent of the province) remains as large-scale commercial operations, including some very large farms, wildlife conservancies and estates in the lowveld that remained largely intact (Moyo 2011b).

Table 2 offers an overview of the socio-economic characteristics of the different sites in our study, presented in relation to the four districts and the four different types of resettlement 'scheme', highlighting the diversity of contexts, livelihood assets and strategies. Land holdings vary significantly between the 'scheme' types, with the A2 small-scale commercial units having the largest land areas. A1 self-contained scheme areas include both grazing and arable land, while A1 villagised and informal areas represent only the arable land. Maize production and sales vary significantly, with A1 self-contained schemes performing best and, with a few notable exceptions, A2 schemes performing poorly. Average crop production levels unsurprisingly decline between the relatively wetter sites of Gutu and Masvingo to the dry area of Mwenezi. Cattle ownership is highest in the A2 ranches of Mwenezi, although with stocking rates still low and cropped area minimal.[10] The overall profile of the new settlers is presented in the table, referring to the age and educational qualifications of 'household heads'. A2 farmers tend to be slightly older and better educated, but

---

[9]The acquisition of whole farms without subdivision to create 'large-scale A2' farms has occurred in recent years and has been characterised by political corruption. In Masvingo province, these farms average over 2000 ha.

[10]Stocking rates on the Mwenezi A2 ranches average 14.4 ha per animal, although herd sizes are building up. Recommended rates are around 10 ha per livestock unit for beef production in this dry area.

Table 2. A socio-economic profile of the study sites (average amounts across survey households).

Cluster	Gutu			Masvingo			Chiredzi		Mwenezi		
Scheme type	A1 self-contained	A1 villagised	A2	A1 self-contained	A1 villagised	A2	Informal	A2	A1 villagised	A1 informal	A2
Age of household head	39	34	43	36	40	43	37	46	n.d.	33	44
Educational level of household head	Form 2	Form 2	Form 3 or higher	Form 2	Form 2	Form 3 or higher	Grade 7	Form 3 or higher	Grade 7	Grade 7	Form 2
Land holding (ha)	35.5	4.0	232.1	33.0	3.9	167.0	6.2	39.1	7.9	8.0	868.7
Area cropped (ha)	5.6	3.1	6.6	8.4	3.4	n.d.	4.6	16.8	3.6	4.0	0.5
Cattle owned (nos)	6.9	5.4	25.2	11.9	4.4	11.7	4.4	14.8	4.7	4.9	60.3
Maize output in 2006 (kg)	2790	2627	3133	7385	3140	6500)	2256	2582	449	104	n.d
Sales (GMB and local) in kg in 2006	1310	1157	896	5283	1766	54563	378	1357	104	0	n.d
% owning a scotch cart	32%	24%	20%	68%	63%	75%	40%	33%	55%	50%	64%
House type (% with tin/ asbestos roof)	43%	40%	25%	45%	44%	100%	42%	78%	96%	100%	100%
% receiving remittances	44%	39%	15%	n.d.	23%	0%	21%	28%	44%	52%	64%

Source: Survey data, 2007–08 (n.d. means no data available, or not applicable) (Scoones *et al.* 2010, 44).

overall, compared to nearby communal areas, this is a relatively well educated and young population.

This radical transformation of land and livelihoods has resulted in a new composition of people in the rural areas, with diverse livelihood strategies. In order to understand more about who was doing what we undertook a 'success ranking' exercise in all 16 sites across Masvingo province. This involved a group of farmers from the area ranking all households according to their own criteria of success. Four broad categories of livelihood strategy emerged from these investigations (following Dorward [2009] and Mushongah [2009]). These are listed in Table 3.

Over a half of all the 400 sample households – across A1, A2 and informal resettlement sites – were either 'stepping up' – accumulating assets and regularly producing crops for sale – or 'stepping out' – successfully diversifying off-farm. These households were accumulating and investing, often employing labour and ratcheting up their farming operations, despite the many difficulties being faced. But not everyone has been successful. 46.5 percent of households were finding the going tough, and were not regarded as 'successful' at this stage. Some were really struggling and only just 'hanging in'; others were in the process of 'dropping out', through a combination of chronic poverty and ill health. Joining the land invasions and establishing new farms in what was often uncleared bush, previously not used for arable agriculture, was not easy. It required commitment, courage and much hard work. It was not for everyone.

Others without start-up assets have been unable to accumulate, and have continued to live in poverty, reliant on the support of relatives and friends. Some have joined a growing labour force on the new farms, abandoning their plots in favour of often poorly-paid employment. Within the 'stepping out' category, some

Table 3. Livelihood strategies in Masvingo province.

Category	Livelihood strategy	Proportion of households
Dropping out (10.0%)	Exit – leaving the plot	4.4%
	Chronically poor, local labour	3.3%
	Ill health affecting farming	2.2%
Hanging in (33.6%)	Asset poor farming, local labour	17.8%
	Keeping the plot for the future	10.3%
	Straddling across resettlement and communal areas	5.6%
Stepping out (21.4%)	Survival diversification	2.8%
	Local off-farm activities plus farming	5.3%
	Remittances from within Zimbabwe plus farming	5.0%
	Remittances from outside Zimbabwe plus farming	4.4%
	Cell phone farmers	3.9%
Stepping up (35.0%)	*Hurudza* – the 'real' farmers	18.3%
	Part-time farmers	10.6%
	New (semi-)commercial farmers	4.7%
	Farming from patronage	1.4%

Source: summarised from Scoones *et al.* (2010, 228-9).

are surviving off illegal, unsafe or transient activities that allow survival but little else. Still others are straddling across two farms – one in the communal area and one in the new resettlement – and not really investing in the new areas, while some are simply keeping the plot for sons or other relatives.

It is not surprising that there have been such variable outcomes. In the period since 2000 there has been virtually no external support. Government was broke and focused support on the elite few, and the non-governmental organisations (NGOs) and donors have shied away from the new resettlement areas for political reasons. Instead, most new farmers have been reliant on their own connections, enterprise and labour. Without support to get going, many have found it difficult, and it has been those with a combination of access to assets, hard work and luck that have really made it.

Overall, in our study sites there is thus a core group of 'middle farmers' – around half of the population – who are successful not because of patronage support, but because of hard work. They can be classified as successful 'petty commodity producers' and 'worker peasants' who are gaining surpluses from farming, investing in the land from off-farm work and so are able to 'accumulate from below' (Scoones *et al.* 2010, 2012; cf. Cousins 2010). This is, as discussed below, having a positive impact on the wider economy, including stimulating demand for services, consumption goods and labour.

New land, new people

One of the most repeated myths about Zimbabwe's land reform is that all the land went to 'Mugabe's cronies'; those with access to elite connections and benefiting from political patronage. This did, of course, happen, and continues to do so. Tackling such extreme excesses of land grabbing through a land audit, as provided for in the 'Global Political Agreement' for power sharing, remains a major challenge. But elite capture is not the whole story of Zimbabwe's land reform; nor indeed the dominant one.

Who got the land and what is the profile of the new settlers? Our study of 400 households across the 16 sites from Masvingo province showed by far the majority of the new settlers are ordinary people (Table 4). While 'ordinary' is certainly a category that lacks clarity, these are essentially people who had little or very poor land in the communal areas or were unemployed or with very poorly-paid jobs and

Table 4. Settler percentage profiles across schemes.

	A1 villagised	A1 self-contained	Informal	A2	Total
'Ordinary': from other rural areas	59.9	39.2	69.7	12.2	49.9
'Ordinary': from urban areas	9.4	18.9	22.6	43.8	18.3
Civil servant	12.5	28.3	3.8	26.3	16.5
Security services	3.6	5.4	3.8	1.8	3.7
Business person	3.1	8.2	0	10.5	4.8
Former farm worker	11.5	0	0	5.3	6.7
N	*192*	*74*	*53*	*57*	*376*

Source: Census data, 2007 (N=376), including all sites (Scoones *et al.* 2010, 53).

living in town. About half of all new settler households were from nearby communal areas and another 18 percent were from urban areas. These people joined the invasions because they needed land, and thought that the new resettlements would provide new livelihood opportunities. As discussed further below, this was not a politically-organised grouping with strong connections to ZANU-PF. The remaining third of household heads was made up of civil servants (16.5 percent overall, but increasing to around a quarter of all settlers in A1 self-contained and A2 sites), business people (4.8 percent overall, but again proportionately higher in the A1 self-contained and A2 sites), security service personnel (3.7 percent overall, employed by the police, army and intelligence organisation) and former farm workers (6.7 percent overall).

Farm workers made up 11.5 percent of households in the A1 villagised sites, with many taking an active role in the land invasions. In one case a farm worker organised and led the invasion of the farm where he had worked. Given that in other parts of the country, farm workers were displaced in large numbers, often ending up destitute, living in camps on the farms (Chambati, this collection), this is perhaps surprising. Yet this reflects the extent and nature of labour on the former large-scale farms in Masvingo province. Unlike in the Highveld farms, where large, resident labour forces existed without nearby communal homes, our Masvingo study sites were formerly large-scale ranches where labour was limited, and workers came, often on a temporary basis, from nearby communal areas and were not permanent residents attracted originally as migrant labour from nearby countries.

Across all of these categories are 'war veterans'. As household heads they make up 8.8 percent of the total population. The category 'war veteran' is however diverse and again perhaps misleading. Prior to the land invasions, most were farming in the communal areas, a few were living in town, while some were civil servants, business people and employees in the security services. At the time of the land invasions in 2000, many indeed had long dropped their 'war veteran' identity and had been poor, small-scale farmers in the communal areas for 20 years since the end of the liberation war. Those who led the land invasions were often able to secure land in the A1 self-contained plots, but many were sidelined in the allocation of larger A2 farms. However, most were not well connected politically before 2000, although through the Zimbabwe National Liberation War Veterans Association, they became so and part of the political drive towards land reform, although with multiple disputes with the party leadership (Sadomba 2011).

Land was allocated unevenly to men and women. In most cases it is men whose names appear on the 'offer letters', the permits issued to new settlers by the government. Yet women were important players in the land invasions, providing support to the base camps during the 'jambanja' period, and subsequently investing in the development of new homes and farms, as wives, sisters, daughters, aunts and so on. However, across our sample only 12 percent of households had a woman named as the land holder on the permit. The highest proportion of such cases was in the informal settlements, as women often saw the land invasions as an opportunity to make a new independent life and escape abusive relationships or accusations of witchcraft.

So who among these groups are the so-called 'cronies' of the party, well-connected to the machinery of the state and able to gain advantage? As discussed above, those able to gain land through patronage included those who grabbed often large farms around 2008, as well as some of the A2 farmers able to manipulate the system. While the A2 farmers in our sample are certainly more 'elite' than those who invaded farms and took on small A1 farms, many could not be described as rich or politically well-connected. Former teachers, extension workers, office clerks and small-scale business people dominate this group. Others however have political connections that have allowed them access to patronage support from the state during the last decade. These are often absentee land owners – so-called 'cell phone farmers' – presiding over often under-utilised land, perhaps with a decaying new tractor in the farmyard. Yet, despite their disproportionate influence on local politics, these well-connected elites are few and far between, making up around five percent of the total population in our study areas (see the categories 'cell phone farmer' and 'farming from patronage' in Table 3, for example). A few gained access to farms with good irrigation infrastructure (such as in the sugar estates), but there is no consistent pattern in their distribution, as others took on dryland ranches. Perhaps because of the distance from Harare, the relatively poorer agro-ecological conditions, the lack of high value infrastructure and the particular local political configurations, in Masvingo province such elite capture is not the dominant story, despite the media assumptions. Masvingo is of course not Mazowe or Marondera, but even in such Highveld areas the situation is much more diverse than what mainstream portrayals suggest, with the new land reform areas dominated by smallholder farmers in A1 schemes, as Matondi (2010) and Sadomba (2011) attest for Mazowe, for example.

How much land did each of these groups get? Table 5 shows this for each of the scheme types, dividing the A2 sites into irrigated and dryland farms. For the A1 villagised and informal sites, the area measurements refer only to arable land, while for the other sites it represents the whole allocation to the households. Land cleared represents that where arable fields had been created and cultivated in 2007–08. The data shows that, for each of the scheme types, so-called 'ordinary' settlers did not receive any less land than other groups; in some cases more. Business people and civil servants were able to clear more land in most instances, due to access to resources to hire labour. Those linked to the security services – the group most likely to be associated with the political-military elite – received marginally more land than the average in the A1 self-contained and informal sites, but less in other sites.

As Table 5 shows, the different 'scheme' types thus create a pattern of differentiation within the land reform areas, with A1 villagised and informal sites receiving the least arable land, although with access to communal grazing (not included in the data). A1 self-contained land (including both arable and grazing) is larger, and where the highest areas of land cleared for production are observed. But it is the relatively small number of A2 farmers who received the largest land areas, including some large ranches in the dryland areas where very limited areas are cleared for arable production, as well as often quite large plots with irrigation potential. Compared to the small-scale farmers on the A1 schemes where intensive mixed farming has taken off, the A2 farmers have found the establishment of new farms much more difficult. With larger areas, the need for equipment and labour, as well as financial investment, and the economic conditions pertaining for much of the study period made getting new enterprises going very difficult. This is reflected in the relatively low areas cleared and low stocking rates, for instance.

The land reform has thus involved diverse people with multiple affiliations. Being influential in the land invasions, war veterans often managed to secure better plots,

Table 5. Land owned and cleared by settler and scheme type.

	A1 villagised		A1 self-contained		Informal		A2 dryland		A2 irrigated	
	Ha owned	Ha cleared	Ha owned	Ha cleared	Ha owned	Ha cleared	Ha owned	Ha cleared	Ha owned	Ha cleared
'Ordinary' – rural	4.8	3.8	31.9	9.3	6.6	5.1	247.0	7.0	–	–
'Ordinary' – urban	4.1	3.5	35.1	9.3	6.6	5.4	194.5	3.5	84.5	31.2
Civil servant	4.1	3.5	35.5	7.3	7.5	5.3	248.3	16.0	67.8	24.4
Security services	3.8	3.5	36.3	9.7	6.7	6.5	–	–	54.0	9.0
Business person	4.0	4.0	36.4	14.1	–	–	272.0	5.0	39.6	32.5
Farm worker	4.4	3.9	29.0	8.5	8.0	4.0	–	–	–	–
Average	4.6	3.8	33.9	8.9	6.7	5.1	232.0	8.7	70.2	27.3

Source: Census data, 2007 (A2 dryland excludes Asveld farm due to lack of area data).

although not always larger ones.[11] We were unable to ascertain party affiliation of those on our sites, but figures from recent elections suggest that there are significant numbers of MDC (Movement for Democratic Change) opposition supporters in these areas, even if they do not admit this publicly[12]. Given the often violent clashes associated with electoral politics, especially linked to ZANU-PF youth militia, many sensible people carry a ZANU-PF party card, even if they vote for the opposition. Many of those who joined the land invasions could not be regarded as 'cronies' in any reasonable sense; many had no party affiliation, they were simply interested in gaining access to land so long denied them.[13] While the land invasions clearly became highly politicised, and the atmosphere of the 'base camps' on the invaded farms was tightly ordered and politically controlled (Chaumba *et al.* 2003), those who ultimately benefited were much more diverse than those with close political ties. Again, as discussed above, who got the land in the A1 sites very much depended on the very particular dynamics of an individual invasion, who was leading it and how contested the farm was.

The large group of civil servants, particularly on the A2 plots – and in our sample especially in the sugar estates – were often teachers, agricultural extension workers and local government officials. While not being poor and landless from the communal areas, most could not be regarded as elite, nor often particularly well-connected politically. Indeed, in simple financial terms many were extremely poor, as government wages had effectively ceased during the economic crisis to 2009.

The net result was a new mix of people in the new resettlements. In the A2 schemes, for example 46.5 percent of new farmers have a 'Master Farmer' certificate[14], while in the A1 self-contained schemes 17.6 percent do. Some 91.6 percent of A2 farmers had at least three years of secondary schooling, while this proportion was 71.6 percent and 44.8 percent in the A1 self-contained and villagised schemes respectively. The new resettlements are dominated by a new generation of farmers, with most household heads being under 50, many born since Independence. A2 schemes are dominated by the over 40s, but often include people with significant experience and connections. That overall 18.3 percent of households came from urban areas (increasing to 43.8 percent in the A2 schemes) is significant too, as connections to town have proved important in gaining access to services and support in the absence of official programmes in the rural areas.

These data from Masvingo province are reflected in other studies from other areas of the country (Moyo *et al.* 2009, Matondi 2010 and contributions to this collection reviewed by Cliffe *et al.*). The overall picture is complex, but a simple

---

[11]War veterans had land areas above the average in the A1 villagised schemes only (at 6.8 hectares). In all other instances their land holdings were actually on average marginally lower than the average.
[12]The MDC-T party won in seven of the 15 constituencies across the four districts of our study area in the 2008 parliamentary election, taking 41.2 percent of the vote, against ZANU-PF's 52.3 percent. ZANU-PF maintained its stronghold in Mwenezi and Chiredzi districts, but lost in Gutu and parts of Masvingo district (Scoones *et al.* 2010, 29).
[13]In this respect, we disagree with Zamchiya's analysis (this collection) which assumes that nearly everyone is a 'crony' and/or a 'party supporter', rather than accepting that people switch allegiance opportunistically to gain strategic advantage, as described by Mkodzongi (2011). Such an alternative interpretation recognises the complexity and contradictions of public and private politics in the highly contested Zimbabwe setting, which of course varies significantly by region.
[14]A quite rigorous agricultural qualification, the result of training by the ministry of agriculture's extension arm.

narrative that land reform has been dominated by grabbing by elites is clearly inaccurate. Land previously occupied by a single farmer, often absent but with a manager and a few workers resident, is now being used by a highly diverse group of people. Overall, the new resettlements are populated by younger, more educated people with a greater diversity of backgrounds, professional skills and connections than their neighbours in the communal areas and old resettlements.

The land reform has resulted in a new social composition of people on the land, with a diversity of people from different backgrounds, with new skills, connections and sources of capital for investing in production. The new resettlements are therefore not a replication of the 1980s resettlement schemes or an extension of the communal areas, nor are they simply scaled-down version of large-scale commercial farms. Instead, a very different social and economic dynamic is unfolding, one that has multiple potentials, as well as challenges.

Investing in the land

One of the recurrent myths about Zimbabwe's land reform is that investment has been insignificant in the new resettlements: the land lies idle, people are not committed to farming and infrastructure is destroyed, neglected or non-existent. Perceptions of a lack of order and poor tenure security have further contributed to this view. Many assume that investment will not proceed without legally enforceable property rights, yet in our sites no leases have been agreed, and in some sites 'offer letters' (permits to occupy issued by the state) have not even been issued. Our studies have shown this narrative of low investment, disorder and lack of development is far from the case. Unlike the old resettlements which were plagued by a sense of insecurity, at least in the 1980s, due to the permit system and top-down imposition of planning requirements (Bruce 1990), it seems that trust in local authorities, combined with political assurances from across the political divide, has been sufficient for many to invest significantly in the new resettlements. This is of course not to say that insecurities do not exist. In the 'informal' settlements, unrecognised by the state, threats of eviction continue, and appropriation by political elites, particularly around the 2008 election period, represented another source of insecurity. Certainly, unstable macroeconomic factors until 2009 added to this and undermined opportunities for capital investment. But, despite this, impressive strides have been made in clearing the land, in purchasing livestock, equipment and transport and in building new settlements.

In developing their farms, most new farmers have had to start from scratch. For the most part the Masvingo study sites were ranches: large expanses of bush grazing, with limited infrastructure. There were scattered homesteads, a few workers' cottages, the odd dip tank, small dam and irrigation plot, but not much else. When groups of land invaders took the land they established 'base camps', under the leadership of war veteran commanders. Surveys of soil types and water sources were undertaken by the land invaders. The new settlers then pegged fields and marked out areas for settlement. Soon, once the official FTLRP was launched, officials from the government arrived and imposed an official plan, based on land use planning regulations, as well as much pressure to accommodate more people. Some had to move their shelters and clear fields anew. But, within a remarkably short time, people began to invest in earnest. There was an urgency: fields had to be prepared for planting, structures had to be built for cattle, granaries had to be erected for the harvests to be stored, and homes had to be put up for growing numbers of people to live in.

A peopled landscape of houses, fields, paths and roads soon emerged. Human population densities increased significantly and livestock populations grew. Stocking densities on beef ranches were recommended to be around one animal per ten hectares; now much larger livestock populations exist, combining cattle with goats, sheep, donkeys, pigs and poultry. Investment in stock has been significant, with cattle populations in particular growing rapidly, especially in the A1 sites.

One of the major tasks facing new settlers has been clearing land. In A1 village sites, on average each household had cleared 6.8 hectares by 2008–09, while in A1 self-contained and A2 sites an average of 13.3 hectares and 23.7 hectares had been cleared. In the A1 sites most of this was being cultivated, while in the A2 sites much less intensive land use is observed (Table 5). In addition, people have constructed numerous gardens, all of which have required investment in fencing. In addition, people have dug wells, built small dams, planted trees and dug soil conservation works. Investment in fields was complemented by investment in farm equipment, with ploughs, cultivators, pumps and scotch carts purchased in numbers.

Building has also been extensive in the new resettlements. Some structures remain built of pole and mud, however, after a year or two, when people's sense of tenure security had increased, buildings using bricks, cement and tin/asbestos roofing increased. Some very elaborate homes have been built with the very best materials imported from South Africa.

Transport has been a major constraint on the new resettlements. With no roads and poor connections to urban areas, there were often no forms of public transport available. This was compounded by the economic crisis, as many operators closed down routes. This had a severe impact. Lack of access to services – shops, schools, clinics – and markets meant that people suffered. Investing in a means of transport was often a major priority. Bicycles in particular were bought in large numbers, but also cars, pick-ups and trucks.

What is the value of all this investment? A simple set of calculations which compute the cost of labour and materials used or the replacement cost of the particular item show that, on average, each household had invested over US$2000 in a variety of items in the period from settlement to 2008–09 (Table 6).

Table 6. The value of investments in the new resettlements[15].

Focus of investment	Average value of investment per household (US$)
Land clearance	$385
Housing/buildings	$631
Cattle	$612
Farm equipment	$198
Transport	$150
Toilets	$77
Garden fencing	$29
Wells	$79
Total	$2161

Source: Scoones *et al.* (2010, 87)

---

[15]The values were calculated using US dollar-equivalent replacement costs for labour, materials, equipment etc., based on an average investment per household across the full sample of 400 cases (Scoones *et al.* 2010, 77–87).

This is of course only a small subset of the total, as such private investment does not account for investments at the community level. Across our sites, churches have been established, schools have been built, roads cut and areas for shops carved out as part of community efforts. Labour and materials have been mobilised without any external help. In the A1 sites in particular this highly-motivated and well-organised pattern of self-help has dominated (cf. Murisa, this collection). While the state has been present, it has not always been helpful. The re-planning of village and field sites was resented by many, as the land use planning models dating from the 1930s were re-imposed, with fields removed from near rivers and streams and villages placed on the ridges far from water sources. Planning laws were also invoked in the destruction of nascent business centres as part of Operation 'Murambatsvina' (Potts 2008).

Extension workers are few and far between and veterinary care almost non-existent. Instead, people have used their own knowledge, skills and connections in developing their agriculture, often relying on those with Master Farmer qualifications which they had gained in their former homes in the communal areas. Without dipping, the explosion of tick-borne animal diseases has been devastating, but many farmers have purchased spray-on chemicals, often organising themselves in groups to tackle the problem.

So without the state and without the projects of donors and NGOs – and significantly without formal title or leasehold tenure – the new settlers have invested at scale. Extrapolating the results from our sample and for the limited set of items assessed to the whole province this adds up to an investment of US$91 million across all new resettlements; a substantial amount by any calculation (Scoones *et al.* 2010, 86). Such a level of investment suggests that in most resettlement sites (perhaps with the exception of the 'informal' sites, where no 'offer letters' have been issued) there is a sufficient sense of security of tenure to allow investment at scale, undermining the claim that what is needed now is a formalisation of tenure regimes and the offering of some form of title (Matondi 2010).

But is this an argument that people can just do it on their own, and should be left to their own devices? Emphatically: no. There are plenty of things that need to be done, and where external support is necessary. In order to get farming moving in the new resettlements a significant investment in infrastructure – roads, wells, dams, dips and so on – will be needed. This is unlikely to come from individual and community contributions, although the considerable entrepreneurial initiative and deep commitment to investment in the new resettlements is an important platform on which to build.

A smallholder agricultural revolution in the making?

A recurrent myth about Zimbabwe's land reform is that it has resulted in agricultural collapse, precipitating widespread and recurrent food insecurity. There is little doubt that the agricultural sector has been transformed, as discussed above, but our data show that there has been surprising resilience in production.

Take maize production on the resettlement farms in Masvingo province. We tracked production on all 400 farms in our sample over seven seasons between 2002 and 2009 (Table 7). The data shows a steady increase in output over time as farms became established, and draught power and other inputs were sourced. The trend was not smooth, however, and the major droughts in this period saw low yields. Availability of seeds and fertiliser was also highly variable across years, with various

Table 7.   Percentages of farmers harvesting greater than a tonne of maize.

District	Scheme Type	2002–03	2003–04	2004–05	2005–06	2006–07	2007–08	2008–09
Gutu	A1 self-contained	18.4	50.0	45.5	75.0	63.4	28.6	61.5
	A1 villagised	13.3	39.1	24.0	79.3	63.3	36.7	78.6
	A2	0.0	0.0	44.4	75.0	66.7	n.d	63.6
Masvingo	A1 self-contained	55.3	63.2	56.4	100.0	100.0	51.3	100.0
	A1 villagised	28.0	38.1	45.8	95.7	91.2	15.8	77.9
	A2	0.0	25.0	25.0	n.d	75.0	75.0	100.0
Chiredzi	A2	14.3	38.5	46.2	50.0	66.7	50.0	88.9
	Informal	18.8	10.2	3.9	86.5	51.0	24.5	62.5
Mwenezi	A1 villagised	26.9	8.0	0.0	4.8	0.0	0.0	0.0
	Informal	11.5	11.5	0.0	0.0	26.7	6.7	0.0

Source: Maize census, 2003–09 (N=400; n.d = no data) (Scoones *et al.* 2010, 108).

government schemes delivering patchily and unreliably. And patterns of differentiation across households were also very evident.

As Table 7 shows, in the better rainfall years of 2005–06 and 2008–09 the proportion of households producing more than a tonne of maize – sufficient to feed an average family for a year – was significant across all sites. For example, following the 2009 harvest between 63 percent and 100 percent of households outside the lowveld sites in Mwenezi produced more than this threshold. If sorghum and millet were added to the tally, more than 60 percent of households, even in the dryland Mwenezi sites, produced sufficient quantities for self-provisioning. Surpluses may be sold or stored, providing a buffer for future years. Around a third of households sold maize, sorghum or millet regularly in this period. For example in 2009 two-thirds of such households in the A1 self-contained settlement sites sold over a tonne of maize, although marketed output was not so high on the A2 farms, by contrast.

A major constraint especially to maize production in this period, however, has been input supply – both of seed and fertiliser. Local production of agricultural inputs declined dramatically from 2000 due to the economic conditions prevailing. Attempts to provide inputs through government programmes – whether the Agricultural Sector Productivity Enhancement Facility, Operation 'Maguta' or the Champion Farmer programme – largely failed. The agricultural policy environment until 2009 was characterised by 'heavy-handed state intervention funded through quasi-fiscal means which distorted markets and incentives and undermined the economy' (Scoones *et al.* 2010, 99). These schemes benefited some, but they also opened up significant opportunities for corruption. Most new resettlement farmers had to source their own seed, reverting to local re-use and imports from South Africa. Fertiliser use dropped dramatically, although the new farmers had the benefits of relatively virgin soils for a period.

By contrast, in some sites, cotton production has boomed. This is particularly so in the 'informal' site of Uswaushava in the Nuanetsi ranch. Here cotton production has increased significantly (Table 8). Cotton sales provide significant cash income for nearly all households. Six different private cotton companies operate in the area, supplying credit, inputs and marketing support – allowing cotton producers to access inputs and other support through other means. New cotton gins have opened up too, creating employment further up the value chain.

Investment in cattle has been particularly important across the sites, but particularly in the A1 schemes, and for certain 'success groups'[16] (Table 9). Contrary to the pattern noted by Dekker and Kinsey (this collection) for the old resettlement areas, cattle numbers are increasing in the new resettlement areas, providing an important source of draught power, milk, meat and cash sales and savings. Particularly dramatic increases in holdings since settlement are seen in the A1 self-contained sites (across all 'success groups'), and in the top success group in the A1 villagised and informal sites. Again, by contrast, the A2 farmers were accumulating less, often because of disease outbreaks and theft.

Markets are key to the resettlement farming enterprises, and these are expanding in new ways around different commodities (see Mavedzenge et al. 2008, for example, for livestock). These are new markets, often operating informally, sometimes illegally. Compared to those that existed before, they have been radically reconfigured by the restructuring of the agrarian economy following land reform and deeply affected by the economic crisis that plagued the country for much of the past decade. Detailed studies by Mutopo (this collection), for example, show the gendered dimensions of such new informal markets, as well as their impressive

Table 8.   Changes in cotton production in Uswhaushava, Chiredzi cluster, 2001–2008 harvests.

	2001	2002	2003	2004	2005	2006	2007	2008
% farmers growing cotton	18%	35%	29%	35%	29%	68%	92%	89%
For cotton farmers, average area planted to cotton (ha)	1	1.5	1.6	1.4	1.4	1.7	1.7	2.1
For cotton farmers, average output of cotton (bales = 200kg)	2.3	1.7	0.8	1.6	0.6	5.8	7.5	6.4

Source: Annual crop census (Scoones *et al.* 2010, 114).

Table 9.   Mean cattle holdings: changes by scheme type and success group.

	Success Group 1		Success Group 2		Success Group 3	
Scheme type	At settlement	2008	At settlement	2008	At settlement	2008
A1 villagised	6.3	10.4	4.5	4.5	1.9	2.6
A1 self-contained	11.2	16.2	1.3	10.9	0.9	3.7
A2	18.9	20.5	13.6	14.8	11.1	4.4
Informal	7.5	12.5	4.5	3.8	0.0	0.5

Source: Survey data, 2007–08 (N=177) (Scoones et al. 2010, 118).

---

[16]As discussed above, this is the local characterisation of 'success' used in the study to differentiate settlers, with Success Group 1 being the most 'successful' according to local criteria. Cattle ownership and accumulation was, unsurprisingly, one of the key indicators.

dynamism. Yet, since they are evolving at such a pace, it is difficult to keep track of how agricultural markets work, and formal data on agricultural production and sales is very shaky indeed; and inevitably highly politicised with different arms of government and different international agencies presenting figures, based on very little ground-truthing, to support a particular view. The statistical basis for assessing the success or otherwise of land reform at a national level thus remains extremely limited, and so detailed local studies of the form of production and the functioning of markets are essential to build a more complete picture.

While across our research sites there are of course some who produce little and have to rely on local markets or support from relatives, overall we did not find a pattern of production failure, widespread food insecurity and lack of market integration. On the contrary, we found a highly differentiated picture, but one which has at its centre smallholder agricultural production and marketing; one that could, given the right support, be the core of a new 'green revolution' in Zimbabwe. By contrast to the previous boom in smallholder production in the early 1980s following Independence, the Masvingo sample suggests a larger proportion of farmers is involved. Around half are succeeding as 'middle farmers' and a third as highly commercialised producers, compared to only 20 percent in the 1980s (cf. Stanning 1989); and of course at a much larger scale than the rather isolated successes of that earlier period (Eicher 1995, Rohrbach 1989).

Dynamic local economies

Of course on-farm success can result in off-farm economic growth, as linkages are forged in local economies (cf. Delgado *et al.* 1998). This is an important dynamic in the new resettlement areas, given the geographical juxtaposition of new resettlement areas of different types with old communal and resettlement areas. Since 2000, the rural economy has been radically spatially reconfigured, with the old separated economic spheres of the large-scale farms and the communal areas being broken down. The result is a shift to new sites for economic activity, connected to new value chains and new sorts of entrepreneurs, linking town and countryside.

The dynamic entrepreneurialism resulting should, we argue, not be under-estimated and represents an important resource to build on. Across our sites, we have small-scale irrigators producing horticultural products for local and regional markets; we have highly successful cotton producers who are generating consider-able profits by selling to a wide number of competing private sector companies; we have livestock producers and traders who are developing new value chains for livestock products, linked to butcheries, supermarkets and other outlets; we have traders in wild products, often engaged in highly profitable export markets; and we have others who are developing contract farming and joint venture arrangements for a range of products, including wildlife. We also have an important group of sugar producers with A2 plots on the lowveld estates who, very often against the odds due to shortages of inputs, unreliable electricity supplies and disadvantageous pricing, have been delivering cane to the mills, as well as other diverse markets, alongside diversification into irrigated horticulture production on their plots.

The new farmers are also employing labour (Table 10; see Chambati, this collection). This is often casual, low-paid employment, often of women, but it is an important source of livelihood for many – including those who are not making it as part of the new 'middle farmer' group identified above. The new resettlements sites

Table 10. Patterns of permanent and temporary labour hiring for cropping and livestock rearing.

	A1 and informal					A2				
	Temporary cropping	Temporary livestock	Permanent both	Permanent cropping	Permanent livestock	Temporary cropping	Temporary livestock	Permanent both	Permanent cropping	Permanent livestock
Percentage of households employing workers	20	13	9	11	9.3	67.6	43.5	44.8	71.9	43.3
% of these female	48	31	26	32	25	27	7	23	26	28

Source: Survey, 2007–08 (N = 177) (Scoones *et al.* 2010, 132).

have become a magnet for others, and households on average have grown by around three members since settlement through the in-migration of relatives and labourers (cf. Deininger *et al.* 2004 for discussion of a similar dynamic in the old resettlement areas). On average, A2 farm households have employed 5.1 permanent workers and regularly employ 7.3 temporary labourers, while those households in A1 schemes and in informal resettlement sites employ on average 0.5 permanent workers and 1.9 temporary labourers. Comparing this level of employment with what existed before on the former cattle ranches, where perhaps one herder was employed for each 100 animals grazed over 1000 hectares, the scale of employment generation afforded by the new resettlement farms is considerable.

There is frequently a sense of optimism and future promise among many resettlement farmers we have worked with. SM from Mwenezi district commented, 'We are happier here at resettlement. There is more land, stands are larger and there is no overcrowding. We got good yields this year. I filled two granaries with sorghum. I hope to buy a grinding mill and locate it at my homestead.' Comparing the farming life to other options, PC from Masvingo district observed, 'We are not employed, but we are getting higher incomes than those at work.' Despite the hardships and difficulties – of which there are many – there is a deep commitment to making the new resettlement enterprises work.

Future policy challenges

Despite the political and economic challenges that Zimbabwe continues to face (Raftopolous 2010), along with outstanding legal challenges and concerns of the international community, there is a broad consensus that Zimbabwe's land reform is not reversible. To move ahead, a sustainable and democratic political settlement is clearly an essential precursor, one that balances rights (of different sorts, and not only those over former private freehold property) and redistribution (and so issues of equity, broad participation in economic activity and redress of historical disadvantage). But whatever new political alliance runs the country in the future, a major challenge remains: what should be done in the new resettlements to build a sustainable, growth-oriented agricultural base? As we found in Masvingo, and others have discovered elsewhere in the country, there is much to build on in terms of basic investment, as well as the skills and knowledge of the new settlers. The challenge is a new one however for agricultural research and development. As the head of extension in the province put it: 'We don't know our new clients: this is a totally new scenario.'[17] Responding to this scenario requires careful thought. As discussed, the new resettlement areas are not a replication of the communal areas, nor are they a scaled-down version of the old commercial sector. These are very different places with new people with new production systems engaging in new markets – all with new opportunities and challenges. The new farmers are often highly educated, well-connected and with important skills. Support for marketing or input supply via mobile phone updates, or agricultural extension or business planning advice offered via the Internet offer real opportunities, for example. If given the right support, we argue, the emergent group of new 'middle farmers' on the new resettlements, both A1 and A2, can drive a vibrant agricultural revolution in Zimbabwe (Scoones *et al.* 2010).

---

[17]Comment by the provincial agricultural extension officer at a workshop in Masvingo in 2006.

This has of course happened before: with white commercial farmers from the 1950s and with communal area farmers in the 1980s (Rukuni *et al.* 2006). But both past agricultural revolutions required support and commitment from outside, something that has been starkly absent in the past decade. Zimbabwe's green revolution of the 1980s has been much hailed, but this was mostly in high potential communal areas and was quickly extinguished following structural adjustment. The nascent green revolution in the resettlement areas potentially has far wider reach, both geographically and socio-economically. But if the new resettlements are to contribute to local livelihoods, national food security and broader economic development, they unquestionably require investment and support. This means infrastructure (dams, roads), financing (credit systems), input supply (fertiliser, seed), technology (intermediate and appropriate) and institutions and policies that allow agriculture to grow.

Getting agriculture moving on the new resettlements through building on existing achievements must be central priority for policy today. What should the top priorities be now? A commentary on our book's conclusions suggested that these were merely 'gestural' and would surely be made irrelevant by the on-going political contests at the national level.[18] Yet the existence of a pervasive, violent, militarised corrupt politics, which entrenches certain positions and dominates a negative cycle of elite capture, does not mean that there are no countervailing forces, driven by other interests. We believe that the political struggle for an accountable, democratic politics must be linked to changing practices on the ground, and to energised policy thinking that takes the new realities into account. With a new progressive narrative on land and rural development, for example, a new vision for Zimbabwe might yet emerge which cuts across currently extreme politically-entrenched divisions. This means, we suggest, that engaging with a future policy agenda is not simply irrelevant hand waving, but a practical means to realise broader, widely shared, goals. Here we identify three inter-related challenges.

First, security of land tenure is an essential prerequisite for successful production and investment in agriculture. Tenure security arises through a variety of means. Existing legislation allows for a wide range of potential tenure types, including freehold title, regulated leases, permits and communal tenure under 'traditional' systems. All have their pros and cons. Policymakers must ask how tenure security can be achieved within available resources and capacity; how safeguards can be put in place to prevent land grabbing or land concentration; and what assurances must be made to ensure that private credit markets function effectively. Lessons from across the world suggest there is no one-size-fits-all solution centred on freehold tenure (World Bank 2003), despite its continued allure in the Zimbabwe debate (Rukuni *et al.* 2009).

Instead, a flexible system of land administration is required – one that allows for expansion and contraction of farm sizes, as well as entry and exit from farming. Informal (actually illegal) land rental markets are already emerging on some sites, allowing land transfers to occur. While the excesses of elite patronage and land grabbing must be addressed through a land audit, a successful approach, overseen by an independent, decentralised authority, must not be reliant on technocratic diktat on farm sizes, business plans and tenure types alone. This will mean investing in land

---

[18]http://anothercountryside.wordpress.com/2011/03/23/potential-of-zimbabwe-land-reform-limited-by-violent-state/

governance – building the effectiveness of local institutions to manage resources, resolve disputes and negotiate land access in clear and accountable ways. Without attention to these issues, conflicts will escalate as uncertainties over authority and control persist. This will have damaging consequences for both livelihoods and environmental sustainability. Support for rebuilding public authority from below must therefore be high on the agenda, linked to a revitalisation of local government capacity. Only with this longer-term effort will a more accountable and democratic approach to land be realised and the depredations of a greedy elite avoided.

Second, as discussed above, land reform has reconfigured Zimbabwe's rural areas dramatically. No longer are there vast swathes of commercial land separated from the densely-packed communal areas. The rural landscape is now virtually all populated. Links between the new resettlements and communal and former resettlement areas are important, with exchanges of labour, draught animals, finance, skills and expertise flowing in all directions. As a result, economic linkages between agriculture and wider markets have changed dramatically.

This has given rise to the growth of new businesses to provide services and consumption goods, many only now getting going. Yet the potentials for economic diversification – in small-scale mining, hunting, cross-border trade and a host of other enterprises – are currently constrained by legal and regulatory restrictions. While a regulatory framework will always be required, it must not be excessively and inappropriately restrictive. Businesses must be encouraged to flourish in support of rural livelihoods, capturing synergies with local agricultural production.

To make the most of the new mosaic of land uses and economic activities, an area-based, local economic development approach is required. This would facilitate investment across activities, adding value to farm production. Today, with a new set of players engaged in local economic activity, many possibilities open up. An area-based approach needs to draw in the private sector, farmer groups and government agencies, but with strong leadership from a revived local government, with rethought mandates and rebuilt capacities. Investing in such capacities and building local economies will in turn improve sources of local revenue beyond patronage and so build systems of local accountability.

Third, reflecting a wide range of interests, the new resettlement farmers are highly diverse in class, gender and generational terms. This diversity has many advantages, adding new skills and experiences, but it is also a weakness. Formal organisation in the new resettlements is limited. The structures that formed the basis of the land invasions – the base commanders and the Committees of Seven, for example – have given way to other arrangements, and there is often limited collective solidarity. There are of course emergent organisations focused on particular activities – a garden, an irrigation scheme, a marketing effort, for example – but these are unlikely to become the basis of political representation and influence. Because politics has been so divisive in recent years, many shy away from seeing political parties as the basis for lobbying for change, and there are few other routes to expressing views.

Building a new set of representative farmers' organisations, linked to an influential apex body, will be a long-term task, and will be highly dependent on the unfolding political alliances in rural areas. As we have shown, the new resettlements are characterised by an important and numerically large 'middle farmer' group. There is also a significant group of less successful farmers with different needs and interests. And there are those elites reliant on political patronage who, despite being relatively few in number, are disproportionately influential.

In contrast to the past when smallholders could easily be marginalised and were courted only at elections for their votes, the new farmers – and particularly the burgeoning group of 'middle farmers' – now control one of the most important economic sectors in the country, and must be relied upon for national food supply. Today, the politics of the countryside cannot be ignored, and organised farmer groups may exert substantial pressure in ways that previously seemed unimaginable.

For this reason, the debate about land, agriculture and rural development in Zimbabwe urgently needs to move beyond the ideological posturing of ZANU-PF, wrapped up in violent nationalist rhetoric, or the startling silences on land issues by the opposition political parties and civil society. A new narrative on land is urgently needed, based firmly on the realities on the ground. How the new configuration of political forces will pan out in the future is a subject of fierce contest, but the role of diverse agrarian interests, including new small-scale farmers, will certainly be important.

Reframing the debate on land in Zimbabwe

The Masvingo study has challenged several recurrent myths about Zimbabwe's land reform: for example, that there is no investment going on, that agricultural production has collapsed, that food insecurity is rife, that the rural economy is in precipitous decline and that farm labour has been totally displaced. Getting to grips with the realities on the ground is essential for reframing the debate. This is why solid, empirical research is so important. Only with these facts to hand can sensible policymaking emerge. Evidence rather than emotion must guide the process. While it remains essential to address abuses of the land reform programme according to strict criteria set by a land audit, it is also important to focus on the wider story, dispelling myths and engaging with the realities of the majority.

Land and politics are deeply intertwined in Zimbabwe. The current impasse cannot be resolved by technocratic measures alone: plans, models, audits and regulations are only part of the picture. A reframed debate must encompass redistribution and redress, as well as rights and responsibilities. The recent divisive debate on land in Zimbabwe has seen these as opposites, creating what has been called a 'dangerous rupture' in Zimbabwe's political discourse (Raftopolous 2009).

The past decade of land resettlement has unleashed a process of radical agrarian change. There are now new people on the land, engaged in multiple forms of economic activity, connected to diverse markets and carving out a variety of livelihoods. Bringing a broad perspective on rights together with a continued commitment to redistribution must be central to Zimbabwe's next steps towards democratic and economic transformation. Only with land viewed as a source of livelihood and redistributed economic wealth, and not as a political weapon or source of patronage, will the real potentials of Zimbabwe's land reform be fully realised.

References

Anseeuw, W., T. Kapuya and D. Saruchera, D. 2011. *Zimbabwe's Agricultural Reconstruction: Present State, On-Going Projects and Prospects for Reinvestment.* Draft Document Prepared for The Development Bank of Southern Africa (DBSA): Pretoria and Agence Française de Développement (AFD): Paris.

Bruce, J. 1990. Legal issues in land use and resettlement: a background paper. *Zimbabwe Agriculture Sector Memorandum.* The World Bank: Washington, DC

Chambati, W. 2007. *Emergent agrarian labour relations in new resettlement areas.* African Institute for Agrarian Studies (AIAS) Mimeograph Series, AIAS, Harare.

Chambati, W. 2011. Restructuring of agrarian labour relations after Fast Track Land Reform in Zimbabwe. *The Journal of Peasant Studies*, 38(5), 1047–1068.

Chaumba, J., I. Scoones and W. Wolmer. 2003. From jambanja to planning: the reassertion of technocracy in land reform in south-eastern Zimbabwe? *Journal of Modern African Studies*, 41(4), 533–54.

Cliffe, L., et al. 2011. An overview of Fast Track Land Reform in Zimbabwe: editorial introduction. *The Journal of Peasant Studies*, 38(5), 907–938.

Cousins, B. 2010. *What is a 'smallholder'? Class analytical perspectives on small-scale farming and agrarian reform in South Africa.* Working Paper 16, January 2010. PLAAS, University of the Western Cape.

Deininger, K., H. Hoogeveen and B. Kinsey. 2004. Economic benefits and costs of land redistribution in Zimbabwe in the early 1980s. *World Development*, 32(10), 1697–1709.

Dekker, M. and B. Kinsey. 2011. Contextualising Zimbabwe's land reform: long-term observations from the first generation. *The Journal of Peasant Studies*, 38(5), 995–1019.

Delgado, C., J. Hopkins, and V.A. Kelly. 1998. *Agricultural growth linkages in sub-Saharan Africa.* International Food Policy Research Institute (IFPRI) Research Report, no. 107, IFPRI, Washington, DC.

Dorward, A. 2009. Integrating contested aspirations, processes and policy: development as hanging in, stepping up and stepping out. *Development Policy Review*, 27(22), 131–46.

Eicher, C. 1995. Zimbabwe's maize-based green revolution: preconditions for replication. *World Development*, 23(5), 805–18.

Gunning, J., J. Hoddinott, B. Kinsey and T. Owen. 2000. Revisiting forever gained: income dynamics in resettlement areas of Zimbabwe, 1983–1996. *Journal of Development Studies*, 36(6), 131–54.

Hammar, A., B. Raftopoulos and S. Jensen, eds. 2003. *Zimbabwe's unfinished business: rethinking land, state, and nation in the context of crisis.* Weaver Press, Harare.

Kinsey, B. 2003. Comparative economic performance of Zimbabwe's resettlement models. *In*: M. Roth and F. Gonese, eds. *Delivering land and securing rural livelihoods: post-independence land reform and resettlement in Zimbabwe.* University of Zimbabwe and University of Wisconsin-Madison, Harare, Zimbabwe and Madison, WI.

Marongwe, N. 2011. Who was allocated Fast Track land, and what did they do with it? Selection of A2 farmers in Goromonzi District, Zimbabwe and its impacts on agricultural production. *The Journal of Peasant Studies*, 38(5), 1069–1092.

Marongwe, N. 2003. Farm occupations and occupiers in the new politics of land in Zimbabwe. *In*: A. Hammar, B. Raftopoulos and S. Jensen, eds. *Zimbabwe's unfinished business: rethinking land, state and nation in the context of crisis.* Weaver Press, Harare.

Marongwe, N. 2009. Interrogating Zimbabwe's Fast Track Land Reform and Resettlement Programme: a focus on beneficiary selection. Thesis (PhD). University of the Western Cape, Cape Town.

Matondi, P., ed. 2010. *Inside the political economy of redistributive land and reforms in Mazowe, Shamva and Mangwe districts, in Zimbabwe*, Ruzivo Trust, Harare.

Mavedzenge, B.Z., J. Mahenehene, F. Murimbarimba, I. Scoones, and W. Wolmer. 2008. The dynamics of real markets: cattle in southern Zimbabwe following land reform. *Development and Change*, 39(4), 613–39.

Mkodzongi, G. 2011. Land grabbers or climate experts? farm occupations and the quest for livelihoods in Zimbabwe. Paper European Conference of African Studies 4, Uppsala.

Moyo, S. 1998. *The Land acquisition process in Zimbabwe (1997/8).* Harare: United Nations Development Programme.

Moyo, S. 2001. The land occupation movement in Zimbabwe: contradictions of neoliberalism. *Millennium: Journal of International Studies*, 30(2), 311–30.

Moyo, S. 2011a. Three decades of land reform in Zimbabwe. *Journal of Peasant Studies*, 38(3), 493–571.

Moyo, S. 2011b. Land concentration and accumulation after redistributive reform in post-settler Zimbabwe. *Review of African Political Economy*. 38(128), 257–76.

Moyo, S., W. Chambati, T. Murisa, D. Siziba, C. Dangwa, K. Mujeyi and N. Nyoni. 2009. Fast track land reform baseline survey in Zimbabwe: trends and tendencies. 2005/6. Harare: AIAS Monograph.

Murisa, T. 2011. Local farmer groups and collective action within fast track land reform in Zimbabwe. *The Journal of Peasant Studies*, 38(5), 1145–1166.

Mushongah, J. 2009. Rethinking vulnerability: livelihood change in southern Zimbabwe, 1986–2006. Thesis (PhD), University of Sussex.

Mutopo, P. 2011. Women's struggles to access and control land and livelihoods after fast track land reform in Mwenezi District, Zimbabwe. *The Journal of Peasant Studies*, 38(5), 1021–1046.

Potts, D. 2008. Displacement and livelihoods: the longer term impacts of Operation Murambatsvina. *In*: M. Vambe, ed. *The hidden dimensions of Operation Murambatsvina in Zimbabwe*. Weaver Press, Harare.

Raftopoulos, B. 2009. The crisis in Zimbabwe, 1998–2008. *In*: B. Raftopoulos and A. Mlambo, eds. *Becoming Zimbabwe: a history from the pre-colonial period to 2008*. Weaver Press, Harare.

Raftopoulos, B. 2010. The global political agreement as a 'passive revolution': notes on contemporary politics in Zimbabwe. *The Round Table*, 99(411), 705–18.

Rohrbach, D. 1989. *The economics of smallholder maize production in Zimbabwe: implications for food security*. International Development Papers, no. 11, Michigan State University, East Lansing, MI.

Rukuni, M., J. Nyoni and P. Matondi. 2009. *Policy options for optimisation of the use of land for agricultural productivity and production in Zimbabwe*. World Bank, Harare.

Rukuni, M., P. Tawonezvi and C. Eicher with M. Munyuki-Hungwe and P. Matondi, eds. 2006. *Zimbabwe's Agricultural Revolution Revisited*. University of Zimbabwe Publications, Harare.

Sachikonye, L. 2003. *The situation of commercial farm workers after land reform in Zimbabwe*. Catholic Institute for International Relations and FCTZ, London and Harare.

Sadomba, W.Z. 2011. *War veterans in Zimbabwe's revolution*. Woodbridge: James Currey.

Scoones, I., N. Marongwe, B. Mavedzenge, F. Murimbarimba, J. Mahenehene and C. Sukume. 2010. *Zimbabwe's land reform: myths and realities*. Oxford: James Currey; Harare: Weaver Press; Johannesburg: Jacana.

Scoones, I., N. Marongwe, B. Mavedzenge, F. Murimbarimba, J. Mahenehene and C. Sukume. 2012. Livelihoods after land reform in Zimbabwe: understanding processes of rural differentiation. *Journal of Agrarian Change* (forthcoming).

Stanning, J. 1989. Smallholder maize production and sales in Zimbabwe: some distributional aspects. *Food Policy*, 14(3), 260–67.

World Bank 2003. *Land Policies for Growth and Poverty Reduction*. World Bank and Oxford University Press, Washington, DC and Oxford.

Zamchiya, P. 2011. A synopsis of land and agrarian change in Chipinge District, Zimbabwe. *The Journal of Peasant Studies*, 38(5), 1093–1122.

Ian Scoones is a Professorial Fellow at the Institute of Development Studies at the University of Sussex, UK.

Nelson Marongwe works with the Centre for Applied Social Sciences Trust at the University of Zimbabwe.

Blasio Mavedzenge works with Agritex of the Ministry of Agriculture in Masvingo and is an A1 farmer.

Felix Murimbarimba is an A2 outgrower sugar farmer in Hippo Valley, formerly with Agritex.

Jacob Mahenehene is a farmer in Chikombedzi communal area, and has a A1 villagised plot.

Chrispen Sukume is an independent consultant, formerly with the Department of Agricultural Economics at the University of Zimbabwe in Harare.

Contextualizing Zimbabwe's land reform: long-term observations from the first generation

Marleen Dekker and Bill Kinsey

In the heat of the discussions about the fast-track land reform programme (FTLRP) in Zimbabwe, little attention is given to the experience with land reform immediately following independence. Understanding these past experiences is useful to contextualize current challenges. Although the farmers resettled in the early 1980s started out in a completely different political and economic environment, the challenges in establishing their farms and communities were, at least to some extent, very similar to those reported today. However, livelihoods developed by the farmers in the old resettlement areas have been severely constrained by the macro-economic context. We argue that any discussion on the success of FTLRP should acknowledge the impact of the devastating macro-economic context on the opportunities for smallholder farmers to establish their farms and become agriculturally productive.

Introduction

In the heat of the discussions about the impact and challenges of the recent fast-track land reform programme (FTLRP) in Zimbabwe, it is often forgotten that resettlement is not new to the country. Understanding the past experiences may be useful in contextualizing current challenges. This paper does exactly that by reviewing existing and new data on the experiences of smallholder farmers who resettled on former commercial farms just after independence in 1980. We document the agricultural and economic outcomes in these schemes and the social cohesion developed in the villages over the past 30 years. We do so by reviewing previously published work on the old resettlement areas (ORAs), experience often ignored in the literature on the FTLRP, and by presenting original data collected in these early schemes after 2007.

It is important at the outset to appreciate some of the ways in which land reform of the 1980s and 1990s differed from the post-2000 land redistribution. First, the earlier experience was designed by agricultural technocrats of the old Rhodesian institutions, and the first blueprints were ready even before independence in April 1980. Implementation, accompanied by a panoply of rules and regulations,

---

The authors wish to express their deep appreciation to the Royal Netherlands Embassy in Harare for support to conduct the 2010 round of the panel study and to Michael Shambare and Nyaradzo Shanaynewako for their assistance in collecting data since 2001. The authors would also like to acknowledge the helpful comments of the anonymous reviewers.

administrative and other supporting staff and with substantial donor backing, began just six months after independence. Farms were purchased on a willing-seller, willing-buyer basis and immediately reconfigured into nucleated villages, arable and grazing land, and administrative and public spaces. The early settlers received strong support at the outset, including free tillage and seed-fertilizer packs, credit, water supplies, 100 percent extension coverage, and market and health infrastructure. But it is a common misperception that these uniform five hectare resettled holdings were meant to constitute the core of a new small-scale class of commercial farms. They were not. The planned income targets for the new holdings – at full economic maturity – were meant to equal the minimum industrial wage in urban areas. Thus an implicit objective was to keep the settlers on the land and avoid a flood of job-seekers into urban areas. Indeed, among the many rules the new settlers had to agree to at the outset was one forbidding any off-farm employment. Other rigidities controlled the way land was used and forbade any subdivision of the land allocated.

In contrast, beginning in 2000, farms were seized through occupation and eviction with little or no prior planning. Indeed, at this stage, claiming the land was far more important than any consideration as to how the land would ultimately be used, hence the designation 'fast-track'. Occupiers were not necessarily entitled to a plot on the farms they occupied but many did eventually receive one. Because of the haste involved however many new settlers have been relocated several times as ex-post planning has evolved and as conflicts have arisen. A key difference from the earlier experience is that, aside from the controversial support programmes targeting mainly the larger-scale, politically favoured new farmers, most FTLRP farmers have received no support whatsoever. With no donor involvement and a rapidly weakening state capacity, they thus had to begin the development of their plots and farming activities in circumstances made even more adverse by accelerating hyperinflation and the effective collapse of all supporting institutions. What this group of settlers has or has not achieved is thus largely a consequence of their own initiative and abilities to make use of the land they have been given.

What both groups of resettled households have in common is that they have been confronted by the extraordinary economic meltdown that began in the late 1990s and accelerated dramatically after 2000. Where they may differ importantly is that the first generation of land reform farmers spent 20 years under a completely top-down, authoritarian administrative system whereas the current generation was resettled in almost an administrative vacuum and, as a result, seems to be developing some grassroots managerial innovations. What we do here is examine the experience of the original settlers, both when they settled initially, and as they struggled through the post-2000 turmoil, in an attempt to understand better the circumstances confronting both groups of smallholders.

The review of previously published work in this paper suggests that although the farmers resettled in the early 1980s started out in a completely different political and economic environment, and the numbers involved were much smaller, the challenges in establishing their farms and communities were to some extent similar to those reported for the small-scale fast-track farmers today.[1] Despite the heterogeneity in socio-economic backgrounds of the settlers and challenges in clearing land and

---

[1]It is important to note that the old resettlement programme mainly involved individual smallholders, labelled A or A1 farmers in both resettlement phases, and no large- or medium-scale A2 farmers – a class of farmers created by the FTLRP.

starting agricultural production in areas with different soils and rainfall patterns, the findings from the ORAs show that many households were able to develop their farms and invested strongly in building their communities. In non-drought years families were able to grow enough food to feed themselves, and most also grew some cash crops. Many farmers increased productivity and invested strongly in cattle wealth. By forming both religious and non-religious community-based organizations (CBOs) and through inter-marriage, households were also able to solve problems of collective action and create new communities. These outcomes were achieved over a considerable period of time, and settlers in the ORAs initially kept social and productive connections to their home areas, suggesting resettlement is a process rather than an event. Evaluating the success of a land reform programme should thus take a long-term perspective.

The new data collected since 2007 presented in this paper documents how the economic crisis that has afflicted Zimbabwe since 2000 has had profound effects on livelihoods in the ORAs; with the limited availability of inputs, cropping patterns have changed and agricultural incomes have generally declined. The opportunities to earn additional income from non-agricultural activities have also dwindled and many households have been forced to disinvest: average livestock holdings decreased significantly between 2000 and 2010 and the number of livestock-poor households significantly increased at the same time. Despite these adverse conditions, a small group of farmers has been able to increase their livestock holdings over the past decade. These more recent developments in the ORAs suggest that any review of the effects of the FTLRP on the livelihoods of smallholders should take such factors into account.

In the next section we describe the research locations and data collection. Section three provides insights into the socio-economic development of the resettlement farmers and their communities in the first two decades after resettlement. Section four discusses the influence of macro-economic changes over the past decade on livelihoods in the study areas. We review migration of household members, changes in agricultural production, the opportunities and constraints for alternative livelihood activities and changes in investment and wealth. Section five concludes.

Data and research locations

This account is based on an unusually rich data set covering a household panel survey, the Zimbabwe Rural Household Dynamics Study (ZRHDS),[2] supplemented with data on social networks and in-depth life histories. The ZRHDS covers three ORAs in three provinces, Mashonaland Central, Mashonaland East and Manicaland, and a smaller group of households in communal areas located adjacent to the resettlement schemes. The schemes were all resettled in 1980, 1981 and 1982, and the three communal areas were chosen for study because they were major sources of the small-scale farmers who elected to be resettled in the early 1980s. In most cases, data are available on the households extending as far back in time as the early 1980s, but for the communal area households the retrospective data begin only in 1997.[3] These

---

[2]The ZHRDS is the longest continuous panel study of households ever undertaken in Africa. The full panel is some 500 households resettled in the early 1980s and 150 households from neighbouring communal areas.

[3]Information on resettled households was collected in 1983/84, 1987 and annually from 1992 through 2001. Households from communal areas were surveyed annually from 1997 through 2001.

data have been the basis for several publications that are utilized in this paper. The post-2000 information presented here was collected in 2010 for a sub-sample of households in 18 communities included in the ZRHDS, some of which had also been surveyed during 2007–09.[4]

The three ZRHDS locations reflect considerably different agro-ecological and economic potential. The 11 Mashonaland Central sites are all in Natural Region 2 (NR2), with comparatively high rainfall in a normal year and with an inherently greater potential for a wide range of cropping. There are pronounced local differences in soils however that influence the crops that can best be grown. Bushu, the communal area (CA), has, as would be expected, soils of lower potential than neighbouring Mupfurudzi Resettlement Area, situated on former commercial farmland. Cotton has been the primary cash crop in the area for decades, but tobacco is increasingly being grown by farmers in Mupfurudzi. With high rainfall, malaria is endemic, along with many other diseases and parasites associated with water vectors. Both adult and child nutrition indicators are typically poor (Kinsey 2010). The nearest towns are Shamva, Bindura and Mount Darwin, and travel connections are relatively easy. In recent years, the closure of Madziwa Mine (gold and nickel) has had adverse effects through a reduction in economic opportunities.

The eight Mashonaland East sites are all on the boundary of NR2/NR3 and possess more restricted agricultural potential than the sites in NR2. The soils are almost universally sandy and the area is ideal for tobacco, but tobacco-growing – despite encouragement by agricultural extension officials – has been undertaken only to a limited extent. The area is cool and malaria-free. Maize remains the major cash crop, and there is considerable market gardening of products that can be sold in the nearby township of Hwedza and several smaller service centres. Transportation is reasonably accessible.

The nine Manicaland sites all lie in areas of relatively low potential in NR4. The low average rainfall means the area is marginally suited to cropping and better endowed for livestock-keeping. In addition, the predominantly light, sandy soils favour drought-tolerant crops. There are however pockets of soil where there is a perched water table and where rice can be grown. Child health and nutritional status have consistently been far better in this area than in either of the other two (Kinsey 2010). These sites are remote, with extremely poor roads, and transportation is a problem. The recent discovery of diamonds in the area has provided a stimulus to a range of non-agricultural activities and has witnessed a boom in small-scale retailing.

Establishing farms, communities and livelihoods in the old resettlement areas

Establishing farms and communities

Despite the application of a common set of criteria for the selection of settlers (Deininger *et al.* 2004), there was considerable heterogeneity among resettled households in the early 1980s. The majority of settlers were farming when they applied for resettlement, either on their own account or with their fathers or brothers. Some (also) had other working experiences as seasonal labourers on the commercial farms or in town where they were employed as drivers, welders, builders, cooks, miners, boiler-makers, storekeepers, clerks or teachers. Not many settlers

---

[4]In pre- and post-2000 comparisons, we often limit ourselves to a comparison over time between the 194 households that were interviewed in both 2010 and the earlier years.

were refugees. Some settlers lived in town prior to resettlement, while the large majority of settlers came from villages in the (then) Tribal Trust Lands. Only a small minority of the early settler households, some four percent, were female-headed (Dekker 2004a).

For many households resettlement was not an abrupt relocation from one place of residence to another. Some women and children would stay behind in the former home area, especially when schools were not yet established in the new areas. For the first years farmers also continued to plough fields back 'home' to hedge the risks of agricultural production. In 1984, 40 percent of the settlers still cultivated fields in their home area – on average on five acres. Similarly, settlers initially kept memberships in community-based organisations (CBOs) in their previous home area (Dekker 2004a).

Because of the procedures used in selecting the new settlers and allocating plots, households in the new communities, often in contrast to their home areas, were typically unrelated to and unacquainted with each other, often not even belonging to the same lineage. Yet, the new inhabitants had to solve various problems of collective action relating to natural resource management, inputs for agricultural production, inadequate access to financial and other services, and the management of risk and uncertainty. To varying degrees, the households addressed these problems by setting up CBOs and through inter-marriage (Dekker 2004a).

Barr (2004) documented the establishment of CBOs[5] in the villages in the first two decades after resettlement and compared the development of memberships with that found in villages in communal areas. Resettled villages have many CBOs and reached the 2000 level of CBOs in communal villages already in the mid-1980s. By 2000, the mean number of group memberships in resettled villages was almost double that of the communal areas (5.43 and 2.84 respectively). As religious memberships are very similar between the CAs and the ORAs, this difference is mainly related to differences in non-religious memberships.[6] Villages in communal areas have far fewer cultural, human development and women's groups, and membership in agricultural and sports groups is considerably lower in communal villages compared to resettled villages.

In a detailed analysis of the establishment of CBOs between 1980 and 2000, Barr et al. (2010) show that the associations in the resettlement villages are fairly homogeneous and inclusive. The authors find no evidence that the poor were excluded or chose to exclude themselves from the groups, and there is no evidence of grouping on wealth (the rich grouping with the rich and the poor with the poor). Actually, the villages in which average wealth was lowest at the time of resettlement had the densest networks of CBO membership throughout the two decades. In these villages, it was the wealthier households who started setting up the CBOs, possibly because clearing the land and building homesteads was easier and faster for them. Initially, the poorer households would have had little time for anything but land clearance and building. However, by 1985 they were just as engaged as the wealthier households and remained so over time. Women were also actively engaged in CBOs.

---

[5]Barr uses a broad definition of CBOs, referring to all the nonpolitical-party groups (clubs, religious groups, unions, revolving savings and credit associations, burial societies, etc.) that were established in the village or that attracted membership from the village.

[6]In communal areas, the average number of non-religious groups that at least one household member belongs to is 1.52, compared to 4.09 in old resettlement areas.

Although female-headed households were less likely to become a member of CBOs at some times, they are not excluded and do not choose to exclude themselves from associating with male-headed households: in fact they are more likely to share memberships with them than with other female-headed households. Finally, it was observed that households that arrived considerably later tended either to be excluded from or chose not to join existing CBOs in the village and appeared instead to set up new CBOs with other latecomers.

The experience of the establishment of farms and communities in the ORAs shows many similarities with the small-scale A1 farmers in the FTLRP. In both programmes the established communities comprised settlers with diverse social and economic backgrounds.[7] Also, settlers invested in various CBOs to meet collective action and production challenges. Murisa (2010 and this collection) documents the importance of farmers' groups in the FTLRP communities, while Scoones *et al.* (2010) stress the significance of religious affiliation for these purposes.

The insecurities surrounding the (re)allocation of land under the FTLRP (Matondi 2011, Scoones *et al.* 2010), were, at least to our knowledge, not a feature of the first phase of resettlement.[8] Yet, ORA settlers kept, and to some extent still keep, links with their former home areas for both social and economic reasons. This suggests that the slow abandonment of links to the communal areas that is documented for the FTLRP schemes (Scoones *et al.* 2010, Matondi 2011) cannot be explained by reference to tenure insecurity alone. Resettlement is a process that involves establishing new farming plots, new homesteads, new community structures, new public services, etc. And this clearly takes considerable time.

Establishing livelihoods

By the end of the 1990s, almost two decades after resettlement, the livelihood portfolios of the ORA households showed positive indicators in terms of welfare gains (Table 1). Compared to similar households in communal areas,[9] mean real household incomes of resettled farmers were significantly higher in 1999: Z$9255 for resettled households, compared to Z$5625 for communal households. The sources of income also differed considerably between the two areas. For resettled households, agriculture was by far the dominant source of income. These households also earned some income from off-farm businesses and livestock growth and reported small contributions from remittances and female income. Although agriculture also constituted a primary source of income for communal households, it contributed only 35 percent to total income. Off-farm business revenues, remittances and off-farm income contributed more to total income compared to the resettled households. Livestock produce, livestock growth and female income are proportionally comparable for resettled and communal households. In this respect, communal households are comparable to other rural households in Africa that earn on average 30–50 percent of total income outside the agricultural sector (Reardon 1997).

---

[7]See for example Scoones *et al.* (2010) on FTLRP communities in Masvingo province and Matondi (2011) for Mazowe District.
[8]Nevertheless, a number of farmers who had been resettled in the ORAs 26 years earlier lost some or all their land in the political intimidation that surrounded the 2008 elections.
[9]Similar households in communal areas are households that applied for resettlement in the early 1980s but were not selected into the programme.

Table 1.   Livelihood portfolios of resettled and communal households, 1999.

	Resettled households	Similar communal households
Household income (Z$ 1995)	9,255	5,625
	(percent)	
Of which:		
Crop income	65	35
Off-farm business revenues	11	20
Livestock produce	1	1
Livestock growth	10	9
Remittances	5	16
Female income	6	4
Off-farm income	2	15

Source: Based on ZRHDS data and adjusted from Deininger *et al.* (2004).

In non-drought years, resettled households produced maize in excess of subsistence needs and most farmers produced cash crops such as cotton and – more recently – tobacco (Dekker 2004a). Maize production, even in subsistence farming, was based on a so-called high-input farming regime; 70–90 percent of farmers used hybrid maize seed in the 1999–2000 cropping season with the highest adoption rate in Mupfurudzi (90 percent), followed by 80 percent of the farmers in Mutanda and 70 percent in Sengezi (Bourdillon *et al.* 2003). Thus most farmers participated in input (seeds, chemicals and fertilizer) and output (crops, cattle) markets, while the labour market, especially on neighbouring commercial farms, offered opportunities to generate income for the poorer households or, in times of stress, to meet cash needs for school fees, medical bills or funeral costs (Dekker 2004a).

Compared to communal farmers that applied for resettlement but were rejected, resettled households performed better in agriculture, partly as a result of the relevant agricultural extension services offered to them (Owens *et al.* 2003). They have higher maize yields and higher crop incomes per unit of land, have larger herds, higher expenditures and higher valued capital stock. Although the differences in crop income, the value of livestock and the value of capital stock persist on a per-capita basis, the differences in per-capita expenditures are lower than expected (Deininger *et al.* 2004) because of the larger household sizes in resettlement areas (Table 2).

The investment in livestock wealth in ORAs has been documented in more detail by Kinsey *et al.* (1998), who stressed the welfare gains of the land reform programme for poor families. Their study showed that some 15 years after their resettlement, 90 percent of the households owned cattle and the average herd size had more than doubled – with a mean herd size of 10 animals. Based on these findings, they emphasized that, 'The problem of zero stocks (as a result of a series of negative shocks) ... is not relevant for our population as a whole' (p. 98). Moreover, they argued that cattle provided one of main mechanisms by which people have been able to make better use of the land they have been given.

This earlier work on the old resettlement areas clearly demonstrates the centrality of agriculture in the livelihoods of the settlers, a finding that is also replicated in the FTLRP, for example in the studies by Scoones *et al.* (2010), Mbereko (2010) and Matondi (2011). At the same time, and as a consequence of the changed macro-

Table 2.   Mean household sizes, 2000.

	Resettlement areas	Adjacent communal areas
All areas	9.2	6.1
Mupfurudzi	10.9	6.5
Sengezi	7.2	5.2
Mutanda	9.6	6.7

Source: ZRHDS data (Dekker 2004a).
Note: All differences statistically significant (0.000).

economic circumstances, livelihoods in some FTLRP communities have been highly diversified.[10]

Interestingly, larger household sizes have been documented for other ORAs as well (Harts-Broekhuis and Huisman 2001) and can be explained by at least two factors. First the need for labour and the relative success of agricultural production attracted additional household members. Second, an oversight in the design of the original resettlement programme meant there was no provision to cater for the demand for land by the second generation, except for the succession by one child after the death of the original recipient of the land. Since it was not easy for (married) sons of the settlers to obtain their own land, either within or outside the resettled areas, many of them remained with their parent(s) and ploughed a portion of their parents' arable fields (de facto subdivision). These difficulties in obtaining land to establish their own farms potentially could have led to a reduction in per-capita benefits from the first generation of land reform. Growing family sizes could have further eroded per-capita expenditure and increased pressure on available land resources, eventually changing the nature of resettlement schemes and making them resemble communal areas. This scenario remains speculative however and has been overtaken by history.

Changes over the past decade

Migration

The opportunities to obtain land under the FTLRP, combined with the macro-economic developments that crippled agricultural production, employment and economic activities more generally, have had a profound influence on the mobility of people (Hammar *et al.* 2010). In the 2010 survey, we collected information on almost 3000 individuals that at one time lived on the homesteads of the respondents interviewed (Dekker and Kinsey 2011). Table 3 indicates that just over half of these individuals no longer live with the respondents, and the large majority (more than 85 percent) of those who no longer live there left after 1999. This pattern is strongest in resettlement areas and in the areas with the highest agro-ecological potential, particularly in Mupfurudzi where so many sons remained living and farming on their parents' plots after they got married and started their own families. This pattern led to considerable changes in household size between 2000 and 2010, as reported in Table 4.[11]

---

[10]See, for example, Mutupo (this collection).
[11]Note that changes in household size reflect net-migration and natural growth, accounting both for household members moving out (including deaths) and household members moving in (including newborns).

Table 3. Percentage of household members migrating after 1999.

Location	Percent
Full sample	52
Old resettlement areas	53
Communal areas	45
Natural Region 2	59
Natural Region 3	53
Natural Region 4	40

Source: ZRHDS data

Table 4. Mean household size in 2000 and 2010.

	Old resettlement areas		Adjacent communal areas	
	2000	2010	2000	2010
All areas	9.2	7.4	6.1	5.9
Mupfurudzi	10.9	6.1	6.5	6.1
Sengezi	7.2	7.5	5.2	4.9
Mutanda	9.6	9.0	6.7	6.8

Source: ZRHDS data
Note: All differences statistically significant (0.000).

Looking more closely at the characteristics of the migrants indeed suggests it is the younger generation that has moved away over the past ten years. The average age of migrants was 26 years, but there is a clear difference in age between those leaving before 1999 (31 years) and after 1999 (25 years). If we look more closely at the relationship of the migrant to the household head, it is clear that after 1999, the majority of individuals leaving were adult sons, daughters-in-law and grandchildren (Table 5). In the full sample, 42 percent of the migrants who left before 1999 were adult children, their spouses and children. After 1999, this figure was 62 percent of migrants. Migrants often left as families, especially when seeking to take advantage of the opportunity to start an independent farm provided by FTLRP.

The importance of new farming opportunities is also reflected in the target locations for which migrants left, presented in Table 6. Other rural areas (most likely FTLRP schemes) were most often cited as the location to which migrants went.

Interestingly, there are some clear differences between settlement types and regions when it comes to the other locations that people went to. Movements within the village (through marriage, occupying vacant plots or moving to newly created residential homesteads)[12] are particularly evident in NR2 and NR4, while Harare and other urban areas are important destinations for communal households and

---

[12]In Mupfurudzi, some village heads decided to create new residential homesteads in their village to allow adult children and their families to occupy their own stands. The agricultural fields belonging to the original household heads were not extended and were still informally subdivided.

Table 5.   Adult children, their spouses and children as a percentage of migrants.

	Period when migration occurred	
	Before 1999 (n = 168)	After 1999 (n = 1157)
	(percentage)	
Full sample	42	62
Old resettlement areas	44	63
Communal areas	15	58
Natural Region 2	40	48
Natural Region 3	62	70
Natural Region 4	33	81

Note: All differences are significant at the 0.10 level, except for NR3 (not significant).

Table 6.   Current place of residence of migrant household members[a].

	Within the village (%)	RA or CA[b] (%)	Other rural area (%)	Harare (%)	Other town (%)	Outside Zimbabwe (%)
Full sample	18	9	43	16	10	4
Old Resettlement	19	8	46	15	9	3
Communal	15	11	31	22	15	6
Region 2	24	11	47	8	9	1
Region 3	8	5	40	29	8	10
Region 4	22	7	40	15	15	1

Notes: [a]Differences between settlement types and natural regions are significant at the 0.05 level.
[b]RA refers to those living in the same resettlement area but not in their original village.

households residents in NR3 and NR4, reflecting a historically stronger orientations towards work in town in these areas. Migration to other countries in the region or further away was strongest in NR3.

Members who left the household to settle in a fast-track scheme acquired on average 13 acres (5.3ha) of land in their new place of residence. The majority of farmers received somewhat less – around 12 acres – while a few received only one acre and six migrants were able to acquire more than 20 acres. More than 30 percent of the migrants who went to the fast-track schemes took cattle with them when they moved to their new farms (on average 1.8 with a range of between 2 and 40), and a similar percentage took farming equipment.

For migrants who relocated to a different stand in the same village, the percentages relocating cattle – and equipment – were much lower and many respondents reported they still share cattle and equipment, and possibly labour, with the intra-village migrants.

The members who left the households to look for work predominantly went to Harare, another town in Zimbabwe, or somewhere outside Zimbabwe. Many of them were successful in finding a job; only 12 percent failed to do so. The occupations most frequently mentioned were piece-worker, domestic worker, vendor, teacher or some other job in the public administration, security forces, skilled manual worker, or farm worker.

Migration in search of alternative livelihoods, either in new resettlement areas or abroad, possibly will have benefited the remaining households in terms of remittances. At the same time the young and able-bodied migrants may have deprived the remaining household members of labour, cattle and equipment to work in the fields. Similarly, the migration has affected the composition of intra-village networks. Changes in informal community networks are also reflected in more formal structures, such as CBOs, with the large majority of those that were operational in 2001 having ceased to exist.[13] Information from the recent surveys suggests this may have also altered mutual assistance patterns, an issue that deserves more attention but is beyond the scope of this paper.[14]

Changes in agricultural production

The multiple dimensions of the economic collapse that Zimbabwe experienced in the decade following 2000 had profound effects on the smallholder farmers studied in the ZRHDS research areas. Most notably, the production and distribution of inputs were hampered and hyperinflation, combined with delayed payments and shortages of money, made the income from agricultural production worthless, especially during 2007 and 2008. Although most input-providing companies are still active in the study areas, and some new ones have arrived, they have not been able to guarantee a sufficient supply of inputs to maintain production levels. They were also unable to cope with hyperinflation, and the supply chain was broken by the consequences of economy-wide failures.

These shortfalls in supply of inputs are reflected clearly in the constraints identified by farmers to their agricultural operations during the 2009/10 season. As shown in Table 7, although a quarter of the constraints mentioned by farmers related to rainfall alone, when combined the various constraints relating to inputs constitute almost two-thirds of the total number mentioned.

Across the past decade, the number of farmers using modern inputs (hybrid seeds, fertilizer and pesticides) has decreased, as have the quantities applied to the crops, as is illustrated by Figure 1. Figure 1 shows for the three resettlement areas we examine here the average fertilizer application rate per hectare planted for selected seasons from 1985/86 to 2009/10. The early annual average application rate was 285kg/ha, whereas the most recent was only 16.5kg/ha.[15]

Other constraints that feature strongly in Table 7 are related to labour and draught power for ploughing. These constraints are, at least partly, related to the post-2000 mobility in the study areas. The precise interplay between these different constraints and their effects requires more detailed analysis in future work.

Together, these constraints have affected cropping patterns, as is evidenced in Table 8. Here, we look specifically at the total acreage planted, the acreage under cash crops (tobacco and cotton), the acreage under food crops and the diversity of crops grown by smallholder farmers in the ZRHDS schemes. The figures show a

---

[13]Interestingly, churches have been much less affected, with 89 percent still functioning today.
[14]Some of the data collected since 2007 suggest profound changes in engagement in and patterns of mutual assistance among smallholder farmers. We find for example a decrease in assistance given for funerals and transfers of food, while tillage services and labour assistance are on the increase (see also Dekker and Kinsey 2011).
[15]While a comparable time series for communal areas is not available, the mean application rates across the CAs in 2008/09 and 2009/10 were 7.9kg/ha and 12.8kg/ha respectively.

Table 7.   Main constraints to cropping operations faced during the 2009/10 season.

Constraint identified*	Proportion of all constraints mentioned ($n = 297$)
	(*percent*)
Rainfall unreliable, erratic, low, late	25.6
Lack of seasonal capital to buy farming inputs	20.5
Lack of cattle/draught power	17.2
Inputs: difficult to obtain/late arrival/poor access/ scarcity/not available	10.4
Inputs: Not enough/none	7.7
Labour shortages: alone/small family/lack of manpower	5.4
Personal circumstances: disabled/ill/aged/can't manage farming operations/blindness	3.7
Crops destroyed/damaged by cattle	2.0
Lack of working capital (to buy livestock drugs/ hire labour/buy & repair equipment)	1.7
Crops attacked by pests	1.3
No/inadequate farming implements	1.3
Security of property, loss of land	1.0
Poor prices/markets	0.7
Other (poor soil, no extension service)	0.7
No constraints	0.7
Total	*100.0*

Source: ZRHDS data.

Note: *Although there is overlap between some constraints, they have been separated here according to the wording used and emphasis given by the farmers.

slight decrease in the acreage under cultivation, a strong and significant decrease in the acreage under cash crops, a marginal increase in the acreage under food crops and a significant increase in the diversity of crops grown. These patterns are very similar for both communal areas and resettlement areas, and are most strongly observed in NR2, where there always was a strong cash crop orientation.

Table 9 further suggests that farmers responded to the lack of labour and inputs by reducing their acreage under maize and cotton, and by planting other crops that require fewer inputs.[16] For groundnuts, beans, bambara nut, sunflower and potatoes, seeds are often obtained informally through local social networks (e.g. from the granary of a neighbour) and the crops are normally planted with no or very little fertilizer.

Shortening our perspective and examining changes only over the past two seasons both strengthens the observations above and uncovers some small signs of positive developments. The growing stability and improvement in the economic climate contributed to a wider availability of inputs. There has been a significant expansion of contract farming, with companies providing inputs for cash cropping under the proviso that the harvest must be marketed through them. And the donor

---

[16]The reduction in cotton cultivation in these areas is in sharp contrast to the national trend reported by Moyo (2009) and the experience of fast-track farmers in Masvingo province documented by Scoones *et al.* (2010) and can be explained with reference to geographical differences in input requirements and availability.

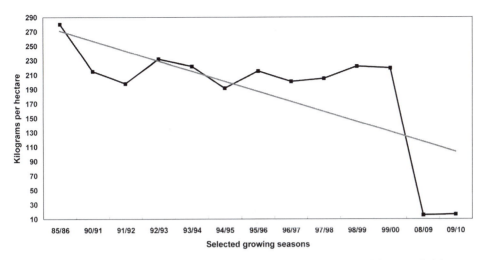

Figure 1. Annual average fertilizer application rates per hectare planted for resettled farmers in ORAs.

Table 8. Changes in cropping patterns, 2000–2010.

	Acreage[a]		P-value of the difference between 2000–2010
	2000	2010	
	(mean)		
Total	5.90	5.51	0.107
Cash crops	1.24	0.79	0.010
Food crops	4.67	4.72	0.567
Crop diversity[b]	3.32	3.93	0.000

Source: ZRHDS data

Notes: [a]Because most of the farmers in the ZRHDS study are of a generation that grew up without the metric system, we use acres here and elsewhere, converting to hectares when it seems helpful to the discussion.
[b]Crop diversity is the mean number of crops grown per household.

community – bilateral and UN agencies as well as NGOs – stepped in and began to import and distribute inputs, primarily seed and fertilizers.

The extent of this aid in-kind can be judged by the fact that well over half the households included in this study received at least some amount of free seed and/or fertilizer during the 2009/10 season.[17] As shown in Table 10, the sources of this aid were diverse.

The inputs provided were in all cases intended to support maize cultivation, and the amounts typically made available were sufficient to permit beneficiaries to plant approximately one acre of maize. At even moderate standards of cultivation, the output from one acre should have been over 1000kg, enough to feed a family of five for a year.

Among the households surveyed, free inputs were not distributed in a completely equitable fashion. While each individual donor undoubtedly had its own criteria for

---

[17]Six percent of the beneficiaries of free inputs received seed only.

Table 9.   Changes in average acreage under cultivation per household (1999/00–2009/10), disaggregated by crop.

Crop	1999/00	2009/10	P-value of the difference between 2000–2010
	(*acreage*)		
Maize	3.65	2.95	0.000
Tobacco	0.14	0.27	0.070
Cotton	0.90	0.25	0.000
Groundnuts	0.87	0.75	0.080
Small grains*	0.49	0.50	0.460
Finger millet	0.08	0.15	0.015
Sorghum	0.38	0.14	0.000
Millet	0.03	0.21	0.000
Bambara nuts (*nyimo*)	0.31	0.21	0.009
All beans	0.13	0.18	0.090
Sunflower	0.11	0.28	0.001
Potatoes	0.004	0.10	0.000
Rice	0.027	0.008	0.024
Other	0.00	0.01	0.030

Source: ZRHDS data.
Note: *Small grains include finger millet, sorghum and millet combined.

Table 10.   Identified providers of free inputs, 2009/10 season.

Provider	Percentage of providers
Africare	1.01
Extension officers	2.02
Catholic Relief Services	17.17
DAPP (an NGO)	21.21
Don't know	2.02
Dhiga Ugute (CADEC/SADC)*	4.04
FAO	42.42
Government	5.05
Named well-wishers	2.02
PLAN International	1.01
Red Cross	2.02
Total	100.00

Source: ZRHDS data.
Note: *The source of the *Dhiga Ugute* (literally 'Dig to fill your stomach') inputs could not be ascertained, but it certainly was not SADC.

determining eligibility for receipt of inputs, there are significant differences in the geographical distribution (Table 11). Most striking is the concentration of inputs according to agro-ecological potential, a finding that perhaps makes some sense when all the inputs provided are in support of maize cultivation. There is also however a striking bias in favour of communal areas.[18]

---

[18]This geographical difference in input provision partly reflects international donor policies on production support. The two priority areas identified by the donors were the communal areas and the old RAs, while the post-FTLRP areas were excluded from this support.

Table 11. Percentage of households surveyed that received free inputs, 2009/10 season, by area.

Area	Households receiving inputs
	(*percentage*)
By Natural Region	
2	93.2
3	34.4
4	16.9
Total	51.3
Significance level across NRs	*0.000*
By tenure regime	
Communal areas	70.0
Old resettled areas	44.8
Total	51.3
Significance level across regimes	*0.000*
By specific locality	
Communal areas	
Bushu	82.4
Chigwedere	100.0
Marange	25.0
Old resettled areas	
Mupfurudzi	96.4
Sengezi	9.1
Mutanda	14.0
Total	51.3
Significance level across localities	*0.002*

Source: ZRHDS data.

At the level of individual locality, the differences are even more pronounced although slightly less significant. All the respondents in one communal area – Chigwedere, for example – received inputs, while less than ten percent in the neighbouring resettlement area – Sengezi – did. Some of the results above can also be explained by the fact that NGOs have particular areas of operation. For example, in the areas surveyed, Catholic Relief Services was operating solely in communal areas, while DAPP was operating solely in a single resettlement area. Government also operated solely in resettlement areas. FAO, in contrast, was active in roughly proportionate terms across the communal and resettlement areas covered.

Nevertheless, across the sample as a whole, there are indications that the provision of inputs through aid, together with the gradually improving economic climate, have led to improvements over the dismal performance in the years following 2000, as is also reported more generally by Moyo (this collection). Table 12 compares cropping patterns and levels of fertilizer utilization across the two most recent seasons – 2008/09 and 2009/10. The focus here is on plots planted with specific crops rather than on the holding as a whole. The reason for this approach is that a single land holding (resettlement plot) often has multiple growers of any given crop. This is especially the case in larger households with resident married children. Thus, for example, one holding in a resettlement area may have four different plots of maize, each cultivated by and under the

Table 12. Changes in cropping patterns and fertilizer application rates between the 2008/09 and 2009/10 seasons, all areas[a].

	Total number of plots[b]		Proportion of total area planted (%)		Mean area per grower (ha)		Proportion of growers applying fertilizer (%)		Mean quantity of fertilizer applied per grower (kg)		Mean quantity of fertilizer applied per hectare (kg)	
	2008/09	2009/10	2008/09	2009/10	2008/09	2009/10	2008/09	2009/10	2008/09	2009/10	2008/09	2009/10
Maize	228	234	53.80	49.45	1.04	0.95	57.85	64.00	112.0	106.1	108.0	109.3
Groundnuts	174	187	13.65	15.62	0.34	0.37	0.00	0.01[c]	0.0	0.5[c]	0.0	1.2[c]
Finger millet	43	18	2.70	1.16	0.27	0.29	0.02	0.06	0.6	0.3	2.1	0.7
Millet	29	28	3.86	4.04	0.52	1.00	0.00	0.00	0.0	0.0	0.0	0.0
Sorghum	41	32	2.44	2.19	0.26	0.54	0.02	0.13	0.5	6.7	1.89	16.6
Bambara nuts (nyimo)	77	85	3.76	4.31	0.21	0.23	0.00	0.01	0.0	1.2	0.0	2.9
Tobacco	20	47	4.81	7.39	1.04	0.39	90.00	99.98	431.1	329.6	413.0	442.2
Sunflower	65	37	4.97	2.42	0.35	0.29	0.03	5.40	1.8	3.0	5.6	10.3
Cotton	26	41	4.46	8.71	0.74	1.04	84.62	85.37	106.0	96.0[d]	142.0	91.8[d]
All beans	65	62	3.32	3.06	0.24	0.22	14.29	14.52	6.0	8.3	26.1	38.0
Rice	6	5	0.14	0.25	0.10	0.02	0.00	0.00	0.0	0.0	0.0	0.0
All potatoes	70	49	1.97	1.23	0.13	0.09	0.05	0.04	0.6	1.8	4.4	16.3
All others	5	8	0.12	0.17	0.13	0.01	0.00	0.13	0.0	0.3	0.0	2.3

Notes: [a] A total of 193 households in both RAs and CAs in each season. [b] There may be multiple plots of any given crop for a household and, because of the frequency of shared responsibilities for many plots, there will normally be more 'growers' than plots. [c] It is believed that applications of gypsum have been misreported as fertilizer applications. [d] Underestimates because of the failure of some growers who had credit for inputs to specify the quantity of fertilizer they received, perhaps because they were diverting it to maize.

responsibility of a different individual, all of whom are members of the same household. Each plot can thus be cultivated according to different standards.[19]

Among the more general conclusions that may be drawn from Table 12 is that average maize plot areas and fertilizer application rates changed very little between the two years, and much the same is true for groundnuts, although the proportion of the total holding area planted with maize declined. The most striking change is the more than doubling in the number of tobacco growers and a strong increase in the numbers of tobacco growers applying fertilizer and in fertilizer application rates.[20] Also striking is the expansion in the proportion of holdings planted with cotton and the mean area per grower, however – despite almost no change in the proportion of growers applying fertilizer – there is a remarkable drop in the rate of application per hectare.[21]

Nevertheless, as noted above, overall fertilizer application rates remain extremely low. Between 2008/09 and 2009/10, resettlement area farmers only managed to increase their average application rate per hectare of planted land from 15.3kg/ha to 16.5kg/ha while communal area farmers increased from 7.9kg/ha to 12.8kg/ha.

The general shortage of inputs almost certainly helps explain why resettled farmers planted only some 52–55 percent of their available land in both 2008/09 and 2009/10 and why communal farmers planted only two-thirds of their available land in 2008/09 and only 57 percent the following season.

It was noted earlier that one characteristic that set resettlement areas off from communal areas during the 1980s and 1990s was the high levels of sales of agricultural produce. Crop sales were the major source of income for old resettlement areas. Crops were predominantly sold through the formal market (Grain Marketing Board, Cottco, Tobacco Sales Floor), especially maize, cotton and tobacco. Some farmers also sold maize and other crops through the informal market in their villages or just beyond. Data on the production and disposal of the harvest from the 2008/09 season, shown in Table 13, indicate that farmers on average produce less, are retaining a greater proportion of their crops and are selling much less (in terms of volume) than in the past. For maize, the percentage of households producing this staple crop has reduced slightly, while the volumes produced have gone down considerably from on average 2982kg per producing household to 1455kg.[22] The proportion of households selling maize went down from 32 percent in 1999 to 23 percent in 2009 and the volumes sold by these households reduced from almost 2000kg in 1999 to 1300kg in 2009. The GMB was not the dominant outlet for sales anymore as many farmers used maize in informal markets to barter for other consumption goods, services such as labour and

---

[19]Over two percent of growers over all areas do not reside on the holding where they are growing their crops.
[20]The decline in mean tobacco area per grower is to be expected since new entrants into tobacco cultivation are inexperienced, and they also have not yet constructed sufficient barn capacity to be able to cure the harvest from a larger area.
[21]A partial explanation for this phenomenon is that growers of cotton on credit reported to the research team that they diverted some of the fertilizer received for cotton to their maize crop.
[22]When compared to other smallholder farmers, these production figures are favourable but, with the larger family sizes in the ORAs, they allow for only a scant surplus – if any. Moreover, the reduction in production is evident.

Table 13. Production and sale of maize, tobacco and cotton in all areas in 1999 and 2009*.

	Maize		Tobacco		Cotton	
	1999	2009	1999	2009	1999	2009
Proportion of households producing	98%	95%	9%	7%	27%	9%
Mean production all households (*kg*)	2919	1380	92	72	458	64
Mean production, producing hh (*kg*)	2982	1455	977	1070	1651	772
Proportion of households selling	32%	23%	9%	7%	27%	7%
Mean sales, all households (*kg*)	642	314	57	72	457	64
Means sales, selling households (*kg*)	1958	1348	647	1070	1651	818
Proportion of producing hh selling	18%	8%	89%	100%	100%	99%

Source: ZRHDS.

Note: The production figures would be more meaningful if the rainfall could also be compared for the two seasons, but one of the effects of the FTLRP was the reduction in the number of rainfall stations from 3000 to well below 100 so valid inter-temporal comparisons are impossible.

productive assets such as cattle.[23] The proportion of households involved in tobacco production and sales has also reduced slightly, but for the households involved, production and sales remained similar or went up. Cotton production has suffered greatly over the past decade. The proportion of households involved went down from more than a quarter to a mere tenth and the volumes produced and sold halved between 1999 and 2009.

There are at least three reasons for these changes in behaviour. One is that a decade of hyperinflation taught farmers that selling (food) crops immediately post-harvest to raise cash and then, if necessary, purchasing staple food supplies later in the season was a losing game. Cash lost its value so quickly that it was far wiser to hold on to any surplus crops in the sure knowledge that physical stocks would lose value much more slowly. The second factor is that, with the collapse of the market for inputs, crop yields declined so that farmers simply had much smaller surpluses than in the past. A third reason is that the prices being offered for certain commodities by state or private market agencies (GMB, Cottco, etc.) were so low in real terms – and payment was often delayed – that farmers withheld crops from the formal market channels. Some farmers did sell small quantities to other farmers in their communities/area or to informal traders passing by.

Opportunities and constraints for alternative livelihood strategies

The reported changes in agricultural production and sales suggest that real incomes from agriculture must have decreased over the past ten years.[24] Moreover, alternative sources of income were less easily found than previously. Before, the adjacent commercial farms often provided opportunities to earn a seasonal income when times were tough (e.g. by working temporary jobs such as weeding or

---

[23]Barter trade was prevalent during the period of hyperinflation, but continues to be important even after the dollarization of the economy in 2009, as hard currency remains scarce in rural areas.

[24]The hyperinflation experienced over the past decade and the associated abandonment of the Zimbabwe dollar make it very hard to substantiate actual reductions in agricultural incomes.

harvesting during peak labour-demand times). With the FTLRP, these opportunities virtually disappeared for the farmers in the areas covered by the ZRHDS. Reports on the non-agricultural activities that farmers often were involved in reveal difficulties in accessing alternative sources of income; farmers did not have the money to invest in such activities (for example, trading) or no materials were available to make items for sale. At the same time, in the context of long-term decline in agricultural incomes, the local demand for products such as sleeping mats, pots, etc. dropped as well.

During the data collection exercise in 2010, households identified 252 non-farming activities that they were engaged in or had engaged in during the previous 12 months. This is an average of some 1.28 activities per household, a figure that compares with the number of activities reported in the early 1990s, but is considerably lower than the two to three activities households were engaged in during the late 1990s.

The types of activities reported are far-ranging, even with the compressed presentation in Table 14. Gardening, predominantly a woman's activity, is the primary activity reported, as has been the case in past ZRHDS survey rounds.[25] Gold-panning also regularly appears, as does petty trading. What appears to have assumed increased importance however is casual labouring. Given the virtual disappearance of commercial farms, it comes as a surprise that more than 20 percent of the sample reported casual labour as a source of income, a clear increase from the 14.4 percent reported by the same households in 1998/99. However only 2.9 percent of all activities took place on a commercial farm. This suggests that many have found opportunities to do piece-work in their own communities. This conclusion is supported by the fact that nearly half of all non-farm activities are undertaken in the same village where the participants live, while another quarter are in a neighbouring village. The increased prevalence of local casual labour may be related to the increased labour needs of households with aging heads and migrant able-bodied children and the cash-constrained economy. Not all the income earned from these activities is in the form of cash. Indeed, the most common form of payment is a combination of both in-cash and in-kind, but exclusively in-kind payments still make up almost 30 percent of the total. While more detailed analysis is needed, this pattern suggests that more people are seeking piece-work to meet specific needs that they cannot access easily in the marketplace, for example, working in a neighbour's field in return for food.

Household investment and wealth

Migration and the general macro-economic hardships have also affected investment and wealth in the old resettlement areas and communal areas studied in the ZRHDS. Previous work on the ZRHDS data has demonstrated that the resettled farmers invested strongly in cattle, and the sale of cattle or other livestock was an important coping mechanism during hard times such as drought, death, illnesses, etc. (Kinsey *et al.* 1998, Kinsey 2010, Bourdillon *et al.* 2003, Dekker 2004b).

---

[25]It should be noted that in Zimbabwe *gardening* is not considered *farming*. The term *farming* is applied only to field crops, whereas most gardening takes place in women's domains – around or near the homestead.

Table 14. Comparison of the types of non-farm income-earning activities undertaken by the same households in 1998/99 and 2009/10, as a percentage of the total activities undertaken.

Activity	1998/99 (n = 402)	2009/10 (n = 252)
	(percent)	
Selling garden produce or fruits	34.1	32.9
Casual labouring/piece-work	14.4	22.6
Gold-panning/diamond mining	3.2	3.1
Petty trading/vending: blankets/clothes/paraffin/knives/etc.	6.2	6.4
Building/carpentry/painting/thatching/brick-making/well-digging	7.7	5.6
Handicrafts/carving yokes/pottery/reed mats	8.4	4.4
Rearing and selling poultry	0.5	3.2
Metalwork: blacksmith/tinsmith/welding	2.2	3.2
Selling milk	3.5	3.6
Formal employment (full- or part-time)	1.5	3.2
Fishing and selling fish	0.0	2.0
Security/neighbourhood police/guard	0.0	1.2
Sewing and selling clothes	0.0	1.2
Repairs: shoes/umbrella/bicycles/clothes	0.3	1.2
Traditional healer/herbalist/treating people or animals	0.3	1.6
Selling firewood or grass	0.5	1.6
Barber/hairdresser	0.0	0.4
Pension	0.8	0.8
Rent: money from lodgers	0.0	0.8
Others: beer-brewing/goat-keeping/selling fruit trees, etc.	15.2	1.2
Total	*100.0*	*100.0*

Source: ZRHDS data

Here, using the data collected in 2010, we present two perspectives on changes in livestock ownership. One looks at the evolution of total cattle numbers and another considers an aggregate measure of cattle wealth. The first perspective is offered in Figure 2, in which we look at the changes in cattle numbers over time for the period 1983 to 2010. From 1983 to 2001, herd sizes grew remarkably steadily – at some nine percent per annum – from one year to the next at the same time that the proportion of families without cattle declined from some 40 percent to under 10 percent (Kinsey *et al.* 1998). The average family herd increased from four animals in 1983 to 10 in 2001 and, until 2000, cattle holdings were a good hedge against inflation.[26]

The years after 2001 however saw a dramatic reversal in the accumulation of cattle for at least two reasons. First, households were compelled to disinvest in order to survive. An extreme example of the way in which herds can be depleted for very little reward occurred in one of the study areas in 2008, when one cow was being exchanged for only 20kg of maize. Second, some migrants took part of the household herd with them, either as the rightful owners of the beasts or under a cattle lending or grazing arrangement, when settling elsewhere to farm. While we are unable to plot the trajectory of cattle numbers over the decade since 2000, by 2010 the average domestic herd was only just over five animals – the same as it had been in 1984, 26 years earlier and just after resettlement began. At the same time, the

---

[26]Because cattle prices moved in parallel with the Consumer Price Index (CPI) or even exceeded the rise in the CPI.

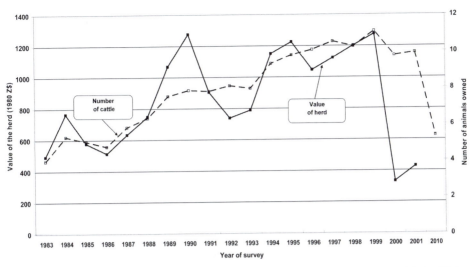

Figure 2.   Mean size and real value of domestic cattle herds, resettled households in ORAs, 1983–2010.

proportion of households without any cattle rose from less than 10 to more than 25 percent.

Another way to look at livestock wealth is to consider the value of livestock in terms of ox-equivalents in order to capture the heterogeneous composition of animal wealth, including goats, calves, heifers, cows and oxen (see Hoogeveen 2001). This is done in Table 15. Household livestock possessions are aggregated using weights based on 1995 market prices.[27] Although this measure is uni-dimensional it does provide a measure to evaluate investments in livestock by households over time. Resettled households had been able to expand their holdings in value terms from an average of a mere 2.7 ox-equivalents in 1980 to just over the equivalent of 7 oxen by 1998, but they experienced a significant decline in the value of livestock holdings to 5.9 in 2000 and to 3.6 in 2010.[28]

These declines were significant for all farmers across all settlement types and natural regions, with the exception of those living in NR3. The contracting herd sizes are visible not only when considering the average number of ox-equivalent beasts, but also when looking at the proportion of livestock-poor households in the resettlement areas. The ongoing economic crisis has in fact resulted in a situation where the problem of zero stocks that was non-existent in the mid-1990s (Kinsey *et al.* 1998) has become relevant indeed, with a doubling of the number of

---

[27]Based on the monetary value of animals sold in 1995, the following weights are used to construct an ox-equivalent value: 1.0 for a trained ox, 0.83 for a bull, 0.71 for a cow, 0.58 for a heifer, 0.59 for young oxen and 0.30 for a calf.

[28]As the decline in aggregate livestock wealth was accompanied by a decline in household size as well (and at least a third of the household members who moved out to the FTLRP areas took cattle and equipment with them), there is only a small net reduction in per-capita cattle wealth between 2000 and 2010. Although this may be interpreted as stability in individual wealth, the increasing proportion of cattle-poor households suggests that an increasing number of farming households in the old resettlement areas lost part or all of their entire resource base, which is an important input for agricultural production.

Table 15.   Average household herd size in ox-equivalents, 2000 and 2010.

	2000	2010	Significance level of the difference in means
	(ox-equivalents)		
Full sample	5.87	3.64	0.000
Communal Areas	4.06	2.42	0.014
Old Resettlement Areas	6.52	4.07	0.000
NR2	6.41	3.21	0.000
NR3	4.33	3.15	0.137
NR4	6.79	4.69	0.058
Full sample			
% of households without livestock	8.25	17.62	0.050
% of households with less than 1 ox.eq.	16.40	37.82	0.000
% of households with less than 2 ox.eq	22.68	47.67	0.000

Source: ZRHDS data.

households that do not have any livestock at all to 18 percent in 2010. The percent of households with just one or two ox-equivalents also more than doubled between 2000 and 2010, from 16 to 38 percent for one ox-equivalent and from 23 to 48 percent for two ox-equivalents. With so few livestock resources left, these households risk being trapped in poverty. Earlier work with the ZRHDS data by Hoogeveen (2001) showed that cattle are indeed an important productive asset: a team of oxen is required to achieve quality and timeliness of cultivation and to obtain a good income. Households without draught animals earn a lower income and, since cattle are expensive, once draught animals are lost or sold, it may be hard to obtain the resources through agricultural production to purchase new ones.

Increasingly, households are also selling cattle to meet their basic needs. While major expenditure priorities remained much the same between 2001 and 2010, households had a much smaller base to draw on in meeting them. In the straitened circumstances of 2010, some expenditure categories vanished altogether, including raising cash to help friends, to repay loans, and for savings. Business investment also declined. In contrast, several of the more wide-ranging expenditure categories – general purchases, paying for services, travel and education – assumed greater importance, suggesting that households' remaining livestock resources were being used more and more as a reserve to finance daily expenditure. The continuing depletion of cattle assets however threatens to reduce household herds below the point of reproductive viability, further reducing tillage capacity and the supply of milk.

The reduction in herd size and increase in the number of livestock-poor households were most strongly felt in Natural Region 2. In this region, the strong outmigration, with the accompanying translocation of cattle, combined with disinvestments reduced the average stock of livestock assets by 50 percent.

Two interesting observations in this context are, first, that livestock ownership by households in ORAs are still significantly higher compared to those in the communal areas, suggesting these households are at least partially able to retain some of the assets they have been able to accumulate. Second, we should be careful in looking at averages only. While most farmers experienced a decline in cattle wealth, some

15 percent of households have been able to accumulate cattle over the past 10 years. The processes leading to this differentiation are at present not entirely clear and cannot unilaterally be linked to participation in cash crop markets, migration processes or livestock endowments in the late 1990s. This group of farmers is potentially comparable to those 'stepping up' in the new resettlement schemes studied by Scoones *et al.* (2010). More detailed analysis is required to understand the determinants of these differential processes of accumulation and impoverishment.[29]

Summary and conclusions

Far too little empirical fieldwork has been done over the past 10 years to understand the full scope of the impact of the FTLRP and Zimbabwe's economic decline on small-scale agriculture and smallholder farmers. We conducted fieldwork to extend a larger, ongoing study that has been in operation since 1983 in three old resettlement areas and adjacent communal areas (from 1997). In this paper, we identify certain important trends affecting the livelihoods of smallholder farmers who were, on average, fairly well established in 2000 and argue that any discussion of the FTLRP should include reference to the earlier resettlement experience in Zimbabwe and acknowledge the impact of the devastating macro-economic environment on the opportunities for smallholder farmers to establish their farms and become agriculturally productive.

The existing data document some similarities between the old resettlement areas and the FTLRP schemes, notably the diverse social mix and investment in new social ties in newly established communities. Similarly, crop production is central in the livelihoods of the settlers, and the pre-2000 livelihoods of old resettlement farmers benefited strongly from extension and input provision. The benefits obtained from land reform were however shared with an increasing number of people since the design of the schemes did not take into account the land needs of the second generation. Although this demographic trend has been partly diverted by the FTLRP, the need to cater to the second generation is also relevant for the FTLRP. The existing data also highlight that resettlement has been an evolutionary process, and newly established farms and communities take time to mature. Thus evaluating the success of a land reform programme should never be done utilizing a short-term perspective.

The new data we collected across 18 selected sites in three different provinces also point to another fundamental conclusion. We reported an overall decline in the size of rural households, especially in resettlement areas, to which people were once attracted because of their agricultural and economic success. Now they are seeking livelihoods elsewhere, and more than half of all household members have migrated away from their original home areas since 1999. Earning a livelihood from farming became extraordinarily more difficult during the past 10 years. Inputs became – and remain – a major problem, and cropped areas – especially cash crops – have been reduced across our sample even though cropping diversity has increased, probably as a risk-management measure. Households have sought to modify their livelihood portfolios through a number of different approaches: by migrating, by

---

[29]Related to this, future work will aim at understanding to what extent the observed average decline of herd size in ORAs is related to the reported accumulation in the FTLRP schemes (for example by Scoones *et al.* 2010).

pursuing a wide range of non-farm activities and by using social networks and transfers. However, we find indications that these alternatives may not be as supportive now as they were in the past. Among the most serious outcomes of the economic turmoil of the past decade has been the depletion of households' livestock assets, even though a small minority have managed to increase their herds. The reduction in herds has not only weakened the overall economic resource base but it also has reduced the ability to manage the land resource through well-timed tillage. Moreover, the divergent trajectories in cattle assets underscore a worrying trend emerging from the broader ZRHDS study: the widening disparity in income and welfare levels among rural households who were originally provided equitable access to farmland.

Finally, the experience of the farmers in the ZRHDS also highlights that well-established smallholder farmers have been and are struggling with constraints and difficulties in pursuing a livelihood, which also affects and has affected smallholder farmers in the FTLRP. Despite these difficulties, and often with no support structure, the new settlers have built up their new farms and communities. An evaluation of the FTLRP should thus acknowledge the broader context in which any of these accomplishments have taken place.

References

Barr, A. 2004. Forging effective new communities: The evolution of civil society in Zimbabwean resettled villages. *World Development*, 32(10), 1753–66.

Barr, A., M. Dekker and M. Fafchamps. 2010. The formation of community based organizations in sub-Saharan Africa: An analysis of a quasi-experiment Leiden: African Studies Centre, ASC working paper, 90.

Bourdillon, M., P. Hebinck, J. Hoddinott, B. Kinsey, J. Marondo, N. Mudege and T. Owens. 2003. Assessing the impact of HYV maize in resettlement areas of Zimbabwe. FCND Discussion Paper No 161. Washington DC: International Food Policy Research Institute.

Deininger, K., H. Hoogeveen and B. Kinsey. 2004. Economic benefits and costs of land redistribution in Zimbabwe in the early 1980s. *World Development*, 32(10), 1697–1709.

Dekker, M. 2004a. *Risk, resettlement and relations: social security in rural Zimbabwe.* Tinbergen Institute Research Series No. 331. Amsterdam: Thela Thesis.

Dekker, M. 2004b. Sustainability and resourcefulness: Support networks during periods of stress. *World Development*, 32(10), 1735–51.

Dekker, M. and B. Kinsey. 2011. Coping with Zimbabwe's economic crisis: small-scale farmers and livelihoods under stress. Leiden: African Studies Centre, ASC working paper 93.

Hammar, A., J. McGregor and L. Landau. 2010. Introduction. Displacing Zimbabwe: crisis and construction in southern Africa. *Journal of Southern African Studies*, 36(2), 263–83.

Harts-Broekhuis, A. and H. Huisman. 2001. Resettlement revisited: land reform results in resource-poor regions in Zimbabwe. *Geoforum*, 32(3), 285–98.

Hoogeveen, H. 2001. *Risk and insurance in rural Zimbabwe*. Research Series No. 247. Amsterdam: Tinbergen Institute.

Kinsey, B., K. Burger and J.W. Gunning. 1998. Coping with drought in Zimbabwe: Survey evidence on responses of rural households to risk. *World Development*, 26(1), 89–110.

Kinsey, B. 2010. Land, food security and livelihoods in rural Zimbabwe: The lost war against poverty. Paper for the 2010 conference of The African Studies Association of the UK, St Antony's College, University of Oxford, 16–18 September.

Matondi, P. 2011. *Zimbabwe's fast track land reform programme*. London: Zed Books, forthcoming.

Mbereko, A. 2010. An assessment of the outcomes of 'fast track' land reform policy in Zimbabwe on rural livelihoods: the case of Gudo ward (Mazvihwa communal area) and Chirere area (A1 Resettlement area). Working Paper 3. Livelihoods after Land Reform Project. South Africa: PLAAS.

Moyo, S. 1998. The land acquisition process in Zimbabwe 1997/98. Harare: United Nations Development Program.

Moyo, S., W. Chambati, T. Murisa, D. Siziba, C. Dangwa, K. Mujeyi and N. Nyoni. 2009. *Fast Track Land Reform baseline survey in Zimbabwe: trends and tendencies, 2005/06.* African Institute for Agrarian Studies (AIAS). Harare.

Murisa, T. 2010. Farmer groups, collective action and production constraints: cases from A1 settlements in Goromonzi and Zvimba: Livelihoods after land reform in Zimbabwe. Working Paper 10. Livelihoods after Land Reform Project. South Africa: PLAAS.

Murisa, T. 2011. Local farmer groups and collective action within fast track land reform in Zimbabwe. *Journal of Peasant Studies,* this collection.

Mutopo, P. 2011. Women's struggles to access and control land and livelihoods after fast track land reform in Mwenezi district, Zimbabwe. *Journal of Peasant Studies,* this collection.

Owens, T., J. Hoddinott and B. Kinsey. 2003. The impact of agricultural extension on farm production in resettlement areas of Zimbabwe. *Economic Development and Cultural Change,* 51(2), 337–57.

Reardon, T. 1997. Using evidence of household income diversification to inform the study of the rural nonfarm labour market in Africa. *World Development,* 25(5), 735–47.

Scoones, I., N. Marongwe, B. Mavedzenge, J. Mahenehene, F. Murimbarimba and C. Sukume. 2010. *Zimbabwe's land reform: myths and realities.* London: James Currey.

Marleen Dekker is a development economist and human geographer and works as a senior researcher at the African Studies Centre in Leiden, The Netherlands. In her PhD research she studied the impact of land reform on participation in social networks and risk coping strategies in Zimbabwe. Her postdoctoral research on Ethiopia focused on the role of social networks in coping with illness and bargaining processes taking place within the household and the wider community. Her current work includes impact evaluations of community-based health insurance initiatives in Togo and Ethiopia. Her research work in Zimbabwe continues to focus on the dynamics in social networks in land reform communities established in the early 1980s.

Bill Kinsey has worked as a development professional for more than 30 years. He is currently a Senior Research Fellow with Ruzivo Trust, Harare, and has an honorary appointment at the Centre for the Study of African Economies at Oxford University. His major research interests centre on land-reform and resettlement programmes in Southern Africa and the effects of government policies on welfare and poverty, particularly child nutrition. These interests have been addressed primarily through an ongoing 29-year panel study of 550 farm households in Zimbabwe across three different agro-ecological zones.

Women's struggles to access and control land and livelihoods after fast track land reform in Mwenezi District, Zimbabwe

Patience Mutopo

Women's access to land and the shaping of livelihoods after fast track land reform should be viewed with a new social and economic lens in Zimbabwe. This paper examines the extent to which negotiations and bargaining by women with the family, state, and traditional actors has proved to be useful in accessing land in one semi-arid district, Mwenezi, in southern Zimbabwe. Based on multi-site ethnography, it shows the complex and innovative ways women adopted in accessing land and shaping non-permanent mobile livelihoods. I challenge the assumption that Western notions of individual rights to land are the best mechanisms for women in Africa; rather it is the negotiated and bargaining processes that exist in patriarchal structures that lead to cultural contracts enabling women's land access. Off-farm activities involving trading in South Africa became a major activity undertaken by the women. Trips to South Africa intensified due to land acquisition, leading to new market searches beyond national borders. The role of collective action and women's agency in overcoming the challenges associated with trading in South Africa is examined within the ambit of the livelihoods analysis.

Introduction

Discourses on land and livelihoods[1] have gained momentum in most Eastern and Southern African countries. Shifts in policy debates result in increasing importance being given to land use and management. Adding the issue of non-permanent mobility[2] to the familiar gender[3], land and livelihoods debates offers a new way to explore gendered mobility influenced by access to fast track farms, which have become the focus of attention, especially since 2000 in Zimbabwe. Attempts at agrarianisation processes with women accessing land have led to intensification of mobile livelihoods. Rural women are increasingly dominating non-permanent mobility, which has traditionally been a male domain. Female mobility with the

---

[1]Livelihood refers to the capabilities, assets and strategies that people use to make a living and to achieve food andincome security through engaging in a variety of economic activities.
[2]Non-permanent mobility is defined as the irregular movement of men and women. It could be either internal orexternal. The aim of the movement that people embark upon is to gain better living conditions or work and earn an income. Non-permanent mobility is also used interchangeably with irregular migration in most studies on migration in human geography.
[3]Gender refers to the socially assigned roles and relations between men and women. Gender attributes are not biological or natural. Gender relationships relate to a range of institutional and social issues rather than a specific relationship between men and women.

aim of securing sustainable livelihoods opens up new gendered geographies of mobility, with rural women participating more in the public space in search of better livelihood pathways. A symbiotic interdependence exists between women's access to land and mobile livelihoods, and this ought to be considered with primacy in rural development thought and practice in Zimbabwe.

Formal and informal processes of land access

The Government's Utete Report on the Fast Track Land Reform Programme (FTLRP) (2003, 40) points out that women were a small minority (18 percent) of those who benefited in terms of formal processes of accessing land, such as the offer letter in the A1 resettlements and leases in the A2 resettlements. A distinction should be observed in terms of some women benefitting from the formal processes and others accessing land as a benefit that emanated from family relations, negotiations, and marriage bonds. I use the the term access to refer to the social and political relations that mediate acquisition and use. Ribot and Peluso (2003) emphasise that different circumstances change the terms and conditions of access and this changes the specific individuals or groups most able to benefit from a resource. Land access from family-level negotiations has not been given much attention in the fast track land reform yet these mechanisms have been very important and have changed the role of women in land access and utilisation in a patriarchal culture.

O'Laughlin (1998, 2002) brought attention to the existence of missing men in Southern Africa's rural areas due to labour migration to South Africa. This created opportunities for women to gain access to land and control production activities. Women-headed households increased as men were missing demographically in rural areas, which could also be a result of urbanisation. O'Laughlin (2009) has termed this phenomenon in which women accessed land as 'relatively secure access to land'. Women-headed households assumed a new form of power over livelihood organisation, and subsistence agriculture remained a key survival strategy. The liberation wars did not even dislodge the women; most of them remained fixed on the land. This reflects that when women become key actors in rural livelihoods, new forms of power trickle down to them, and that power leads to new configurations in land use and livelihood options. Pelizolli (2010) demonstrates that women have taken over the land in a Chokwe irrigation scheme in Mozambique, competing with the few men left, as most men have left to look for full-time employment.

In this work my intention is to show the complex and often hidden means of access to and control over land in Mwenezi district that are adopted by the women. Moyo (2011) asserts that 14 percent of women benefited countrywide; Scoones *et al.* (2010, 55) note that women were beneficiaries 'for eight percent of the A2 plots, 14 percent of the A1 villagised resettlements, 13 percent of the A1 self-contained sites and 15 percent of the unofficial sites, which were not formally registered under fast track and are so prevalent in Masvingo Province'. It is in the unofficial settlements that most women received land in Masvingo Province. Matondi (forthcoming) reports a figure of 14 percent of women acquiring land in Mazowe district. In Zvimba district, 25 percent of A1 beneficiaries were women and 22 percent of A2 beneficiaries were women (Murisa 2007). These figures are actually higher than those found by the Presidential Land Review Committee (Mazhawidza and Manjengwa 2011).

Variations in gender disaggregated data reveal the fact that women in rural set-ups had different participatory mechanisms and motivations for participating in the fast track land reform processes. In as much as mobilisation campaigns were held in the different districts, in Mazowe women played a leading role in the mobilisation campaigns during the land occupations as compared to other districts (Sadomba 2008). I disagree with Matondi and Sanyanga (forthcoming) who argue that the gender statistics hide more than they reveal the extent to which women were ostracised during the fast track process. What should be borne in mind is that the quantitative data present women's entry point into the new farms and also the role they played as individuals in acquiring land.

The fast track presented a life opportunity for most women that had never happened in the history of land relations in Zimbabwe. In comparison to the land resettlement programme in the 1980s and 1990s, the number of women gaining access to land has escalated under fast track, which may be attributed to changes in the way the societies are evolving. In the aftermath of economic structural adjustment programmes, land has emerged as an asset that appreciates in value such that both men and women value the importance of its ownership.[4]

Presentation of the study begins in the next section with a discussion of the methodological and conceptual approach. An introduction to the situation before and after fast track land reform at the study site in relation to how land and livelihoods were constructed is presented as this helps in understanding the case study. I then spell out the methodology of ethnographic observation on which the study is based, including a description of the locality and the terrain to which it had to be tailored. This is followed by a short discussion, building on comparative literature, to spell out the problems women have in getting access to land, and relating this to the structural contexts from customary social relations and from changing colonial and post-colonial practices in southern Africa, and Zimbabwe in particular. Section 5 then reports findings from the literature and from the case study about the complex manoeuvres whereby women acquired land within and beyond FTLRP. The next three sections present the key findings of the investigation on women's role in farming, their incorporation of a pattern of migration to expand their livelihood options which is made possible by the prior access to land, and finally the new livelihood strategies pursued and how they depend on women cooperating.

The case study

This study takes a livelihoods approach (Hebinck and Shackleton 2011, Scoones 2009, Bebbington 1999, Ellis 1998) that is informed by a rights-based analysis that centres on a non-western construction of rights (O'Laughlin, 2009, Sen 1999, Gready and Ensor 2005). It focuses on a detailed ethnographic study of one farm, Merrivale in Mwenezi district of Masvingo. The research focused on Tavaka village, Ward 4, where the beneficiaries have six hectares of farm land and shared grazing lands intended to provide between 30 and 50 hectares[5] per household. The area falls under

---

[4]This line of thinking has been been put forth by Moyo (2011), Jiririra and Halimana (2008) and Jacobs (2010).

[5]However it should be noted that these land coverage measurements differ between households, and are based on the government's policy of what constitutes the A1 model of

Natural Regions IV and V and receives an average rainfall of 450–650mm, (National A2 Land Audit Report, Ministry of Lands, Land Reform and Resettlement 2007, 24). The agricultural season of 2008–2009 was unusual as the area received 700mm of rainfall after almost a decade of dry spells (Government of Zimbabwe Meteorological Report 2009, 35). The major farming activities are livestock production, crop production and wildlife conservancies. The crops most commonly grown by the new farmers are groundnuts *(nzungu),* round nuts/bambara nuts *(nyimo),* maize, cotton, sorghum, millet, leafy vegetables and sunflowers.

New space, new owners and livelihood styles at Merrivale

Before fast track land reform Merrivale farm was home to one white farmer who had control over approximately 4000 hectares of land that was used for cattle ranching and wildlife farming. There were 50 permanent workers at Merrivale who included foremen, a domestic servant, a cook, a gardener and cattle herders. Female workers were mainly doing the household chores and the looking after cattle was done mostly by men. During wet periods there was need for extra labourers and 30–35 casual workers were employed.[6] In total there were at most 85 workers at the farm. All these workers came from the surrounding communal areas of Lundi, Neshuro, Maranda and Chimbudzi. During the week they stayed on the farm as there were accommo-dations built for the farm workers. Some permanent farm labourers stayed with their families on the farm and others left their families at their communal areas in Mwenezi.

However with the advent of FTLRP they all left the farm and either retreated back to their homes or sought labour on the few remaining ranches in Mwenezi East such as Mukumi estate, on other farms still operated by white farmers, or on A2 plots such as those at Moriah, Qapane and Sweetwaters.[7] In recent debates on the livelihoods of former farm workers, Chambati (2011) stresses that before FTLRP only an estimated 71 percent of farm workers' wages conformed to the gazetted statutory requirements, but even so only 30 percent could meet their household needs from their wages. To supplement their incomes they were involved in other income generating activities, such as farming on plots made available by their employers or at their communal homes.

However since 2000, Merrivale has been home to approximately 300 families,[8] providing livelihoods for three times as many households as formerly. These families came from the surrounding communal areas, and their livelihoods are based not just on cattle rearing but also rain-fed agriculture, specialising in cotton, sunflowers, maize, ground nuts, round *(bambara)* nuts and sorghum. A key finding of this study is that additional smallholder livelihoods have been created. Women have been manoeuvring to create mixed livelihood portfolios. Women undertake diversified livelihoods through different pathways and styles and are cognizant of their own initiatives and capabilities in an area that has been historically seen as unsuitable for crop production. This refutes Chimhowu and Hulme's (2003) contention that the state had sought to create livelihoods in an unplanned way that is unrealistic in a dry

fast track farms in Zimbabwe. For a detailed account of the debates on the dynamics of farm sizes under fast track land reform see Matondi(2011), Moyo (2011) and Marongwe (2009).
[6]Interview with Margaret Spencer and Karen Caister, original owners of Merrivale before fast track, Pietermaritzburg, Kwa-Zulu Natal, South Africa, February 2010.
[7]Interview with village head, Tavaka village, January 2010.
[8]Mwenezi Rural District Council files on farm populations, accessed in February 2010.

area. The land at Merrivale has developed into a mixed farming livelihood zone that supports more people than the previous arrangement in which one owner had control over vast tracts of land.

Cousins and Scoones (2010), Scoones *et al.* (2010), and Wolmer (2008) have pointed out that livelihood enhancement and the different livelihood pathways that have emerged in the south-eastern low veld of Zimbabwe should be understood within the differing notions of natural and social landscapes in Southern Africa. I used a multi-sited ethnographic method of inquiry in order to understand the issues of women, land, non-permanent mobility and livelihoods at Merrivale Farm. Fitzgerald (2006) notes that in ethnography the 'field' is not simply a geographic place waiting to be entered, it is a space whose boundaries are constantly negotiated and constructed by the ethnographer and the members of the community. The resulting methodological mandate 'to follow the people' (Marcus 1995) as they travel between different localities takes seriously the mobility processes. Thus I had to do research in the sending (Zimbabwe) and the receiving area (South Africa). I resided with an elderly woman in one of the households at Tavaka village for a period of 16 months (April 2009 to August 2010). I undertook an in-depth study of 20 households based on interviews and close observation of the activities undertaken within these households. Other data were collected primarily by participant observation and in-depth interviews; individual life histories were important in understanding the land acquisition modes and livelihood pathways that women pursued. Desktop reviews of the existing literature on land rights, interviews, focus group discussions and transect walks were all used to investigate women's land-based livelihoods at Merrivale.

Out of the 20 households that were under study, 10 had mobile women travelling to South Africa frequently (mostly monthly) to sell pumpkin leaves, *nyevhe* (an indigenous vegetable), *ivhu repachuru* (termite mound), groundnuts, bambara nuts (known as round nuts in Zimbabwe), *mopane* worms and craft items made from wood and reeds. The other 10 households had varied characteristics: in five of them neither spouse was involved in non-permanent mobility. The other five had interesting patterns, with the women going at least twice in three months because their husbands were doing menial jobs in South Africa and they had to be at home to look after the children and farm the land. Selection of the study site and the households was purposive; this enabled me to understand the different coping mechanisms adopted by the different types of households that were under study. Snowball sampling was also employed as some respondents helped in recruiting and identifying others. As households in the village were interviewed during the initial stages of the research they would suggest other households with similar characteristics that suited my research theme.

Transect walks with the farm owners and the village head enabled me to find out information on agricultural production, cropping patterns, and sizes of the plots. Carrying out the transect walks with the villagers helped in identifying the land that was under conflict over use of resources, such as grazing lands. In theory these were to be subdivided, but in practice the grazing lands, according to my observations, were more of an open pool resource as the villagers believed in sharing the grazing and watering points for the livestock at the *mativi*[9] or Chatagwi dam. Management

---

[9]*Mativi* is a Karanga term which refers to natural water sources that do not dry up. They are associated with certain specific symbolism such as the sacredness of the place where they are found in Karanga society.

of grazing lands and water points as common pool resources was a practice that the settlers imported from their former communal homes. Interviews were undertaken with the district administrator of Mwenezi; an extension officer from the Agricultural Extension Office; officials from the Ministry of Agriculture, Lands and Rural Resettlement; and officials from Care International.

Focus group discussions were held at Chatagwi primary school. They were conducted with three separate groups; one group was comprised of men only, another of women and the other was a mixed group. Each group had a maximum of 12 individuals and a minimum of eight. I made sure that the number would not exceed 12 so as to enable orderly discussions. I prepared focus group discussion guides that I gave to the facilitators. I engaged independent facilitators who were known to the community and with whom the groups were comfortable in discussing all the issues so that I could acquire more information. I became an observer myself as I wanted to take notes on my own and not influence the process of the group interviews. At the end of each session I gave out an evaluation guide in which I asked the participants to express their views of the sessions and their usefulness and also bring to my attention the weaknesses of the discussions. The three focus groups enabled me to understand the different perceptions about how women accessed land: those who gave out the land, the way tenure arrangements were shaped and by whom. It also gave me insights into how women responded to questions in the presence of men, especially in the mixed group. Focus group discussions buttressed the information gathered on transect walks as well.

The research was multi-sited in that I also undertook trading trips to South Africa with these women. Participant observation was important in acquiring first hand information on how non-permanent mobility was organised by the women. As I accompanied the women on trading trips to South Africa, I also engaged in the selling of the products, so as to gain understanding of what happens in the domain of non-permanent mobility and trading by the women from Mwenezi. In-depth interviews were held with the women; some of these were carried out in Johannesburg's Bree Street with some of the women who had gone to sell their produce. I analysed the data using the focused coding method, where I built themes after extensive reviewing of my field data notes and personal observations. The major themes that developed out of these methods are the ones that I explore in this work to make sense of women's struggles to realise their right to land and build livelihoods in a new fast track resettlement village.

Women's access to and control over land

A World Bank Report (2003, 23) notes that 'if women have access to and control over land then family livelihood patterns improve. Most women-headed households have better management policies in terms of farming practices, marketing of produce and use of the income'. In Zimbabwe most women do farm their husband's land but they do not have any form of title (deed or customary acknowledged right) to that land. These take a form limited to 'land offers' from the Government of Zimbabwe in the fast track resettlement areas. Similar patterns also existed in the communal areas where land use and control has been under the domain of men and traditional leaders, making land a male-controlled resource. However in communal areas it should be understood that the interplay of customary, codified and colonial laws has disadvantaged women with regards to land access.

Makura–Paradza (2010), Cheater (1986) and Gaidzanwa (1994) note that in the communal and resettlement areas of Zimbabwe land access has become a field of contestation based on legal pluralism and semi-autonomous social fields that are constantly remodified by traditional authorities and local governments so they may continue controlling land. This creates a difficult social environment for women trying to assert a right to land, as much as their livelihood is tied to ever-changing rules governing this critical resource (Jacobs 2010, Tsikata 2003). Rural peasant women face exclusion. particularly from different political and traditional regimes that control land, in spite of their immense contribution to food production. This treatment of women has been accompanied by 'gender specific discursive justification' in Zimbabwe (Goebel 2005, 147). Land is a preeminent political resource for the state and for different actors at different geographic levels, and at this stage competing assertions of legitimacy and territoriality are always at interplay (Alexander 2003).

The Utete report treats women as a homogenous group, yet within fast track it is important to bear in mind that women were just not discriminated against 'as women'. Women, like men, are not a homogenous category since they belong to different classes, ethnic groups, races, political parties and professions. Derman and Hellum (2004) and Mazhawidza and Manjengwa (2011) note that under FTLRP most women from urban or rural areas were discriminated against along political, social and economic lines. This had more to do with women being treated as minors in everyday life in most cases. This has constantly been challenged by non-governmental organisations (NGOs) such as the Women and Land Lobby Group, which tirelessly advocated that women, in spite of their biology and marital status, had a right to land during fast track (Moyo 2011).

There was also differentiation between wives and women heads of households. Jacobs (2010), O'Laughlin (2009) and Goebel (2005a and b) have pointed out that the social importance associated with women's marital status is always an important factor in household relations in Africa. O'Laughlin (2002) insists that most households in Southern Africa have oscillating periods during which women sometimes assume headship of households, making social differentiation difficult. Married women have a better social standing than non-married women in most patrilineal societies in Africa. During the fast track process, the state and some traditional actors wanted to capitalise on using culture to exclude women heads of households from accessing land, but in practice women heads of households emerged as victors as compared to their married counterparts (Mazhawidza and Manjengwa 2011, Scoones *et al.* 2010).

Most married women could not engage with the fast track process as they had to continue managing their communal homes while the husbands participated in the mayhem phase. In as much as the situation tended to disadvantage women, in A1 schemes the single, divorced and widowed capitalised on the use of social networks and political party affiliations and gained new power forms that enabled them to acquire land. O'Laughlin (2009) and Agarwal (1994, 2003) insist that such complexities require the appreciation of communitarian approaches to rights based on respect for the family and evolving living traditions and laws, as these have always been critical in facilitating women's access to land empirically. In spite of the social customs the different categories of women assumed new land acquisition modes that were built on resilience, agency and changing the traditional land control spheres.

Chingarande (2008) argues that patterns of land ownership after FTLRP, as well as other rights such as tenancy, resettlement permits and leases, show that nearly a fifth of women have independent rights of ownership, access or control over land. Women in general lack the bargaining power to negotiate their right to land and livelihood. This is made clear by Article 23(3) of the Zimbabwean Constitution, which allows discrimination against women in matters involving family issues, and land is regulated as a customary law matter in most cases in Zimbabwean societies. Most male attitudes towards land redistribution and security of tenure with regards to women have not been positive due to socialisation processes which view land as a resource associated with men, and women as the primary providers of labour. Gender stereotypes affect women's public and private lives. Moyo (2007, 78) notes that, 'women received the least resettlement land even though their skills and labour tend to be critical to food production and rural livelihoods'.

In some cases women gained access to land through land invasions, as these subverted formal forms of patriarchal traditional or administrative authority (Scoones *et al.* 2010, Sadomba 2008, Moyo 2011). This gave the opportunity to some women, often widows, divorcees, and those ostracised from their communal area communities as they were able to join the invasions and gain access to land (usually in A1 villagised schemes). Scoones *et al.* (2010) noted that the women were valued in the invasion process and in the base camps as independent and able to help with a range of gendered domestic tasks such as cooking and singing in the base camps; they did not have important positions like base commanders but at times some assumed posts of secretaries and treasurers. For them it was a liberation, and an escape from other settings where as women they would not gain access to such rights. This explains in part the higher number of women having access to land through offer letters in their own right compared to communal areas where traditional patriarchal lineage authorities allocate land or the old resettlements where a bureaucratic administrative authority that is equally patriarchal in many ways allocates land. The 'quotas' for women in some 'Seven Member Committees'[10] earlier on also helped women manoeuvre access. While this may have changed in the decade since, it was an important feature of fast track, and worth mentioning.

Strategies used to acquire land by the women at Merrivale

In their quest to access land, women at Merrivale used various mechanisms to circumvent the male-dominated pattern of formal allocations. The land accessed was mainly virgin land on the former commercial ranch, which made it more agriculturally productive in the short term as it had not been farmed at all but had been heavily manured. Most of the women were ordinary rural women who had been raised in the communal areas and depended on farming, but they had different educational backgrounds. Toulmin (2007) and Berry (2002) point to general trends in Africa of access to land being 'negotiated'. Most of the married women pointed out that they would use their negotiating skills, for instance employing the right language to speak with their husbands, especially in their bedrooms. They would also know the right times to start the discussions when their husbands were in good

---

[10]These were semi-formal planning groups that were enacted during the fast track process so that order could be asserted in the beneficiary selection criteria and land demarcation exercise.

moods. Jackson (2003) emphasises the need for reflexive ethnographic methodologies in understanding gender as social relationships, subject positions and subjectivities, on meshing of shared and separate interests within households and on the power residing in discourse that shapes land access and security by women.

Sen (1999) offers three interrelated concepts for analysing gender and production relations: 'negotiative conflict, rights appreciation and cooperative bargaining'. Sen's analytical framework based on rights and livelihoods entitlements sheds light on the household as a dynamic site where various actors negotiate diverse spaces and strike bargains as part of efforts to position themselves for more equitable gains. In these processes, some members' rights are subjugated while others, particularly those of male members, are respected and asserted. Access to land by these women reflects the need to create a bargaining site where power dynamics erupt and have to be managed as much as possible to the benefit of the women. Tsikata (2009, 13) argues that, in their full ramifications, taken up in the literature on intra-household gender and inter-generational relations, 'questions of sexuality and sexual politics open a large arena of contestations and insights which have the potential to transform our understanding of livelihoods'. This implies that sexual relationships influenced land negotiations and bargaining processes in terms of land access by the women.

Access to land for most women involves them being in some form of relationship tied to men. This could either be a husband or brother. One woman said, 'the only land that we have as ours are the water gardens,[11] but men still control their use; the problem is our names do not even appear on the offer letters for the bigger portions of the land'.[12] Women are slowly realising that the information on documents has an impact on their tenure security. If the permits are to be handed over and their names do not appear on them, their right to inheritance in situations of divorce or death of a spouse is compromised. Outreach programmes undertaken by Women and Law in Southern Africa across the national electronic media soon after FTLRP in 2004 could help explain the women's awareness. My own assessment is that there are clear unresolved tenure issues in A1 settlements that apply to both women and men leading to insecurity among both groups; this leads to a subsequent fear of dispossession by the farmers. In the focus group sessions, the men and women kept pointing out that in as much as they believed in the occupation of the land by being there on the land physically, they still feared eviction as they lacked papers that proved secure ownership. I interpreted the papers they referred to as proof of the formal ownership of land such as title deeds. This view is further supported by Clause 7 of the offer letter which states that the offer 'can be withdrawn at any time, with the government having no obligation to compensate for any improvements that might have been made'.[13]

Fear of losing plots was raised by a number of men and women as they said how uncomfortable they were without codified tenure papers. Women do not have the independent right to use the water gardens and this affects the way they can manage and even plan the crops to grow as there is almost always male influence. I observed this myself at the gardens: as we watered the vegetables men would constantly visit

---

[11]Water gardens *(magadheni)* refer to the vegetable gardens that women control, which are also known as hand irrigated gardens in some instances.
[12]Interview with Mbigi, Tavaka village, May 2009.
[13]Offer letter for Mrs Ziracha, Tavaka village, June 2009.

and give suggestions with regards to management and vegetables to be grown and the way the water gardens had to be maintained. In their quest to develop the right to independent use of the water gardens, the women would adopt the strategy of appearing to be listening but would decide as a group to deviate from the expected rules laid down by the men. Deviation from the expected norm of accepting male advice, which is deeply rooted in rural Zimbabwean societies, became an important survival and coping strategy for the women trying to control water gardens.

At Merrivale, some of the women pointed out that they now had their own fields after being married for a number of years, as this established trust. These notions of trust were based on the number of years one had been married and also the interactions that one had with the community that had developed into relations that were considered important. Marriage offered established trust which could not be easily eroded, and women felt it was important to build the trust and earn it from their husbands. Observing cultural practices, an assessment and respect of traditional authorities, was also a key feature of building trust for the women. Such that the notion of security based on offer letters and legal courts becomes less important to the women as the state's offer letters are not crucial with regards to settlement in the community; what is important is physical presence on the land, hence this does not affect their daily lives.

The land they referred to as theirs constitutes the areas where they plant women's crops, which are traditionally bambara nuts, ground nuts and vegetables from the water gardens. These are classified as women's crops because they are considered as more labour intensive and requiring patience which can only be provided by women. The crops are seen as being of low market value and do not fetch a lot of money as compared to cash crops that are dominated by men. This has become traditionally accepted in households but with some reservations from most men. Men argued that women tended to invest more time in planting bambara nuts and ground nuts that if invested in cotton and maize would yield more production. Based on conversations that I had with the women, limitations on owning land by women are linked to a myriad of factors including cultural notions that support the belief that 'when married you do farm your husband's land and you cannot have your own separate fields until after a number of years have passed'.

The need to have individual income emerged as a factor that made women realise the importance of accessing and controlling land. The married women had now established close relations with their husbands and the clan and so could now safely ask for land. Most of these women had been married for more than five years and had stayed with their husband's families before coming to settle at Merrivale. For instance, Mrs Chimbari is married to a member of the village ward development community. She had managed to negotiate with her husband that she has three fields of her own where she can grow the crops she sells in South Africa. The negotiations took some time: 'nekuti vane rubato mumatongerwo eminda nenzvimbo ndaibvira kure kusvika zvafamba, bhedhuru rinoshanda sisi kutaura nyaya dzakadai [as he is involved in land governance within the village head I had to start from a social distance, as you know that the bedroom is the place to discuss such issues which are important such as land]'. She continued:

> The land still belongs to my husband, though it would be good for me to have ownership too. Especially if he dies, I am afraid of being chased away and losing

everything during asset distribution, yet I have been putting all my effort into this plot. (Field notes, June 2009)[14]

I realised that marriage has a deeper and more symbolic meaning for access to resources by women than one might think. Schmidt (1992) and Bourdillon (1992) assert that the Shona have always believed in marriage and that it carries important symbolic, social, political and economic weight for the couples and their families. After some time in the marriage the husband may delineate some fields to the wife or he may decide not to do so. Women had to maintain a certain level of social distance with their husbands as they negotiated for land access. Even so there was some form of tenure insecurity as the wives felt that they would lose the land in the event of the deaths of their husbands. Bebbington (1999, 2034) points out that, 'there is a conjunction between place and the reproduction of cultural practices that are important inputs and outputs of livelihood strategies'.

Marriage enabled the women to acquire land and have access to their own fields within the husband's family after a certain period of time; these are referred to as 'tsewu' in the Karanga and Shangaan traditions found in southern Zimbabwe. Jeater (2007) and Schmidt (1992) have observed that differences exist in the Karanga and Shangaan cultures in terms of language, rituals and marriage systems. In terms of their marriage set up the women have differing levels of power and influence in the household, for example, during one of my field visits I went to a Shangaan home and the wife just greeted me and continued talking to me while the husband cooked and served me, something which is seen as totally against the culture and disrespectful to one's husband in the Karanga clan. Most of the cultural practices of the Shangaan have been influenced by the Karanga traditions due to intermarriages and sharing land spaces such that they are now more mixed communities adhering to Karanga beliefs and mixing with their own cultural system.

The fields are of much importance to women; when I grew up my mother had a tsewu.[15] Bourdillon (1992) points out that the Shona mythology of agriculture should be understood from the fact that young and older women were able to survive as agriculturalists because they had a portion of their own individual land where they could grow crops and vegetables that fed the household all year round. However with the colonial system of governance and the reframing of women's rights to land, women have now lost this right to land in most instances: due to the imposed colonial norms of proper and respectable marriages there have been constant reinterpretations of culture such that men have often been able to remodify traditional cultural norms willy nilly.

The concept of the tsewu field is problematic, because it is also tied to the reproductive status of the wife in the Karanga tradition and not as much in the Shangaan culture.[16] In the Shangaan tradition if a woman is not married she could

---

[14]This extract is based on discussions I had with the women as I interacted with them either in their fields or as I helped them look for firewood in the village. These discussions were useful as they unravelled the critical dimensions of women's access to land at Tavaka village, Merrivale farm.

[15] The issue of tsewu (the female field acquired from the husband after giving birth to at least two children) was critically explained to the researcher by Mrs Elisah Chauke, the Assistant District Administrator of Mwenezi.

[16]Interview with Assistant District Administrator, Mwenezi Rural District Council Offices, Mwenezi, February 2010.

acquire the *tsewu* from her parents so that she could help look after aging parents and the extended family. In some Shona customs it is only upon the birth of the first child that a woman has to be given her own field. This makes it rather difficult for a woman to assert her rights to land as cultural considerations have to be appreciated in African societies. Deviating from the norm and being inquisitive are frowned upon.

At Merrivale, most of the women pointed out that such processes still happen and sometimes it was after the birth of the second child that they were given their own huts (*kupihwa imba yerumuro kubva mavamwene*) accompanied by a field. I interacted with polygamous households and from my discussions of their marriages, I found the *tsewu* field was given to every wife after giving birth to children but this depended on the availability of land. This had also been another factor in why the families had moved from the communal areas of Chimbudzi, God Star, Lundi and Gudo and settled at Merrivale: there was more land which could accommodate these cultural practices. Peters (2010) in a matrilineal study examined how daughters had the right to land and sons had to source and acquire rights of use to that land through marriage, negotiation and bargaining. Seur (1992) and Jackson (2003) has observed that in matrilineal societies, over time men's interests in the conjugal unit have come into conflict with those of the matrikin group because in order to command their allegiance as sons, they require reassurance of their inheritance rights in the parental property. The conflicts have been exacerbated by the scarcity of land in these societies, such that negotiations to solve the conflicts had to include both men and women as in patrilineal societies. I argue that gender relations are not set in any society but rather they are an outcome of negotiations and bargaining in a context of a particular cultural and political history.

Mushunje (2001) supports the contention by noting that the fact that a woman was allocated a piece of land after she had given birth to her first child implies that those who were so unfortunate to be childless were never considered to be worthy of a piece of land. Childlessness condemned women to landlessness and lack of sustainable livelihoods on the land they worked hard on. The vegetable garden was probably given to women as a piece of land that was meant to provide for the immediate consumption needs of the family.

However in as much as the system of allocating women their access to the *tsewu* was shaped by all these contradictions, most men pointed out that they would abide by the norm because they feared that if they did not it would have a bad effect on their children since the mother's wrath upon her death would affect their marriages or lives. 'I have given my wife the field, according to custom, because I do not want this to form part of the basis of an avenging spirit that will affect my children and family, I have to respect it since it has always been a custom, that even my father abided by'.[17] O'Laughlin (2009) argues forcefully that the 'rights' approach assumes a particular individualistic and liberal stance that does not appreciate the social and cultural norms of African societies – hence the failure of the neoliberal agenda in solving African land conflicts. I have demonstrated that gaining access to land and the right to use the land may not happen through such individualistic routes, but through more collective negotiation and bargaining (within lineages, families, marriages). These gender relations are highly skewed and power laden but it has often been overlooked that the involved people understand and know how to

---

[17]Interview, G.C.M.,Tavaka village, July 2009.

manage the cultural complexities. From my findings it emerged that women can have remarkable bargaining power in certain domains (kitchens, gardens, some crops, certain trade and market routes). Individual liberal rights approaches that are emphasised by the language of rights and World Bank policies and simple statistics of 'ownership' by an individual do not necessarily capture lived realities.

Tsikata (2009, 14) asserts that, 'after all, the conjugal unit remains an important site of negotiation of access to land and labour even in situations where it is embedded in more extensive kinship and residential units'. Questions of marital residence practices, the physical and social demands of child-bearing and -rearing, and the hetero-normativity, pro-natalism and son preference of many African societies have ramifications beyond personal freedoms and status. These are important in the structuring of men's and women's access to and control over land and over their own and others' labour and always prove to disadvantage women in the daily struggle for their right to access land and create a sustainable livelihood source in a Western liberal notion of individual rights, but within the communitarian systems of rights and livelihoods that exist in Africa, these negotiated settlements are appreciated and tend to be very effective in facilitating women's access to land.

Land in some instances was also acquired through the concept of the economy of affection,[18] particularly if one was related to the village head or some influential person like a war veteran. One woman from the village was able to acquire land for her family using this approach but the problem was that she did not have an offer letter as the plots had been given to them after the land reform process in 2003. The land reform process might seem as if it has ended; my own understanding is that it is still ongoing. This is buttressed by Article 5 of Zimbabwe's Global Political Agreement, which states that those who were disadvantaged in the exercise, particularly women and members of other political parties, can still have access to land since 69 percent of the land is still underutilised countrywide.[19] Women relied on the village head for support in staying on and farming their land. Though they stayed with their families, they had to make sure that they were always on good terms with the village head and never crossed him, as they feared that their land could be repossessed. Nevertheless, negotiation of reproductive rights resulted in women's access to some land and this led to their having the capacity to engage in other livelihood systems tied to land use.

Twenty-five percent of women in my sample pointed out that they were also renting land from some of the new farmers who were no longer interested in the resettlement areas and had gone back to their original homes in Maranda, Neshuro, Chimbudzi or Gudo communal lands in Mwenezi District. Out-migration of these farmers back to their original homes demonstrates that some people accessed land and felt that they had no security of tenure since they had offer letters, which are not recognised as title deeds that prove legal ownership. Hence, many felt insecure with the new plots. One person said that people felt 'disconnected' from the resettlement lands and so preferred to return to their villages. This could either refer to lack of security of tenure or lack of connection with one's ancestral spirits. Another noted in

---

[18]This refers to the importance of family and village ties in resource distribution in Africa. This is also known as the politics of villagisation in which distribution of resources is done with the family members and relatives of the distributor benefitting first before the other villagers that are not relatives of the distributor.

[19]See Matondi (forthcoming) for further critical reflections on the ongoing land reform process in Zimbabwe and what it means in the land and livelihoods revolution in Zimbabwe.

a focus group discussion, 'Some people feel that they do not own the land as we do not have any form of real proof except these offer letters, which others do not have. They fear that the owners might come and repossess the land'. Retreat to the communal areas was seen as a viable option since the traditional authorities know who owns which portion of land. The people felt that it was much better to be in their former homes where they really had a right to the land as they had farmed on it for most of their lives as opposed to the new spaces where they were not sure about whether they were the legal owners of the land due to the contestations over fast track land reform processes.

Thirty percent of the women pointed out that they employ young men from around the communal areas and plant their crops on their parents' farms, since it was easy to negotiate with your own kith and kin for farming land. One woman said in an interview, 'If you share the same totem[20] you will not go wrong; that person has to be your father'. Some of the women said that they had used the formal channels of applying for land for resettlement but they were still awaiting responses from the district and provincial authorities.

At Tavaka village, there are only five female plot holders out of the overall 40 plot holders, two of them are war veterans and another two inherited the land because their husband and mother-in-law respectively had died. Female inheritance in this case was not contested by the relatives as the son of the deceased woman was working as a teacher and so the wife was asked to take over the land since she had participated in the land occupations with her mother-in-law and father-in-law. This case was peculiar because most of the relatives of this particular family had remained in the communal area and had better lives so they were not interested in land in Tavaka village.

From the discussions that I had with the men and women about inheritance matters, I found out that there are two scenarios depending on how inheritance laws are applied. The death of a young or middle-aged male farm owner automatically meant that land had to be transferred to the surviving spouse, mostly the wife but under the guidance of a male relative. However in Karanga tradition the wife does not assume total control of the land as the husband has brothers and male children who also assume heirship. The wife would remain on the plot but decisions could be made by the brothers even if they were in communal areas in cases in which there were no male children or the children were still young. However in cases of elderly widowed women, they had full control of the land and livestock. They made decisions regarding the farming activities on their own. In such cases elderly women have full rights to land just like men as they are considered to be mature.

The other woman had acquired the land through kinship relations with the war veterans in the area. The women war veterans[21] had used their ties to access the land

---

[20]A totem is equated with a specific animal in Shona culture. Every person has a totem, this is done to respect family relations (*ukama*) and avoid intermarriages among people of the same totem. At Merrivale the most common totems are fish, elephant, rat, zebra and lion. For instance the men who belong to the fish (hove) are addressed as *muzikani* or *save* and the women as *sambiri*. People belonging to the fish totem are not supposed to eat fish as doing so leads to loss of teeth or sickness. Information on totems and the resultant effects of not adhering to them is based on my own cultural knowledge and my personal communication with the people at Merrivale farm.

[21]War veteran refers to a person who joined the war in 1962 or before and underwent military training, Sadomba (2008). I treat this group as a semi-formal structure because they accord

and so they had some form of 'quasi-control' over the land. I argue that it is only 'quasi-control' since they only had offer letters and had no formal title to the land such as a permit or title deeds. Local perceptions elsewhere may differ from my perspective. If the war veterans are perceived to be able to ward off any attempted seizure of the land they farm, that, to them, constitutes control. Legality and the realities on the ground may differ and hence access to and control over land by women means in practice the capacity to use the land without inhibitions and the community knowing that it is your land. However, such control is difficult for women in their own right to attain, particularly without negotiated relationships with men, especially through husbands.

Social capital[22] emerged as a prime factor that enabled women to acquire land at Merrivale in the form of networks based on kith and kin, church or women's clubs, where women would encourage each other to try and access land from these and various other institutions dealing with land reform issues. Social networks created by the women led to the building up of livelihood networks that led to land acquisition in the village. Social relationships relevant to land such as group associations and networks are an important conduit for realisation of women's land rights. Meinzin-Dick and Mwangi (2008, 37) have noted that

> groups and social networks comprise collectives aware of their common interests and may include loose networks, patron-client ties, kinship, friendship, religious groups, while gender and age are important categories. Through these social ties individuals and groups have laid legitimate claim to land for productive use, whether cultivation or grazing, whether seasonal or non-seasonal, whether for fruit trees on individual land, or for water on state land. These rights may be derived from very localized relationships and longstanding practices or norms or they may be allocated through projects, programmes or state authority.

Livelihood construction and the activities that men and women undertake in giving meaning to land in any agro-ecological zone depend on what land means to the people who use it on a day-to-day basis. Makura–Paradza (2010) has pointed out that single, divorced or married women will always adopt livelihood pathways that will not jeopardise land-based asset accumulation. This is one crucial reason that land redistribution will always lead to multiple livelihood systems that redefine questions of agrarian political economy, gender, rights and livelihood issues in relation to land access and utilisation. Land reform is not just about land per se, it is also about creating new livelihood portfolios, which was necessarily complex in the difficult political, economic and social environment that emanated during FTLR. A 'policy options' report to the United Nations Food and Agriculture Organisation (FAO 1986) on agrarian reform in Zimbabwe had already in 1986 made the point

---

themselves status that they think is useful in the dynamic nature of governance and societies, although this is controversial in the Zimbabwean context. Sadomba (2008, 265), himself a war veteran, in his PhD thesis based on participant observation of the fast track process in Mazowe argues that war veterans are a governing body and hence a formal structure as they form the central pillar of governance in the country.

[22]Nemarundwe (2003) notes that social capital refers to standardised networks or networks of variation or flux that people enter into so as to regulate their livelihoods. Standardised networks for example are burial societies which have constitutions, and networks of variation or flux are the ones loosely created just for a specific time and continuously change. People can negotiate in networks of variation and they do not have formal rules.

about the logic for assessing impact in terms of livelihoods created and not just productivity.

My results are in consonance with Moyo (2011) as fast track led to ordinary women increasingly controlling land in as much as this is not recognised and appreciated. I challenge Matondi (forthcoming) who posits that it was only women from special interest groups, such as war veterans and those who had political connections, who got access to land, and when numbers are examined it is the powerful women who make up most of the numbers. However Scoones *et al.* (2010) and Moyo (2011) argue that it was mostly ordinary women who accessed land in large numbers and not so much the politically connected, refuting the political elitism of women's access to land. Accounts of land acquisitions through bargaining and use of negotiations have not been researched much and if such studies are given primacy, the figures could actually increase. Whitehead and Tsikata (2003) are of the notion that a re-examination of the local level studies on land use by women in sub-Saharan Africa can safely lead one to conclude that a re-return to custom might be useful in solving women's land problems in Africa.

Sixty percent of the people said their lives had improved, compared to 40 percent who argued that their lives had deteriorated after land reform – because there have been recurrent droughts and their land is not so fertile, requiring lots of fertiliser they could not afford to buy. When I asked them about the government's Operation 'Maguta'[23] they were quick to point out that the crops had been affected by the drought and they did not yield much. I could observe through the gestures they were making that their responses were not necessarily the whole truth and perhaps they sold the seeds and the diesel so as to earn some money. An official for Care International said that, 'Masvingo province is generally a very dry area such that growing leafy vegetables is sustainable. That is why most women are involved in horticulture as the vegetables take a short time to mature and will not be affected by dry spells due to lack of moisture'.[24] Falco *et al.* (2010) reports that crop biodiversity has in recent years been affected by climate change and soil-related deficiencies resulting from different land use systems. Farmers should therefore be encouraged to use crop adaptability measures that suit the changing climatic conditions.

Vulnerable agricultural livelihoods in dry land areas lead to diversification strategies that enable households to survive. Non-permanent migration strategies emerge as key livelihood pathways that are pursued by the farmers. This has been a long term pattern in Southern Africa (Murray 1980, Scoones *et al.* 1996, Brycesson 2002). The last three decades have seen some notable changes, with women increasingly involved in migration beyond the national borders, which is changing domestic relations (Dodson 2008). Previously men were involved more in migration to South Africa and women stayed at home engaging in the household and agricultural activities. Non-permanent mobility strategies became increasingly important in Zimbabwe post-2000 due to the challenges that the new farmers faced in establishing new farms, in pursuing agricultural activities with insufficient state support, and with the collapse of the formal economy within Zimbabwe, which affected agribusiness operations. This also brought about the need to search for

---

[23]Operation 'Maguta' was a programme of the Government of Zimbabwe during fast track to provide the newly resettled farmers with farming machinery and fuel so that they could boost agricultural production.
[24]Interview with Care International Programme Officer, Masvingo Province, April 2009.

alternative markets and hence South Africa, Mozambique and Botswana became new market sites for agricultural commodities.

Non-permanent mobility as livelihood strategy tied to access to land

Non-permanent mobility[25] has taken centre stage in livelihood and migration debates internationally. In this part of Zimbabwe before FTLR, discussion of women's livelihoods centred mainly on farming, and few women embarked on cross border trading trips to supplement their income due to the economic structural adjustment programme (Muzvidziwa 1997, 1998). They lived in the communal areas of Maranda, Lundi, Gudo, Chimbudzi and God Star. Trading of items such as peanut butter, green mealies and chickens was also undertaken by the women within their former communities and the towns of Masvingo, Gweru, Harare and Bulawayo. However, FTLR ushered in a new trading regime that led to the need to move from the old livelihoods to new ones based on cross border trading in food products (Mutopo 2010). The past 10 years in Zimbabwe with the difficult economic and political terrain led to a new political economy centred on the need for creating a viable economy and livelihoods such that even rural women had to devise new livelihoods niches based on land acquisition. Changes in the social, economic and political spectrum have an influence on rural land-based livelihoods. My findings corroborate those of Scoones et al. (2010) who note that cross border trade among women in rural Masvingo had increased, especially on the fast track farms due to the proximity of South Africa, creating new commodity chains for agricultural and livestock products.

Migration in Southern Africa has been marked by complexity that has increased as a result of recent changes in migratory patterns, which have become more inter-regional. As the mining industry in South Africa, which was the traditional destination of male migration, has declined, migratory strategies have been redefined and other sources of income have been identified involving cross border trade and the informal sector. The Southern African Development Community report on Gender, Remittances and Migration (SADC 2007) has shown that 47 percent of women in the Southern African region are involved in the informal sector and cross border trade. Non-permanent mobility as a strategy is gaining more currency in rural areas. The SADC report (2007) highlighted that compared with other countries in the region, Zimbabwe has more women irregular migrants than other countries.

The case of Mrs Maidei Chigushe (56 years) gives an indication of the kinds of strategies involved. Mrs Chigushe is from Merrivale resettlement area, Mwenezi. She and her husband acquired an A1 plot and practise subsistence farming and animal husbandry; they mainly grow sugarcane, maize, groundnuts and sorghum, since their plot is in Natural Region IV. She also goes to Dick Farm, where the A2 newly resettled farmers have irrigation and so they have plentiful vegetables year round, to buy and order more vegetables. She began a marketing enterprise in 2005, and travels to Johannesburg weekly to sell her horticultural products, in Berea Street. She arrived in Johannesburg one Sunday morning having boarded a J.R. Choe bus,

---

[25]I define non-permanent mobility as the periodic movement of women to South Africa to trade their agricultural commodities normally for periods ranging from three days to two weeks. The women do not settle in South Africa but travel every month and maintain their livelihood base at Merrivale in Mwenezi, Zimbabwe.

which costs 400 Rand. She also pays R100 for the luggage on the bus. She packs her vegetables in sacks and since she has to declare them at the South African border, she normally puts R100 aside to pay the South African immigration official; she just puts the money in her passport and does not say anything. The Zimbabwean side is not a problem because as she is in transit she is not searched and does not have to declare her goods. She brought pumpkin leaves, salted ground nuts, powdered okra, small reed baskets and mounds of soil from vleis *(ivhu repachuru)*, which she was selling for relatively small amounts of money. Before beginning this business, she used to do farming and sell sugarcane and other commodities at Ngundu station. She was able to get a passport in 2005 and she decided to cross the border into South Africa to sell her produce. In order to apply for the visa she had to ask her husband to sell one of their beasts so that she could raise the R2000 required by the South African Embassy; thus the husband in this case was supportive of his wife's enterprise. On a good trip she makes R1000, which she says is enough for her family to survive. Her customers are Zimbabweans and other foreigners. She told me,

> I am satisfied, my child, the plot that we acquired has enabled us to send our siblings to school and some have even gone to University like you. We are now able to eat and dress properly. This time we did not have problems with planting seed because I just had to put the bags on the bus. The maize seed from South Africa does not need lots of fertiliser, even though the *sadza* (maize porridge) does not satisfy much, it is better than nothing. (Field notes 2009)[26]

This demonstrates that Mrs Chigushe was stepping out of poverty as a result of access to land and the capacity to grow crops and to trade them in South Africa. The different livelihood pathways she undertook demonstrated that by engaging in farming she was able to realise her right to a livelihood, which was enabled by her right to free movement to South Africa to search for markets. The construction of such new agrarian dispensations has led to greater independence for many women through the capacity to engage in regional trade. Makura-Paradza (2010) has noted that women were able to build on their rights to land by constantly negotiating different pathways and means in order to meet sustainable livelihoods. Rural areas are undergoing enormous changes through such processes, and women's role in smallholder farming is increasingly gaining prominence in international agricultural discourses.

This evidence serves to provide a starting point for appreciating how women negotiate the agricultural spaces around them even in complex set-ups, and also the importance of land as a strategic resource for women in Zimbabwe. Mrs Chigushe's case demonstrates that land is an important asset that can advance women's rights and helps to empower them. My observations and interviews at Merrivale led me to realise that land as an asset is imbued with multiple meanings, of which home is an important one for many people. I argue that the nexus of land, farming, home and access to markets is critical for the well-being of women, in particular. At the same time, I also agree with Goebel (2002) and Vijfhuizen (2002) that the meaning and role of land for women changes not only in relation to local, national, and international trends, but according to the particularities of women's lives (for

---

[26]The case history developed gives an indication of how access to land after fast track land reform has changed the lives of the women as they develop other agro-commodity chain platforms where they can trade their agricultural products.

example age, marital status, whether or not they have dependent children, and so on). Assets are neither fixed nor static; their value changes. Different assets are important at different times.

Engaging in mobile livelihoods enabled women farmers to gain income that was used to enhance the status of the farm, showing how a right to independent use of land leads to accumulation of monetary and non-monetary assets that are crucial for the sustenance of the household. Some of the income Mrs Chigushe made was used for purchasing farming implements, paying school fees for children and looking after relatives in the former communal area. Use of the income was negotiated with her husband as he played a role in the family budgetary process. During discussions with her on the trading trips to South Africa, she highlighted that she also kept part of the income for herself as she had to continue with her trading activities into South Africa. Sometimes she would conceal the exact amount of money from her husband because she still wanted to retain that sense of self-ownership of the money.

Cheater (1974) observed that farming has a relationship with mobility strategies. She goes further to point out that entrepreneurial strategies driven by mobility contributed to better outcomes for farmers in the high purchase areas in Musengezi. Conceptualising rural women's access to land in terms of human rights in this way means acknowledging that right solely on the basis of their existence within a community, rather than because of women's specific role as providers of food, or the need to increase the efficiency of agricultural production, or to increase the welfare of their daughters and sons (Monsalve Sua'rez 2006). Questions of the rights of women to land and sustainable livelihoods are integrated and should be understood as based within the cultural systems that govern women's daily lives and other non-farm activities.

The women were faced with quite a number of vulnerabilities during their trips to South Africa. The logistical issues raised by some of the interviewed women reveal instances when their produce was taken and thrown away at the border, as there are global and national rules that determine entry of agricultural produce into South Africa and these include specified limits on quantities. Any importation of agricultural produce also requires permits from the Zimbabwean government, which the women found to be a hassle; they were not interested in acquiring trading permits that necessitated them travelling to Harare. In as much as these rules were there and the women were aware of them, they always wanted to take more quantities than were allowed since social capital[27] was important in the movement of their produce into South Africa. Sometimes due to delays at the border, leafy vegetables such as rape would spoil quickly.

During one of the trips with the women I observed that as we got to the border these women had sad faces. One woman said to me that the business of going to South Africa was difficult. At one point all the produce she had was taken away at the South African border because they asked her to produce a cross border permit, which would allow her produce entry. Instead she proceeded to South Africa with a few packets of *mopane* worms which she had in her bag with her clothes. She had to find a piece-work job so that she could supplement the income from the *mopane* worms so as to buy a few groceries for her family.

---

[27]This refers to the different forms of social networks that the women were involved in their logistical trips to South Africa, which involved different actors such as bus drivers, courier syndicates that involved groups of men, and women at the border posts.

The women all feared South African immigration officials. They were subjected to body searching and verbal abuse, violations of their right to free movement. For instance, I witnessed one police officer shouting to one of the women during one of my trips with them, 'Body search! Women this side with women police officers; men that side with male police officers. In case you are hiding some stuff'. Because they wanted to go and sell their produce, however, the women could not complain as they feared being sent back to Zimbabwe. Such treatment really undermines their humanity. The multi-layered strategies of trying to survive for Zimbabweans have been marked by xenophobia in South Africa. Hammar, McGregor and Landau (2010) reinforce that, 'In the popular imagination, this melee was triggered in no small part by the perceived "human tsunami" of migrants flooding in from Zimbabwe, who were seen as adding to unemployment, crime, the lack of housing and services for ordinary South Africans. Other interpretations stressed a long history of ethnicised violence in South Africa, negrophobia and self-hate'. The xenophobic attacks should also be understood as issues embedded within ethnocisation politics and unresolved state formation processes in Southern Africa (Njamnjoh 2002). These forms of violence did not deter the women from trading, showing the agency they possess despite such an unpleasant reception in South Africa.

The tensions and threats involved in international border crossings were evident in some of the informants' testimonies. For instance, a Zimbabwean bus driver warned female passengers: 'We are now approaching the South African border. Those who need help with declaring their goods should see me. I guess you are all aware of the way we do it. The prices keep increasing. Also beware of thieves, as you know that they are now targeting women going to sell like you'.

It was routine for the women to face requests for 'extra' or extortionate payments, for instance, from a South African border official, 'No mama, you have to leave your goods ahaha; it is too much [to] put them on the scale. It is R500 for you to cross with all these bags of round nuts'.

Other threats and vulnerabilities arose from the need to sleep outside once in South Africa, at Safari Motors in Louis Tritchardt (the town now named Musina), at the Butterfield bakery and at Park Station and Bree Street in Johannesburg. This necessity raised other concerns like being mugged. All these involved threats to the right to life and the search for better livelihoods. In order to deal with these threats the women would sometimes pay the attendants at the spots where they slept so that they could be protected against thieves. The women I travelled with sometimes would even sell on credit a R5 bundle of vegetables; this demonstrates how the recent global financial crisis is impacting the small-scale traders who deal with food products, as consumers' buying patterns also change. The buyers and sellers have to find ways of cushioning themselves against the global economic forces that have pushed food prices beyond the reach of many. Adepoju (2008) notes that the feminisation of mobility is now about creating livelihoods and hence new coping mechanisms for both rural and urban women in sub-Saharan Africa.

My sample from Merrivale also contained women who were not involved in non-permanent mobility. Thirty-five percent of this group of women were engaged in farming and would sell their agricultural produce locally in the surrounding communal areas of Chivi or in the urban centres of Masvingo, Bulawayo, Beitbridge and Harare, especially processed groundnuts(made into peanut butter), maize, *mopane* worms and vegetables. From my personal observations and interaction with the women who traded their agricultural produce locally, it seemed they had regular buyers from the towns who would come to purchase their maize and *mopane* worms, particularly for resale in the urban areas. The women who travelled to the urban areas would also buy goods for resale back at Merrivale and the surrounding areas, such that competition was increasing during August 2010 as most women who either traded locally or in South Africa were competing to bring back the most sought-after goods. This evidence reveals that acquisition of land by the women led to the right to search for markets independently in a quest to reorganise and reorient rural livelihoods.

This differs from Jacobs (2000), who argued in the 1980s that resettled women lost their material and economic niches, trading, markets and personal contacts. My study reveals that in the new communities the women are building new contacts and markets, unlike in the old resettlement areas where movement of settlers was strongly discouraged as the aim was to promote and enforce full-time farming. In the old resettlement schemes, land had been allocated using different means – planned and regulated by resettlement officers – as opposed to the new fast track farms (Kinsey 2002). The fast track farms are very different as they are more open and allow for greater negotiation in most livelihood activities. The non-permanent mobile women also cited that their aim was to travel to South Africa or Botswana where their produce would fetch more money than locally and the buying of goods for resale was much more viable since there are competitive prices and more shops to chose from, unlike in Harare, Bulawayo or Masvingo where goods such as clothes and food items are still expensive.

Agrarianisation, livelihoods styles and collective action

Aside from trading in agricultural produce, when in South Africa the women would also venture into other trades; they did not rely on selling produce only. However, it should be noted that these other income-earning activities were secondary; selling of agricultural produce remained the main activity. The menial jobs were mainly done after they had finished selling the fresh produce. Some plaited hair. Others cooked thick white maize porridge (*sadza*) in backyard restaurants to raise money. One woman that I travelled with would be plaiting hair while another woman would market her produce, showing the cooperative work and collective action[28] that women employ when they do business together. Collectives emerged as a useful coping strategy as most of the women do not travel individually. They are normally in groups and might take different routes when selling in the same suburb, but will always devise a meeting place so that they keep track of each other. The cooperative efforts were based on the spirit of working together (*mushandira pamwe*) that is much emphasised in the Shona culture on working relations and land-based livelihoods.

By helping each other, the women promoted a spirit of solidarity in managing and improving livelihood options. They developed bonds that emanated from working together, promoting the spirit of collective action that is emphasised in African culture, through the spirit of '*ubuntu*'. I observed that cooperation was also a social learning process. Kruissjen *et al.* (2009) notes that it is 'the process through

---

[28]This entails cooperation and cooperative efforts that regulated the women's agricultural commodity trading activities in South Africa.

which groups of people (or stakeholders) learn: together they define problems, search for and implement solutions'. I realised that it emerged as a process where the individuals developed and agreed on marketing routes, prices, and means of dealing with difficult situations that could arise during the trading trips in a foreign land. Hence collective action emerged as coordinated behaviour of groups towards a common interest, and assessment and adoption of solutions if there were any mishaps during the trading (Kruijssen *et al.* 2009). In South Africa it should be noted that collective action efforts were about safety and security in a difficult and often dangerous environment in which foreigners were not welcome and subjected to violent xenophobic attacks.

Sixty percent of the women from my sample, and other women not part of my sample who constituted 70 percent[29], who embarked on non-permanent mobility reported that they would also purchase some products for resale either in Masvingo or in the surrounding farms. These products ranged from sugar, cooking oil, flour, and vegetable seeds to shoe polish, hair extensions, *davita* drinks, and clothes. The United Nations Development Report on Mobility and Development (UNDP 2009) notes that as more and more women from all sectors of society are engaged in seasonal migration, in search of markets, jobs and better lives, it should be understood that women are slowly becoming an important component in global trade development. There is now a new resurgence of the feminisation of mobility and land-based livelihoods influencing mobility as evidenced by this work.

Conclusion

My analysis reveals that the FTLRP allowed access to land for women. A few women managed to acquire land individually, the majority via men and particularly through marriage and cultural contracts. I demonstrated how despite the contradictory patterns in patriarchy, women had the ability and mechanisms to make use of these contradictions in accessing land and shaping mobile livelihoods. Traditional patriarchal/patrilocal systems meant that women gained access to land largely through intra-household negotiations, and often hidden, informal means (weapons of the weak), often restricted to taking place after the birth of their first child as demonstrated in the paper through the *tsewu* concept. Land access is not formalised, but based on negotiation and trust relations. This brings some insecurity, but for most of the women investments continued based on this arrangement. This work has challenged the Western liberal notion that emphasises individual human rights to land as the most effective strategy to address the issue of African women's access to land. Negotiation and bargaining have proved to be more useful in helping women gain access to land, even in the most disputed land reform such as the Zimbabwe

---

[29]It should be noted that as I travelled with the women, I also met and interacted with different women from Harare, Bulawayo, Masvingo, Mutare, Chiredzi, Beitbridge and Gutu. They all sold many different items and they would purchase different items for resale, with urban women purchasing electrical goods and clothes as compared to myself and the rural women I travelled with purchasing mostly food items. This was mainly due to the different market spaces and consumers. During August 2010 women from Merrivale also started buying cell phones for resale, demonstrating how trading also opens up the global world of the technology revolution since in Mwenezi Econet cell phone lines now have reception and a network after major upgrades by Econet wireless company.

case. The interactive model of women's land access that includes the state, family and traditional authorities has proved to be effective at Merrivale.

I also unravelled how access to land was critical and shaped women's economic independence and became the basis of non-permanent mobility with the aim of trading in South Africa. Merrivale is located in a (very) dryland area, where agriculture is risky and other off-farm opportunities are necessary. This was especially important given the collapse of the formal economy in Zimbabwe until February 2009. Due to the location of Merrivale farm, cross border trade and non-permanent mobility emerged as key livelihood options for women and income earning. Intensification of trading activities in South Africa was linked to land acquisition in the new fast track farms. This independent source of off-farm income, linked to land access through negotiated relations in a patriarchal setting, offered new opportunities for women. The new entrepreneurial activities of the women, as much as they were lucrative, exposed them to new risks and vulnerabilities. However through notions of agency the women never gave up their livelihood pathways.

Acknowledgements

I want to thank the anonymous reviewers who made constructive comments that led to the outcome of the paper. I wish to thank also Professor Lionel Cliffe, Dr Prosper Matondi, Dr Eveson Moyo, Professor Robin Palmer, Professor Paul Hebinck, Professor Michael Bollig and Professor Rudo Gaidzanwa for useful suggestions. The research grant from the Volkswagen Stiftung (Germany) under the Project, Human Mobility, Institutions and Access to Natural Resources in Contemporary Africa is sincerely appreciated.

References

Adepoju, A. 2008. *Migration in sub-Saharan Africa*, Issue 37). Nordiska Africa Institute Uppsala, Sweden.

Agarwal, B. 1994. A field of one's own: gender and land rights in South Asia. Cambridge: Cambridge University Press.

Agarwal, B. 2003. Gender and land rights revisited: exploring new prospects via the state, family and market. *Journal of Agrarian Change*, 3 (1 and 2) January and April Issue, 184–224.

Alexander, J. 2003. *The unsettled land: state making and the politics of land in Zimbabwe 1983–2003*. Oxford/Harare/ Athens, James Currey Publishers/ Weaver Press/ Ohio University Press.

Bebbington, A. 1999. Capitals and capabilities: a framework for analyzing peasant viability, rural livelihoods and poverty. *World Development*, 27 (12), 2021–2044.

Berry,. S. 2002. Debating the land question in Africa. *Comparative Studies in Society and History*, John Hopkins University Press, California.

Bourdillon, M. 1992. The Shona peoples: an ethnography of the contemporary Shona, Gweru, Zimbabwe: Mambo Press.

Brycesson, D. 2002. The scramble in Africa: reorienting rural livelihoods. *World Development*, 30(5), 725–739.

Chambati, W. 2011. Emergent agrarian labour relations in Zimbabwe's new resettlement areas, African Institute for Agrarian Studies Monograph Series.

Cheater, A. 1974. Aspects of status and mobility among farmers and their families in Musengezi African purchase land. Paper presented at the 3rd Rhodesian Science Conference, Salisbury.

Cheater, A, 1986. The role and position of women in pre-colonial and post-colonial Zimbabwe, *Zambezia*, X111, (ii), University of Zimbabwe publications, Harare, pp. 65–77.

Chimhowu, A. and Hulme D. 2003. Livelihood dynamics in planned and spontaneous resettlement in Zimbabwe: converging and vulnerable. *World Development*, 34(4), 728–750.

Chingarande 2008. Gender and the Fast Track Land Reform in Zimbabwe, African Institute for Agrarian Studies, Monograph Series, Harare.

Cousins, B. and Scoones, I. 2010. Contested paradigms of viability in redistributive land reform: perspectives from Southern Africa. *Journal for Peasant Studies*, 37(1), 31–66.

Derman, B. and Hellum, A. 2004. Land reform and human rights in contemporary Zimbabwe: balancing individual and social justice through an integrated human rights framework. *World Development*, 32(10), 1785–805.

Dodson, B. J. 2008. Gender, migration and livelihoods: migrant women in Southern Africa. *In*: N Piper, ed. *New perspectives on gender, migration, livelihoods, rights and entitlement*. NewYork and London: Routledge.

Ellis, F. 1998. Livelihood diversification and sustainable rural livelihoods. In D. Carney, ed. *Rural sustainable livelihoods*: What contribution can we make? London: Department of International Development.

Falco, S. *et al*. 2010. Seeds for livelihood: crop biodiversity and food production in Ethiopia. *Ecological Economics*, 69, 1695–1702.

FAO 1986. Policy options for |agrarian reform in Zimbabwe: a technical appraisal. Rome: UN Food & Agriculture Organisation.

Fitzgerald, D. 2009. Towards a theoretical ethnography of migration. *Qualitative Sociology*, 29(1), 1–24.

Gaidzanwa, R.B. 1994. Women's land rights in Zimbabwe. *Opinion*, 22(2), 12–16.

Goebel, A. 2002. Men these days are a problem. Husband taming herbs and gender wars in rural Zimbabwe. *Canadian Journal of African Studies*, 36(3), 460–489.

Goebel, A. 2005a. Zimbabwe's fast track land reform. What about women? *Gender, Place and Culture*, 12(2), 145–172.

Goebel, A. 2005b. *Gender and land reform: the Zimbabwe experience*. Montreal and Kingston: McGill & Queen's University Press.

Government of Zimbabwe, 2003. Report of the Presidential Land Review Committee under the Chairmanship of Dr. Charles Utete. Harare.

Gready, P. and J. Ensor, eds. 2005. *Reinventing development: translating rights-based approaches from theory into practice*. London: Zed Books.

Hammar, A., McGregor, J. and Landau, L. 2010. Introduction. Displacing Zimbabwe: crisis and construction in Southern Africa. *Journal of Southern African Studies*, 36(2), 263–283.

Hebinck, P. and Shackleton, C. 2011. *Reforming land use and resource use in South Africa: impact on livelihoods*. London: Routledge.

Jackson, C. 2003. Gender analysis of land: beyond land rights for Women. *Journal of Agrarian Change*, 3(4), 453–480.

Jacobs, S. 2000. The effects of land reform on gender relations in Zimbabwe. *In:* Bowyer-Bower, A.T.S and Stoneman, C., eds. *Land reform in Zimbabwe, constraints and prospects*. London: Ashgate.

Jacobs, S. 2010. *Gender and agrarian reforms*. London and New York Routledge.

Jeater, D. 2007. *Law, language and science: the invention of the native mind in Southern Rhodesia, 1890–1930*. London: Greenwood.

Jirira, K.M. and Halimana, M.C. 2008. A gender audit of women and land rights in Zimbabwe. Desk review final report. Prepared for the Zimbabwe Women Resource Centre and Network (ZWRCN)'s Project for the European Commission and Action Aid UK on women, land and economic rights.

Kinsey, B. 2002. Survival or growth? Temporal dimensions of rural livelihoods in risky environments. *Journal of Southern African Studies*, 28(3), 615–629.

Kruijssen, F. *et al*. 2009. Collective action for small scale producers of agricultural biodiversity products. *Food Policy*, 34, 46–52.

Makura-Paradza, G.G. 2010. Single women, land and livelihood vulnerability in a communal area in Zimbabwe, African women leaders in agriculture and the environment, AWLAE SERIES, No. 9, Wageningen Academic Publishers.

Marcus, G. E. 1995. Ethnography in/ of the world system: The emergence of multi-sited ethnography. *Annual Review of Anthropology*, 24, 95–117.

Marongwe, N. 2009. Interrogating Zimbabwe's fast track land reform and resettlement programme: a focus on beneficiary selection. PhD Thesis. University of the Western Cape, Cape Town, South Africa.

Matondi, P.B., ed. 2012. *Zimbabwe's Fast Track Land Reform Programme*. London: Zed Books.

Matondi, P.B. and Sanyanga, R. (forthcoming). *'Revolutionary Progress' without Change in women's land rights in fast track farms in Mazowe. In*: P. B. Matondi, ed. Zimbabwe's Fast Track Land Reform Programme, Zed Books, London.

Mazhawidza, P., and Manjengwa, J. 2011. The social, and political transformative impact of Fast Track Land Reform Programme on the lives of women farmers in Goromonzi and Vungu-Gweru Districts of Zimbabwe. International Land Coalition Reseach Reports. Available from: http://www.landcoalition.org/publications/social-political-and-economic-transformative-impact-fast-track-land-reform-programme-li [accessed September 2011].

Meinzin-Dick, R. and Mwangi E. 2008. Cutting the webof interests: pitfalls in formalizing property rights. *Land Use Policy* 26, 36–43.

Meteorological Report, 2009. Department of Meteorology, Ministry of Agriculture, Zimbabwe, Harare.

Monsalve Suárez, S. 2006. Gender and land. *In*: P. Rosset, R. Patel and M. Courville, eds. *Promised land: competing visions of agrarian reform*. Oakland, CA: Food First Books.

Moyo, S. 2007. The land question in Southern Africa: a comparative review. *In: The Land question in South Africa: the challenge of transformation and redistribution*, edited by Lungisile Ntsebesa and Ruth Hall, HSRC Press, Capetown.

Moyo, S. 2011. Three decades of agrarian reform in Zimbabwe: changing agarian relations. *Journal of Peasant Studies*, 38(3), 493–531.

Murisa, T. 2007. Social organisation and agency in the newly resettled areas of Zimbabwe. The case of Zvimba district, Monograph Series, Issue, No. 1/07. Harare: African Institute for Agrarian Studies.

Murray, C, 1980. Migrant labour and family changing structures in the rural periphery of Southern Africa. *Journal of Southern African Studies*, 6(2), 139–157.

Mushunje, M. 2001. Women's Access to Land in Rural Zimbabwe. University of Zimbabwe School of Social Work, Working Paper Series.

Mutopo, P. 2010. Women trading in food across the Zimbabwe-South Africa border: experiences and strategies. *Journal for Gender and Development*, 18(3), 465–477.

Muzvidziwa, V.N. 1997. Rural urban linkages: Masvingo's double rooted female head of households. Zambezia, XXIV (11), University of Zimbabwe publications, Harare, 97–123.

Muzvidziwa, V.N. 1998. Cross border trade: a strategy for climbing out of poverty in Masvingo, Zimbabwe. Zambezia, (XXV) (1), University of Zimbabwe publications, Harare, 29–58.

National A2 Land Audit Report, 2007. Prepared by the Ministry of Lands, Land Reform and resettlement and The Informatics Institute, SIRDC.

Njamnjoh, F.B. 2002. Local attitudes towards citizenship and foreigners in Botswana: an appraisal of recent press stories. *Journal of Southern Africa Studies*, 28(4), 755–775.

Nemarundwe, N. 2003. Negotiating resource access. Institutional arrangements for woodlands and water use in Southern Zimbabwe. D.Phil thesis, Department of Rural Development Studies, Uppsala, Sweden.

O'Laughlin, B. 1998. Missing Men? The debate over rural poverty and women headed households in Southern Africa. *Journal for Peasant Studies*, 25(2), 1–48.

O'Laughlin, B. 2002. Proletarianisation, agency and changing rural livelihoods : forced labour and resistance in Mozambique. *Journal for Southern African Studies*, 28(3), 511–530.

O'Laughlin, B. 2009. Gender justice, land and the agrarian question in Southern Africa. *In*: A, Akram-Lodhi and C. Kay, eds. Peasants and globalisation: political economy and rural transformation and the agrarian question. London: Routledge.

Pellizoli, R. 2010. ' Green Revolution' for whom? Women's access to land in the Mozambique Chookwee irrigation scheme. *Review of African Political Economy*, 37(124), 213–220.

Peters, P. 2010. Our daughters inherit the land and our sons use the land. Matrilineal and matrilocal land tenure and the new land policy in Malawi. *Journal of Eastern African Studies*, 4(1), 179–199.

Ribot, J.C. and Peluso, N. 2003. A theory of access. *Rural Sociology*, 68(2), 153–181.

Sadomba, W. Z. 2008. War veterans in Zimbabwe's land occupations: complexities of a liberation movement in an African post- colonial settler society. PhD Thesis, Wageningen Agricultural University.

Schmidt, E. 1992. *Peasants, traders and wives: Shona women in the history of Zimbabwe, 1870– 1939*. Portsmouth: Heinemann, United Kingdom.

Scoones, I. with Chibudu, C. Chikura, S., Jeranyama, P., Machanja,W., Mavedzenge, B., Mombeshora, B., Mudhara, M. Mudziwo, C., Murimbarimba, F., Zizera, B. 1996. *Hazards and opportunities: farming livelihoods in dryland Africa: lessons from Zimbabwe.* London: Zed Books.

Scoones, I. 2009. Livelihoods perspectives and rural development. *Journal of Peasant Studies,* 36(1), 171–196.

Scoones, I., Marongwe, N., Mavedzenge, B., Murimbarimba, F., Mahenehene, J., and Sukhume, C., 2010. *Zimbabwe's Land Reform: Myths and Realities.* Oxford, London: James Currey, Harare: Weaver Press.

Sen, A. 1999. Development as Freedom. Oxford University Press, Oxford.

Seur, H. 1992. Sowing the good seed: the interweaving of agricultural change, gender relations and religion in Serenje District, Zambia. PhD thesis, Wageningen University, the Netherlands.

Southern African Development Community (SADC) 2007. Report on gender, remittances and migration, 2007. South African Institute of International Affairs.

Toulmin, C. 2007. Negotiating Land Access in West Africa, *In:* Derman, B., Odgaard, R., and Sjaastad, E. eds. *Conflicts over land and water in Africa.* Oxford: James Currey.

Tsikata, D. 2003. Securing women's interests within the land debate in Tanzania: recent debates in Tanzania. *Journal of Agrarian Change,* 3(1), 149–183.

Tsikata, D. 2009. Gender, land and labour relations and livelihoods in Sub-Saharan Africa in the era of economic liberalisation: towards a research agenda, Feminist Africa 12, Institute of Gender Studies, University of Cape Town.

United Nations. 2009. Development Report on Mobility and Development.

Utete, C.M.B. 2003. Report of the presidential land review committee on the implementation of the fast track land reform programme, (2000–2002), under the chairmanship of Dr Charles M.B.Utete.

Vijfhuizen, C. 2002. *The people you live with: gender identities and social practices, beliefs and power in the livelihoods of Ndau women and men in a village with an irrigation scheme in Zimbabwe.* Harare: Weaver Press.

Whitehead, A. and Tsikata, D. 2003. Policy discourses on women's land rights in sub-Saharan Africa: the implications of the re-return to the customary. *Journal of Agrarian Change,* 3(1), 67–112.

Wolmer, W. 2008. *From wilderness vision to farm invasions: conservation and development in Zimbabwe's south east Lowveld.* London: James Currey, and Harare: Weaver Press.

World Bank. 2003. Land policies for growth and poverty reduction: a World Bank policy research report. Oxford: World Bank and Oxford University Press.

Patience Mutopo is a PhD research fellow with the Cologne African Studies Centre, University of Cologne, Germany. She is also a research associate with Ruzivo Trust in Harare, Zimbabwe. Her research interests focus on gender, land rights, biofuel production, agricultural value chains, health and agro ecosystems. She is a member of the Legal Empowerment of the Poor coordinated by the Department of Environment and Development at the University of Oslo and the Norwegian Centre for Human Rights. She is also a member of the Land Deal Politics Initiative, a research consortium run by the Institute of Social Studies, The Hague, the Institute of Development Studies, University of Sussex, the Polson Institute of Development Studies, University of Cornell and the Institute for Poverty, Land and Agrarian Studies, University of the Western Cape.

Restructuring of agrarian labour relations after Fast Track Land Reform in Zimbabwe

Walter Chambati

The fast track land redistribution programme generated new agrarian labour relations altering the tying of labour on the large farms to tenancy, supplemented by casual labour from the communal areas. Job losses and displacement occurred, but this is not the whole story as new and diverse sources of rural employment have emerged, including high levels of self-employment on small farms (A1) supplemented by casual employment. Large farms dependent on wage labour experience labour shortages despite the mechanisation drive. However, communal areas and A1 farmers continue to provide labour to large farms, although labour supplies are negotiated on new terms.

Introduction

One effect of the recent land reforms in Zimbabwe – the generating of new agrarian labour relations – has been overlooked in most literature after 2000. The extensive redistribution of over 80 percent of white-owned large-scale commercial farms (LSCFs) under two resettlement models (A1 and A2) – the majority to peasants from the Communal Areas – under fast track land reform (FTLRP) has reformed the unequal agrarian structure into one more broadly based (Moyo and Yeros 2005, Moyo 2011a). The new agrarian labour relations that emerged are not yet adequately understood, and require an analysis of the relation between changing patterns of agricultural production and labour utilisation, and their effects on different classes and segments of society. These linkages are missing from several studies,[2] but some (see Moyo *et al.* 2009, Scoones *et al.* 2010) do provide this kind of analysis.

This piece examines the transformation of agrarian labour relations in Zimbabwe after the FTLRP since 2000, including the emergence of a new agrarian employment structure which is undermining the social relations based on residential labour

[1] I wish to acknowledge the advice received from Sam Moyo and support of AIAS researchers (Ndabezinhle Nyoni and Steven Mberi) in data collection. Helpful comments were also received from Paris Yeros, Lionel Cliffe, Ben Cousins and two anonymous reviewers.
[2] e.g. Masiiwa and Chipungu (2004), Richardson (2005), Hellum and Derman (2004), and Scarrnechia (2006).

tenancy,[3] and it addresses what this has meant for former farm workers' social reproduction. The new types of farm and non-farm labour that have emerged, and the social and economic conditions under which this labour is employed, are discussed based on detailed empirical evidence.

The dominant perspective in analysing agrarian labour relations after 2000 has tended to be steeped in modernisation narratives, in which formal wage labour in the capitalist LSCFs is treated as superior to self-employed forms of labour in the 'backward' peasant sector (Freund 1984), assuming that returns to wage labour are greater than those of self-employed peasants. Thus (rural) employment is narrowly conceived as largely consisting of wage employment on LSCFs, primarily because self-employment by peasants does not fit the formal employment criteria used by neo-classical economists (Leavy and White 2003). The redistribution of LSCFs to peasants is thus equated with the 'end of modernity' (Worby 2003), with unemployment as the sole consequence.

In Zimbabwe, many contributions to the debate on agrarian labour relations after the FTLRP focus only on impacts on former farm workers[4] – loss of jobs and displacement. It could be argued that this is a conservative orientation towards implicitly defending the poorly remunerated wage labour in the former LSCF. Displacement of former farm workers addresses two groups: those physically displaced from LSCF compounds and those 'displaced *in situ*' (Hartnack 2007, Magaramombe 2010). The latter are still resident in the compounds but are out of 'work'. In this analysis 'work' is narrowly conceived as farm wage work, which ignores other forms of petty commodity production. The analysis of the physical displacement of from the farm compounds does also not adequately examine the insecure residential tenure that farm workers faced before 2000. The notion of 'physical displacement' does not acknowledge the extent to which the FTLRP has re-established self-employed peasant jobs that were displaced by colonial land dispossession or investigate whether this represents losses or gains in the overall scheme of social life.

Another strand of studies has followed a narrow 'human rights' approach limited to civil and political rights to the exclusion of socio-economic rights, including the new livelihoods of former farm workers and peasants. In this approach farm workers and white farmers are portrayed as passive victims of violence at the hands of war veterans and peasants during the land occupations (ZHRF and JAG 2007, JAG/RAU 2008). These studies tend to be silent on the new structure of agrarian labour and how farm workers were mobilised against land reform by land owners and their trade union, the General Plantation and Agriculture Workers Union of Zimbabwe (GAPWUZ), emphasising job protection rather than a new livelihood after redistribution. Nor do they examine farm workers' occasional mobilisation by war veterans and peasants to join the land occupations (Sadomba 2008), and in turn their recruitment by new farmers as labourers.

By focusing on former farm workers, methodologically, these studies limit their analysis to one aspect of agrarian labour (e.g. Sachikonye 2003, Magaramombe

---

[3]The form of labour tenancy existing in Southern Africa does not entail rental payments but the residential rights of 'labour tenants' in the farm compounds were tied to labour supply (see Neocosmos 1993).

[4]See Sachikonye (2003), Hartnack (2007), NRC (2003), Magaramombe (2003), and Rutherford (2004).

2003), to the neglect of new farm workers, land beneficiaries as employers, labour unions, and peasants as labourers and employers. They tend to use single-district-based surveys whose results cannot be generalised (e.g. FCTZ 2002) and even single farm case studies (e.g. Hartnack 2007, Magaramombe 2010). Others have hardly undertaken any empirical assessments and their analysis of the labour situation is mostly based on media reports (e.g. NRC 2003). Absence of empirical data is thus a limiting factor in many analyses, which fail to explore the differentiated types of labourers and the relationships that exist between family, wage, and non-farm work.

Agrarian labour relations need to be understood in their historical context, and in former settler colonies such as Zimbabwe these were based on specific land-labour utilisation relations created by land dispossession. This labour regime was also accompanied by economic and extra-economic factors which shaped labour supplies (Arrighi 1970, Neocosmos 1993).The debate after the FTLRP eschews the historical evolution of agrarian labour relations that separated producers from their means of production and forced them into cheap wage work in white farms and industries, creating a peculiar form of 'bonded' agrarian labour relations based on agricultural workers' residential tenure on the farm compounds being linked to employment (Arrighi 1970, Clarke 1977, Moyo 2011a). The political power conferred by land ownership was used to entrench total control over the landless workers (Clarke 1977).

The repossession of previously dispossessed land or 'expropriation from the expropriators' (Shivji 2009) affects agrarian labour relations that were structured around land alienation and private property. The political power of the white farmers over labour relations through compulsion of wage labour in return for residency was undermined in the 2000s by redistribution and vesting of land tenures under the state, and the allocation of land through public leases and permits, leading to the transformation of the labour residential tenancy relationship (Moyo 2011a). Land redistribution provides peasants wider access to means of production for autonomous social reproduction through self-employment.

This research seeks to present an evidential base for these transformations, and in particular is informed by empirical studies in newly redistributed areas conducted by the African Institute for Agrarian Studies (AIAS) since 2003, as well as other secondary sources. These include district-based surveys (Chambati 2003, Sunga 2003), a national baseline survey in six districts covering 2089 land beneficiary and 761 farm worker households (Moyo et al. 2009) and follow-up fieldwork and visits between 2007 and 2010 (Chambati 2009, Chambati and Moyo forthcoming).[5] The overall survey design is detailed in Moyo (2011a). Agrarian labour relations after 2000 are examined from the perspective of land beneficiaries, and new and former farm workers. The 2005/06 farm labour surveys interviewed 414 former farm workers that were still resident in the redistributed lands and 347 new farm workers. In the absence of population figures on farm workers, convenience sampling was utilised to select respondents, who were asked about their labour history in the LSCFs, current employment patterns, socio-economic conditions, land tenure and land access, organisation and access to social services.

---

[5]The national baseline survey was conducted in six provinces: Chipinge, Chiredzi, Goromonzi, Kwekwe, Mangwe and Zvimba districts. Follow-up visits entailed interviews with land beneficiaries, workers, extension officers and district councils in Zvimba and Goromonzi districts to get updates on developments.

This article first examines how colonial land dispossession and policy shifts after 1980 shaped agrarian labour relations in Zimbabwe. Second, the restructuring of agrarian labour relations since 2000 is examined generally, while the new labour processes that have emerged are detailed. What the new labour processes mean for the social reproduction of agrarian labourers is then discussed, followed by assessments of labour struggles. The conclusions summarise our argument.

Land dispossession/repossession and agrarian labour relations

Forced land expropriations from the indigenous people created a dual agrarian structure, with white minority control of large tracts of freehold agricultural land and overcrowded Tribal Trust Lands, which formed the basis for a cheap labour strategy (Palmer 1977). The proletarianisation of African peasants evolved gradually, as some were initially allowed to retain their ancestral holdings on alienated land in the 1890s as tenants and producers of food and raw materials required by mining and tertiary industries, resulting in labour shortages (Arrighi 1970). Various taxes (hut and poll) introduced to force peasants into wage labour did not resolve labour shortages as some met their taxes from producing a greater surplus (Arrighi 1970, Neocosmos 1993). Thus, the labour supplies on LSCFs were initially built on the basis of migrant workers from Nyasaland, Northern Rhodesia and Mozambique (Palmer 1977, Clarke 1977).

After LSCFs became fully developed around 1920, most tenants were forcibly relocated to the reserves (Arrighi 1970). Land alienation, taxation and repression of peasant farming[6] forced locals to join the foreign migrants in wage labour (Amanor-Wilks 1995). This transformed their participation in the labour market from 'discretionary' to a 'necessity' (Arrighi 1970, Moyo 2011a). The share of migrant workers declined from 60 percent in 1956 to 43 percent in 1969 and to 34 percent by 1974 (Clarke 1977).

A master-servant relationship evolved in LSCFs that was based on a residential labour tenancy (Rubert 1997, Tandon 2001) aided by repressive racial legislation.[7] This was used as bargaining power by employers (Clarke 1977). White farmers used intimidation, racial abuse, arbitrary dismissals and violence to manage labour (Neocosmos 1993, Rutherford 2001). Farm wages were paltry and remained stagnant between 1940 and 1970 (Arrighi 1970).

Farming in the LSCFs was diversified from the labour intensive maize/tobacco production system since the 1970s to include other outputs[8] utilising high levels of modern technologies[9] and capital intensive production systems[10] mainly geared towards exports (Loewenson 1992). The LSCFs' capital intensification increased the number of workers from 225,455 in 1969 to 311,913 in 1974, but the share of permanent workers declined from 82 percent to 75 percent (Clarke 1977).

---

[6]During the 1930s a two tier maize and livestock marketing system was introduced in which peasants received lower prices (Arrighi 1970).
[7]These included the *Master and Servants Act of 1899,* which treated black workers as the property of land owners, and the *Native Juveniles Employment Act of 1926,* which forcibly bonded unemployed blacks in the towns to LSCFs (Amanor-Wilks 1995).
[8]Sugar, cotton, wheat, soyabeans, coffee, tea, beef and dairy
[9]Hybrid seeds, pesticides and fertilisers
[10]e.g. irrigation.

After 1980, the state introduced new labour regulations that included farm workers and abolished the *Master and Servants Act of 1899* (Kanyenze 2001). These included the requirement for state consent in worker dismissals and minimum wages that increased real wages of farm workers by over 50 percent (Amanor-Wilks 1995). Farm worker trade unionism and workers' committees also emerged, but their impact on improving worker welfare was limited (Loewenson 1992, Rutherford 2001). Trespass laws restricted trade unions' access to LSCFs, and GAPWUZ only managed to recruit a third of the permanent workers by 1999 (Kibble and Vanlerberghe 2000). There was also a growth in non-governmental organisations (NGOs) on LSCFs that ran welfare improvement projects for farm workers (Moyo *et al.* 2000).

The Economic Structural Adjustment Programme (ESAP) in 1991 reduced state intervention in the labour markets and eroded earlier gains. Collective bargaining replaced minimum wages, resulting in a decline in the average real wages of farm workers from 60 percent of the Poverty Datum Line (PDL) in 1990 to 24 percent in 1999 (Kanyenze 2001). Farm workers supplemented their wages with other income generating activities[11] and petty commodity production on small gardens provided by 20 percent of LSCFs and in Communal Areas (Vhurumuku *et al.* 1998). The LSCFs intensified their diversification into exports during ESAP (Moyo 2000). Wage employment in the LSCFs grew from 218,172 in 1983 to 334,521 by 1996, and by 1999 there were 314,879 workers (CSO 2001). The share of casual labour increased from 25 percent to 50 percent (CSO 2001), meaning that half of the labour force was part time by 1999.

Land reforms prior to 2000 only redistributed 3.5 million hectares of land to 73,000 households by 1998 (GoZ 2001). Farm workers were not considered as a specific land beneficiary category by policy until 1998, but fell into the broader category of 'poor and landless' (Moyo *et al.* 2000). Some farm workers occupied mostly abandoned farms in the early 1980s, with the support of the Zimbabwe African National Union Patriotic Front (ZANU-PF), and were later regularised by official resettlement (Moyo *et al.* 2000). Farm workers were marginalised by the shift in land reform beneficiary selection policy to focus on the resettlement of 'productive' farmers in a context in which they were not viewed as independent producers (Moyo 1995). Furthermore, some nationalist discourses perceived them as 'foreigners' not qualified for resettlement (Moyo *et al.* 2000). With the slow land reforms and absence of data on the numbers of farm workers who accessed land during this period, it is fair to conclude that very few benefitted.

Although social policy shifts after 1980 improved the lot of farm workers, they did not significantly alter agrarian labour relations. The farm compound persisted as an institution tying residency to employment. Many studies done before the FTLRP have exposed the appalling living conditions (housing, health, education, malnutrition, poverty) of farm workers.[12] Many white farmers continued to institute some of the elements of the masters-servant relationship to repress labour. This has been termed 'domestic government' (Rutherford 2001).[13] The spatial dispersion of the

---

[11]e.g. petty trading and gold panning.
[12]See e.g. Loewenson (1992), Amanor-Wilks (1995), Tandon (2001), Rutherford (2001).
[13]The white farmers resolved disputes internally and transcended the employment contract by resolving marriage and other social disputes in the farm compound (Rutherford 2001). Many used violence in labour management (Kanyenze 2001).

LSCFs limited the reach of state labour officials, and government provided minimal social services to farm workers as they largely considered them as the responsibility of farmers (Loewenson 1992).

The unequal agrarian structure meant that in the Communal Areas, labour intensive farming of cheap food crops (mostly maize) on small plots of land was the basis of social reproduction for subsistence and surplus sales in domestic markets (Moyo 1995). Labour was mostly provided by self-employed female workers, although a few hired wage labour (Adams 1991). Despite the growth in agricultural production and productivity (maize and cotton) in the Communal Areas after 1980 (a result of increased government support), the declining farm sizes and land quality caused by demographic growth meant that many could not meet their social reproduction needs (Moyo 1995). Agricultural productivity declined from the 1990s as the state reduced input subsidies, forcing many households into cheap wage labour (Oni 1997). By 1999, migrants constituted around 30 percent of the LSCF workers (Sachikonye 2003). Deepening poverty,[14] growing landlessness and massive retrenchment of urban workers after ESAP fuelled demands for land redistribution by 1997 (Moyo 2000, Moyo and Yeros 2005).

Restructuring of agrarian labour relations after the FTLRP

The FTLRP restructured Zimbabwe's agrarian labour relations in a number of ways. The first was the expansion of peasant livelihoods, as 63 percent of the land beneficiaries were from Communal Areas (Moyo et al. 2009, 22). Various forms of self-employment on farms and in non-farm activities, as well as farm wage jobs, have been generated, as will be discussed later.

Since not all LSCF land was acquired, the remaining plantations and LSCFs (Moyo 2011b) retained an estimated 100,000 full- and part-time wage workers as of 2003 (Chambati and Moyo 2004, Chambati and Magaramombe 2008).

The second restructuring pertains to the displacement of an estimated 200,000 formal farm jobs nationally (Chambati and Magaramombe 2008), of which about 50 percent were part-time. Between 30,000 and 45,000 workers are estimated to have been displaced from the former LSCFs to Communal Areas, towns, informal settlements (Chambati and Moyo 2004, Chambati and Magaramombe 2008) and neighbouring countries (Rutherford and Addison 2007).

Nationally it is estimated that over two thirds of the former farm workers remained on the former LSCF land (Chambati and Moyo 2004, Moyo et al. 2009, Magaramombe 2010). Nearly 69 percent of the former farm workers indicated that the majority of their colleagues were still resident in the compounds (AIAS farm worker survey 2005/06). There were variations in the displacement, as in Chiredzi, Chipinge and Zvimba districts over 60 percent of the former farm workers reported that most of their colleagues had stayed put, compared to 29 percent and 9 percent in Kwekwe and Mangwe respectively (AIAS farm worker survey 2005/06). Overall there has been a net gain in livelihoods, as 45,000 farm workers and 4,000 physically displaced farmers have been replaced by 170,000 farm households (see Moyo 2011a) plus new types of employment.

---

[14]By 1995, 62 percent of the national population were poor, the majority of whom resided in the communal areas where 80 percent of the people were poor in comparison to 46 percent in urban areas (GoZ 1998).

Few former farm workers remaining in the redistributed lands have transitioned to self-employment on FTLRP land allocations. There were competing demands for land from landless peasants, the urban working class, farm workers, semi-proletarians, and an emerging middle class bourgeoisie. Different arguments have been mobilised on who was supposed to benefit from land redistribution. Some would argue that farm workers should have been the first to get land since their livelihoods are dependent on LSCFs (Magaramombe 2003), while others are in favour of preferential treatment for landless peasants (Mutingwende 2004). Gender equity in land allocations is urged by others (see Moyo 2011a). The FTLRP policy document emphasised landless peasants as targeted land beneficiaries, followed by other groups in need of land, and offered a 20 percent quota of all land allocations to liberation war veterans (GoZ 2001). Former farm workers were neither listed as targeted land beneficiaries nor was a specific land quota offered to them in the FTLRP policy document.

Many former farm workers were excluded from land access, despite their preference for resettlement,[15] constituting only 8.1 percent of the land beneficiaries (Moyo et al. 2009, 22).[16] However, former farm workers deployed multiple strategies to gain access to land, including registration with traditional leaders in Communal Areas, participation in land occupations, and official applications with district land committees without revealing their status as former farm workers (Chambati and Moyo 2004). They did this in fear of victimisation and being left out of the resettlement exercise as perceived 'anti-land reform' reactionaries and MDC (Movement for Democratic Change) supporters (Chambati and Moyo 2004). Only 38 of the 70 former farm workers who got land disclosed their status when they registered for land (AIAS farm worker survey 2005/06).

Chipinge had the highest of proportion of former farm workers among land beneficiaries (17.7 percent), followed by Mangwe (12.4 percent), while the proportions were 6 percent and 7 percent in Kwekwe and Zvimba respectively (Moyo et al. 2009). In Chiredzi, they constituted less than one percent of the land beneficiaries (Moyo et al. 2009). These patterns indicate that former farm workers were not entirely disregarded as a category of beneficiaries (see Alexander 2003, ZHRF and JAG 2007) but in practice were included. Some former farm workers (9.7 percent) informally accessed small pieces of land (between 0.04 and 2.0 hectares) parcelled out from among land beneficiaries, while others (26.8 percent) have Communal Area plots (see Table 1).

Those who did not gain land continue to sell labour for farm and non-farm activities. Overall, 32.1 percent and 22.9 percent of the former farm workers were employed as permanent and casual farm labour respectively (Table 1). Twenty-two out of the 133 former farm workers employed as permanent workers also worked as casual labourers (AIAS farm worker survey, 2005/06). In total, 49.7 percent of the former farm workers in the surveyed districts performed permanent or casual farm work or both (AIAS farm worker survey 2005/06). Their residency in the farm compounds is insecure, and is sometimes considered as 'squatting' as some seek to

---

[15]Many former farm workers (73 percent) preferred resettlement (GoZ/IOM 2004).

[16]In Masvingo Province, they were seven percent of the land beneficiaries (Scoones et al. 2010). The farm worker land allocations are also presented as a proportion of former farm workers as group. The GoZ/IOM (2004) estimates that 15 percent of the former farm workers received land.

Table 1. Employment of former farm workers in agriculture.

Type of employment	Chipinge		Goromonzi		Chiredzi		Kwekwe		Mangwe		Zvimba		Total	
	No.	%	No.	%	No.	%	No.	%	No.	%	No.	%	No.	%
Permanent employment	28	38.4	27	23.9	40	44	11	24.4	11	40.7	16	24.6	133	32.1
Casual employment	23	31.5	25	22.1	24	26.4	3	6.7	2	7.4	18	27.7	95	22.9
Self-employment on:														
FTLRP land allocation	19	26.0	15	13.3	14	15.4	6	13.3	4	14.8	12	18.5	70	16.9
Informal land allocations by employers	7	9.6	6	5.3	10	11.0	7	15.6	4	14.8	6	9.2	40	9.7
Communal Area plots	19	26.0	19	16.8	44	48.4	15	33.3	4	14.8	10	15.4	111	26.8
N	73		113		91		45		27		65		414	

Source: AIAS Inter-district Farm Worker Survey (2005–2006).
Note: Percentages do not add up to 100 percent as some former farm workers are not employed in agriculture, while some permanent workers have multiple jobs.

perpetuate the LSCF labour tenancy, although the state permits them to reside on and farm small plots (Moyo *et al.* 2009), as discussed later.

New structure and forms of agrarian employment

Land redistribution has generated a new agrarian labour structure in which there is competition for labour resources. The first line of competition pertains to the deployment of labour among a set of competing activities that include self-employed farming, other petty commodity production (e.g. natural resources exploitation) and hiring out labour to farm jobs. The second line of competition relates to the competition among the different farming sectors on the former LSCF area, which includes 146,000 peasant households, 22,700 mid-sized farms and 217 large farms on redistributed land, alongside remaining LSCFs and plantations (Moyo 2011a).

Self-employed labour deployed to farming averaged 3.6 persons per household in newly redistributed areas (Table 2). Of the small peasant (A1) and larger scale (A2) farms, 26 percent and 17.1 percent, respectively, were exclusively using family labour for farming (Figure 1). By 2006, the newly redistributed areas had generated an estimated 570,301 new self-employed farm jobs nationally (Table 5). Men accounted for 55 percent of the family labour (AIAS Baseline Survey 2005/06) in contrast with Communal Areas, where women are dominant (Adams 1991). This reflects the slightly male-dominated population in newly redistributed areas (Moyo *et al.* 2009).

The majority of self-employed labourers (74 percent of households) also hired farm wage labour (Figure 1). Within the different resettlement sectors, 71.6 percent and 82.9 percent of the A1 and A2 farms hired wage labour respectively. The use of wage labour on farms tended to be higher in districts with high agro-ecological potential; in Chipinge, Goromonzi and Zvimba over 75 percent used it compared to 40–60 percent in drier Mangwe and Kwekwe districts (Figure 1).

In the small A1 farms, most of the work is performed by the family as 21.3 percent and 57.1 percent of A1 households hired permanent and casual workers respectively. Mid-sized to large A2 farms rely more on full- and part-time wage labour, which was engaged by 50.5 percent and 63.6 percent of these farms respectively (Moyo *et al.* 2009). Agricultural wage labour is being mobilised in small batches in the downsized farms, as few hire more than 10 workers (Moyo *et al.* 2009), compared to an average of 50 workers in the LSCFs (CSO 2001). Another feature of the new agrarian employment is that some land beneficiaries also sell their labour to other farmers (Table 2). Men still dominate the permanent workforce as they did before 2000, comprising 70 percent, but women have substantially increased their share from 10 percent in 1999 (CSO 2001) to 30 percent (Moyo *et al.* 2009).

The farm wage labour is recruited from different sources including Communal Areas, the former farm worker population, unemployed urban people and new land beneficiaries. Many land beneficiaries (42 percent) recruited wage labour from among their extended family and others in the Communal Areas (AIAS Household Baseline Survey 2005/06). The recruitment of wage labour from urban areas was more common in districts close to these areas such as Goromonzi and Zvimba, in which households hired unemployed youths from Mabvuku suburb and Banket town (Chambati 2009). Former farm workers were employed by 36 percent of the land beneficiaries and their recruitment was higher in Chipinge, Chiredzi, Goromonzi and Zvimba districts which had higher concentrations of former farm workers before 2000 (Moyo *et al.* 2009).

Table 2. Emergent structure of rural labour in new resettlement areas.

Level of labour use	A1							A2						
	No. of hh[4]	% of hh	Average labour use					No. of hh	% of hh	Average labour use				% of Total
			FT[5]	PT[6]	Family	Hired out				FT	PT	Family	Hired out	
Low[1]	1134	68.7	0.0	4.22	3.65	0.14		217	49.5	0.0	12.3	3.38	0.02	64.7
Medium[2]	140	8.5	1.0	5.95	3.56	0.17		55	12.6	1.0	12.1	3.45	0.05	9.3
High[3]	377	22.8	7.2	12.75	3.77	0.11		166	37.9	8.2	13.1	3.76	0.06	25.9
Total	1651	100.0	1.74	6.31	3.67	0.14		438	100	3.2	12.6	3.53	0.04	100

Source: Moyo and Chambati (2008) from AIAS Baseline Survey (2005/06).
Notes: [1] Household utilises family labour in combination with part-time labour; [2] Household hires in one ful -time worker; [3] Household hires in at least two full-time workers plus part-time; [4] hh – household; [5] FT – full time; [6] PT – part time.

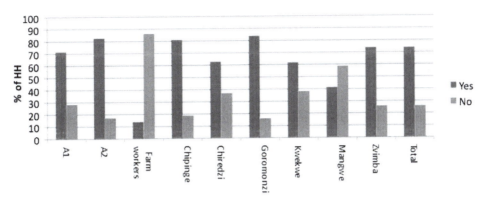

Figure 1. Hiring of farm labour in newly redistributed areas
Source: AIAS Household Baseline Survey (2005/06), N=2089.

New forms of social patronage have been introduced in newly redistributed areas through hiring extended family members into the wage labour force. In these forms, work relations are defined by kinship ties between the employer and the employee (Chambati and Moyo 2004). These social patronage labour relations entail qualitatively different social relations than the remnants of the master-servant relationship in LSCFs, as discussed later. The social patronage labour relations are more common in the A2 farms (16 percent) compared to the A1 farms (11.5 percent) (AIAS Household Baseline Survey 2005/06).

There exists a differentiated pattern of agrarian labour use, in which some land beneficiaries have access to more economic resources – which translates to larger cropped areas and utilisation of more wage labour. Agrarian labour utilisation is one of the critical factors in explaining broader social differentiation in newly redistributed areas (Moyo *et al.* 2009). Most new land beneficiaries (64.7 percent) were low-level labour users who relied mostly on family labour with occasional hiring of part-time workers, while 26 percent were high-level labour users, hiring more than seven full-time workers (Table 2). More A2 farmers (38 percent) were high-level labour users compared to A1 farmers (23 percent) (Table 2). As expected, the number of workers hired increased together with cropped area (Moyo *et al.* 2009).

Small farms had higher agrarian labour intensities (labour employed per unit of farm area) as the average labour intensities decreased as the farm size increased (Table 3) because larger farms tend to have access to labour-substituting machinery (Moyo *et al.* 2009). Thus small farms generate more employment in comparison to the larger farms. In the 1990s, LSCFs had the lowest labour intensity, averaging 0.7 workers per cropped hectare, compared to over 3.5 in Communal and Old Resettlement Areas (Chambati 2009). The permanent labour utilisation per unit of cropped area in newly redistributed areas averaging 0.62 is slightly less than the combined full and part time labour utilisation per unit of cropped area that prevailed in LSCFs before 2000 (Moyo *et al.* 2009). This implies greater capacity for labour absorption when land utilisation reaches optimal levels in the new farms.

Land preparation capacity represents one of the critical constraints for land beneficiaries and which when available at the right season creates the opportunity to crop more land, enhancing productivity and potentially increasing the demand for labour. Only 36 percent of large A2 farms had access to a tractor, while in the small

A1 farms, 52 percent had access to animal-drawn ploughs and 6.2 percent had access to tractors (Moyo *et al.* 2009). The 'fast track' mechanisation programme initiated by the state in 2003 was launched to address this and labour shortages that arose after 2000 (GoZ 2007). Cattle, which provide an important source of draught power, especially for small producers, were owned by 42.8 percent of the land beneficiaries (Moyo *et al.* 2009).

Farm labour utilisation rates were influenced by the capital stock possessed by the households (Table 4). There was a direct relationship between farm machinery and equipment endowments and scale of labour establishment. The high-level labour users constituted the majority of the households with high capital intensity. This relationship tended to be influenced by the areas cropped as, for example, in Zvimba low-capital-intensity households cropped 3.5 hectares in comparison to 17.6 hectares in high-capital-intensity households (Chambati 2009). Meanwhile, those involved in labour-intensive crops, such as tobacco, had high capital stocks (Moyo *et al.* 2009). Thus high labour utilisation rates are directly influenced by access to capital, which in turn determines commodity choice and land use capacity.

Land redistribution and tenure reforms after 2000 have also opened up self-employment opportunities in natural resource exploitation to land beneficiaries, those in nearby communal areas and former farm workers. The LSCF freehold

Table 3. Labour intensities by farm size.

Labour type	Farm size				
	1-19	20-49	50-99	100+	Total
No. of workers per unit area					
Permanent workers	0.33	0.88	0.33	0.16	0.19
Casual workers	1.20	0.37	0.10	0.02	0.71
Family labour	0.60	0.15	0.06	0.21	0.35
Family + permanent	0.94	0.23	0.10	0.04	0.54
N	962	620	191	208	1981

Source: AIAS Household Baseline Survey (2005/06).

Table 4. Level of labour use vs. capital intensity.

Level of Labour Use	Capital Intensity (No. and % of households)							
	Low[1]		Medium[2]		High[3]		Total	
	No.	%	No.	%	No.	%	No.	%
Low	822	68.4	432	63.8	97	46.2	1351	64.7
Medium	93	7.7	80	11.8	22	10.5	195	9.3
High	287	23.9	165	24.4	91	43.3	543	26.0
Total	1202	100	677	100	210	100	2089	100

Source: AIAS Household Baseline Survey (2005/06).
Notes: [1]Low capital intensity – did not own any power-driven implements and relied on animal drawn implements and hiring services. [2]Medium capital intensity – owned at least one power-driven implement, not including a tractor. [3]High capital intensity – owned at least three items from a set of power driven equipment that includes tractors.

property rights excluded other population segments from accessing natural resources through their security protection systems in the form of farm guards and other security people (see Loewenson 1992, Amanor-Wilks 1995, Rutherford 2001). Land beneficiaries with state tenures are less equipped to exclude others from natural resources on their lands. In Kwekwe district, for example, over 20 percent of the land beneficiaries confirmed the presence of informal gold miners on their lands (Moyo *et al.* 2009). Land beneficiaries blame former farm workers for the destruction of natural resources (Chambati and Moyo 2004). Many former farm workers (81 percent) confirmed accessing natural resources on new farms (AIAS farm worker survey 2005/06).

About 21 percent of the former farm workers were involved in the extraction of natural resources on a full-time basis (AIAS farm worker survey 2005/06). Alluvial mining (especially gold and diamonds) has created an alternative source of income for farm workers and others in various provinces. In Goromonzi and Kwekwe districts 30.7 percent and 21, percent respectively, were self-employed in gold panning, while 27 percent in Chipinge panned for diamonds (AIAS farm worker survey 2005/06). Firewood, fisheries and thatching grass sales also provide self-employment (Chambati 2009).

Among some land beneficiary households, farming also competed with these non-farm activities for self-employed labour. Overall, six percent indicated that they were involved in the sale of natural resources (AIAS Baseline Survey 2005/06). This was more common in the drier districts such as Mangwe and Kwekwe where 46.9 percent and 11 percent were involved in gold panning respectively (Moyo *et al.* 2009). Natural resources also provided building materials for homesteads, constructed by 73 percent of the land beneficiaries, and firewood was the major source of energy for 81.3 percent. Petty trading activities such as vending of clothes have also grown as result of unrestricted movement of people and goods with over five percent of the land beneficiaries involved (AIAS Baseline Survey 2005/06).

A key development in the new agrarian labour regime areas has been the shortage of labour reported by 38.4 percent of the land beneficiaries (Moyo *et al.* 2009, 107). Given the high prevalence rate of HIV/AIDS among former farm workers compared to other segments of the population (UNDP 2003), some argue that HIV/AIDS effects explain farm labour shortages (Rutherford 2004). Similarly, food aid targeting former farm workers around 2003 has also been blamed by land beneficiaries for reducing their incentive to work (Chambati and Moyo 2004). However NGOs such as Farm Community Trust of Zimbabwe (FTCZ) have argued that farm labour shortages are not related to food aid, as these were also prevalent where there was no distribution (Chambati and Moyo 2004).

These shortages of labour are primarily influenced by the wage differentials between farm labour and other non-farm sources of income. The shortage of labour was more acute in alluvial-gold-rich Goromonzi and Kwekwe districts where it was experienced by 47.6 percent and 42.5 percent of the land beneficiaries respectively, while in the other districts less than 30 percent faced labour shortages (Moyo *et al.* 2009). In this situation, land beneficiaries with access to capital are able to pay better wages and attract labour better than those without. Farm labour shortages were also faced by the remaining plantations and LSCFs, such as Tanganda Tea Company, which attributed reduced output in 2006 to labour shortages (USAID 2010). The farm labour shortages also arose because of the expansion of self-employment among the potential agrarian labour force from the Communal Areas who gained access to land.

Notwithstanding the labour shortages and production constraints (Moyo *et al.* 2009, World Bank 2006), the current labour utilisation rate indicates an expansion of farm jobs greater than that of the former LSCFs. Extrapolating the district surveys' farm labour utilisation rates nationally, we estimate that by 2006, permanent jobs had grown to over 454,281 in comparison to 167,851 jobs in 1996 (Table 5). Self-employed farm jobs grew by more than 90 fold during the same period (see Table 5).

The FTLRP has thus broadened the role of agriculture as a key source of employment to alleviate Zimbabwe's unemployment problem, which has been growing since ESAP. Tobacco production is estimated to have employed 112,000 people, accounting for 20 percent of the current formal employment, in the 2009/10 season, whereas in 1998 it employed 172,000 people (PWC 2010), or 12.7 percentof the formal employment (CSO 2000).[17]

Social reproduction of agrarian labour

Land redistribution has altered the ways in which agrarian labourers realise their social reproduction. The diversified farming structure provides the agrarian labour force with broader choices for wage work than before. Agrarian political power has been diffused among many smaller employers. Quite crucially, the new dispensation *delinks* the employment rights of former farm workers from their land rights.

Furthermore, the analysis of farm worker's social reproduction needs to take into account the socio-economic context during the last decade. Adverse economic conditions especially between 2002 and 2008 induced production constraints,

Table 5. Estimated growth in full-time and self-employment, 1996–2006[1].

Farm sector	No. of employers	Average farm size	Average no. of employees per farm		Total number of employees	
			Permanent	Self-employed	Permanent	Self-employed
Past Scenario, 1996						
LSCF	6,600	2,000	25.4	0.9	167,851	5,898
New Scenario, 2006						
A1	141,656	40	1.7	3.6	240,815	509,962
Small-medium A2	14,072	71	6.6	3.6	92,875	50,659
Large A2	1,500	600	15	3.6	22,500	5,400
Remaining LSCF						
White LSCF	700	871	25.4[2]	2	17,640	1,400
Black LSCF	1,440	625	25.4[2]	2	36,576	2,880
Corporate estates	874	1,874	50.2	n.a	43,875	n.a.
Total					454,281	570,301
New jobs created					286,430	564,403

Sources: CSO (2001), AIAS Household Baseline Survey (2005/06), World Bank (2006).
Notes: [1]Estimates exclude part-time workers to avoid double accounting as some permanent workers also perform part time work. [2]Employment in remaining LSCFs is assumed not to have changed.

---

[17]Agriculture was the largest formal employer, accounting for 26 percent of the wage labour force in 1999 (CSO 2000).

including price controls on some commodities and shortages of agricultural inputs such as fertilisers, agro-chemicals, farm machinery and agricultural finance (World Bank 2006, Moyo *et al.* 2009) and these have in turn shaped agrarian labour relations. Hyperinflation, which set in around 2006, eroded the average real earnings of both rural and urban workers (World Bank 2006). Since 2009, economic stabilisation through 'dollarisation' and re-liberalisation has begun to reverse some earlier production declines in crops such as tobacco (see Moyo 2011a).

Nonetheless, farm wages were inadequate to meet labour's cost of social reproduction – as before 2000 – as land beneficiaries pressed for low wages in the collective bargaining process in the context of an economic crisis and their just 'starting up'. The gazetted farm worker wages which a few new land beneficiaries pay are well below the rural Poverty Datum Line. In November 2005, the gazetted wage constituted 11 percent of the PDL, which led to the introduction of payments in kind (Chambati 2009) that were received by 50 percent of the farm workers around 2006 (Chambati and Moyo forthcoming). In 2011, prevailing wages (US$30 to US$50) still remain far below the rural PDL estimated at over US$300 (ZIMSTAT 2011). Indeed, 51 percent of the farm workers noted that poor wages were a major challenge to their social reproduction (AIAS farm worker survey 2005/06). Nor were all farm workers able to receive their statutory benefits, as only 40 percent and 36 percent of the farmers provided annual leave and protective clothing respectively (Moyo *et al.* 2009, 110). There were however cases of non-super exploitative[18] wage payments, especially among producers of export crops such as tobacco which require specialised skills (Moyo *et al.* 2009, Moyo 2011a). In this context, some farm workers sold their labour to more rewarding non-farm jobs.

Although farm workers have diversified their sources of employment, the social reproduction of many of them remains poor. About 53 percent of the farm worker households were managing to consume three meals per day (breakfast, lunch and supper) around 2006, while 4.3 percent and 42.6 percent had one and two meals per day respectively (AIAS farm worker survey 2005/06). This was in contrast to land beneficiary households in which over 75 percent were able to consume three meals per day during the same period (AIAS household baseline survey 2005/06). Moreover, proteins were glaringly missing in farm worker diets dominated by the staple *sadza*[19] and vegetables (AIAS farm worker survey 2005/06). Similar to the situation before 2000, malnutrition and food insecurity remain major challenges for farm workers, as noted by 31 percent of them in the survey.

Many farm worker households were not able to send their children to school, as only 22.2 percent of the school-age children (6 to 18 years) were in school (AIAS former farm worker survey 2005/06). Most of the youths were employed in the new farms, with 30.9 percent and 15.8 percent of them as permanent and casual labour respectively. Land beneficiary households presented a different picture as 77 percent of the school-going age were in school (AIAS household baseline survey 2005/06).

There were however some differences in the social reproduction of agrarian labour between kinship and non-kinship work relations. Although the wages paid in kinship work relations were not significantly different from those paid to non-

---

[18]To clarify, non-super exploitation refers to a situation in which wages can meet the costs of social reproduction, while in super exploitation the wages cannot meet workers' costs of social reproduction.
[19]Thick porridge made from maize meal.

relatives, in the former case more workers received additional benefits than in the latter (Moyo *et al.* 2009). For instance, 84 percent of related employees received food rations in comparison with 54 percent of non-related employees (AIAS farm worker survey 2005/06). This was also reflected in the number of meals consumed in related employee households; 67.3 percent of these households consumed three meals per day in comparison wtith 46.4 percent of non-related employee households. Annual leave was also enjoyed by more kin employees (58.9 percent) compared to non-related employees (48.9 percent).

New A2 land beneficiaries favour a residential labour tenancy, contrary to policy that allows former farm workers temporary residence irrespective of employment (Magaramombe 2003, Chambati and Moyo 2004). The implementation of this policy was varied across the districts and in some areas conflicts have emerged as land beneficiaries seek to evict those workers refusing to work for them.[20] About a third of the former farm workers indicated that insecure access to residency and related land conflicts had been problems for them during the FTLRP (AIAS farm worker survey 2005/06). Some A2 land beneficiaries complain that they incur electricity and water charges on behalf of former farm workers who they do not employ (Utete 2003). While the residential rights of some former farm workers are as insecure as they were before 2000, the land beneficiaries are less able to compel labour to work for low wages, often leading to labour shortages. Some farm workers view the FLTRP as an opportunity to dismantle labour tenancy even in the face of eviction threats, which 15.8 percent face (Moyo *et al.* 2009). Other former farm workers also argue that even if they are not working on the new farms, their residency is justified since they are yet to receive retrenchment packages; only 56 percent had received part or the full amount (AIAS farm worker survey 2005/06).[21]

The labour tenancy is also being prolonged by the recruitment of landless relatives and other new workers. Eighty-four percent (or 292) of the new farm workers surveyed were resident in the homes provided by their employers (AIAS farm worker survey 2005/06). Of these, 277 new farm workers (including relative-employees) indicated that they stayed there because of their employment. This labour tenancy arrangement is qualitatively different from the overcrowded compound which was located far away from white farmers' luxurious mansions,[22] as we observed that new farm workers in most cases were housed at the land beneficiaries' homestead and shared several amenities with them including water, sanitation and energy. Labour tenancy also continues in the remaining plantations and LSCFs, where some of those employed continue to reside on such farms under insecure tenure (Moyo 2011b).

Although, the data presented above suggests the entrenchment of functional dualism, whereby farm workers continue to subsidise farming for the emergent mid-sized and large capitalist farms through underconsumption of necessary social

---

[20]The High Court of Zimbabwe has heard several eviction cases of former farm workers (Chambati and Magaramombe 2008).

[21]Permanent workers losing their jobs were entitled retrenchment packages deducted from the LSCF compensation for land improvements (*Statutory Instrument No. 6 of 2002*). However disagreements on the compensation process and delays in the legal aspects of land acquisition meant that few LSCFs had been valued (Moyo *et. al.* 2009).

[22]Between 1996 and 2000, 30 percent of the funds invested in housing in the LSCFs were for farmers when they constituted only 0.2 percent of the population and 70 percent was devoted to the 2,000,000 people that included farm workers and their families (CSO 2001).

requirements that cannot be met by the wage incomes (Moyo and Yeros 2007), after 2000 it persists in a different form. The working class or semi-proletariat have increased their bargaining power or ability to withdraw labour from agriculture and sell it elsewhere.

Furthermore, land redistribution has reduced the *necessity* for some peasants to engage in farm labour markets. However, in the context of the economic crisis during the FTLRP, some small producers who could not mobilise adequate resources to fully utilise their lands also engaged in farm wage labour (see Table 2). This suggests that the semi-proletarianisation process that existed before 2000 with limited access to land (Moyo and Yeros 2005) has been altered as paid work is sought with autonomy of land access.

Land redistribution has also paved the way for non-racial labour relations systems in the new farms. The inhumane treatment that farm workers suffered at the hands of 'superior' white farmers declined significantly after 2000. Only 0.8 percent of the farm workers reported the use of physical beatings by employers in the accomplishment of tasks (AIAS farm worker survey, 2005/06), indicating a shift in some of the struggles farm workers face in improving their material conditions.

Struggle for autonomy among agrarian labour in redistributed areas

Beyond farm-level workers' committees, self-organisation of farm workers before 2000 was restricted on LSCFs through the labour tenancy system. After the land redistribution and tenure reforms farm workers continue to organise themselves with limited assistance from external agents such as political parties, NGOs, and their lead trade unions. During the land occupations around 2000, some former farm workers viewed their future as independent producers and formed alliances with liberation war veterans and peasants in occupying LSCFs (Sadomba 2008). Indeed 18 of the 70 former farm workers (25.7 percent) who got access to land participated in the land occupations and got their occupied lands regularised under the FTLRP (AIAS Farm Worker Survey 2005/06). They also asserted their autonomy in the struggle for land through independent land occupations (Musungwa 2001, Sadomba 2008). Former farm workers were identified as the third dominant group in perpetrating 'human rights' violations against land owners after liberation war veterans and ZANU (PF) members (ZHR and JAG 2007).

Some former farm workers were also mobilised by white farmers and GAPWUZ to defend existing freehold land rights and their jobs, and to oppose the new constitution that contained a clause on compulsory LSCF acquisition (Chambati 2009, Sadomba 2008). The alliances forged during the land occupations also tend to define the current relationships between former farm workers and land beneficiaries. Overall about 18 percent of the former farm workers reported being involved in violent confrontations with A1 land beneficiaries over farm compound land rights and natural resource access. These violent confrontations were reported by over 20 percent in Chiredzi, Goromonzi and Zvimba, where GAPWUZ had a strong membership base and fewer farm workers who participated in land occupations, in comparison to less than seven percent in the other districts. In Mangwe district, which had the highest proportion of former farm worker participation in the land occupations, only 3.7 percent reported these violent confrontations (AIAS Farm Worker Survey 2005/06).

Former farm workers' struggles also relate to them being wrongly perceived as 'foreign' migrants (Moyo *et al.* 2000). In fact 26.5 percent of the former farm workers are descendants of migrant workers who were born in Zimbabwe, while 90.5 percent were born in Zimbabwe (AIAS Farm Worker Survey 2005/06). This is recognised by the Citizenship Amendment Act of 2004 that entitles people born in Zimbabwe of Southern African Development Community (SADC) parents to citizenship, although state administrative capacities are weak in offering registration services (Chambati and Moyo, 2004). Most of the former farm workers of foreign origin consider themselves de facto citizens of Zimbabwe as only 0.5 percent stated a preference to return to their countries of origin after the FTLRP (Chambati and Moyo, forthcoming). By late 2004, the Repatriation Unit of the Ministry of Public Service, Labour and Social Welfare and the International Organisation of Migration (IOM) had not handled any requests from former farm workers for repatriation (Chambati and Moyo 2004).

Farm worker trade unionism has further weakened after the FTLRP. Few farm workers (4.4 percent) were aware of the existence of a labour union in their area and only 3.0 percent were paid up members of a labour union (AIAS Farm Worker Survey 2005/06). Trade unionism among farm workers was more common in Goromonzi (9.6 percent) and Chipinge (4 percent) compared with less than one percent in the other districts. Workers' committees that handle worker grievances exist on few farms, as only 12.0 percent of the workers had such structures at their places of employment (AIAS Farm Worker Survey 2005/06).

The absence of trade unions weakens the voice of farm workers seeking the enforcement of their labour rights. Indeed wage bargaining of farm workers is weak at the National Employment Council for the Agricultural Industry (NECAIZ) level, as land beneficiaries have refused to endorse some collective bargaining agreements (e.g. July 2003) arguing unaffordability, with a weak response from GAPWUZ (Chambati and Moyo 2004). At the local level, farm workers have however deployed their bargaining power to sell their labour in non-farm jobs. The realisation of farm worker labour and social rights is also negatively affected by the limited presence of state labour officials in the newly redistributed areas (Chambati and Moyo 2004).

Furthermore, a few NGOs still remain working in new farms (e.g. FCTZ, Kunzwana Women's Association and Farm Orphan Support Trust [FOST]) from a peak of about 12 in the late 1990s (Chambati and Magaramombe 2008) due to funding constraints (Moyo *et al.* 2009) and state perceptions of them being MDC supporters who utilise welfarist projects for political campaigning (Helliker 2008).

Farm workers are increasingly organised independently in different ways to provide labour services to a range of farmers and others, with the power to withdraw labour. Some formed groups of skilled former farm workers or specialised labour to provide consultancy services to new land beneficiaries as a resistance to the low wages, as they tend to demand higher payment rates than permanent and casual workers. The specialist services being provided by these groups include overall farm planning activities, machinery operations, tobacco operations, and livestock disease management and were utilised by 11.2 percent of the land beneficiaries (AIAS Household Baseline Survey 2005/06). Unskilled former farm workers groups also perform general tasks such as weeding, harvesting and stumping that were used by 24.4 percent of the land beneficiaries (AIAS Household Baseline Survey 2005/06).

Former farm workers are thus engaged in wider autonomous struggles to improve their material conditions and social reproduction.

Conclusion

Land redistribution has generated a new agrarian labour regime comprising many self-employed peasants in farm and non-farm jobs. Agrarian wage employment has been broadened as some A1 and many A2 land beneficiaries also hire wage labour in a differentiated manner. The compulsion to engage in cheap farm labour markets has been lessened for sections of the peasantry who gained access to land and have some autonomy in meeting the requirements for their social reproduction.

New employers have less total control over farm workers than was the case under freehold title as they only have land user rights and cannot enforce the residential labour tenancy. The political power conferred by land ownership is now diffused among many smaller farmers who compete for labour, while de-racialisation of the agrarian labour relations has diluted remnants of the master-servant relationship of the past.

Although functional dualism persists in the new agrarian structure, the semi-proletarianisation process resulting from landlessness and land shortages before the FTLRP has been altered. The freeing of labour from residential tenancy tied to provision of farm labour means that farm workers can sell their labour in competing non-farm jobs. Farm workers resist poor working conditions in the farms through self-organisation in autonomous groups to provide high-wage demand-driven consultancy services to land beneficiaries.

Nonetheless, the social reproduction of many farm workers remains precarious, suggesting the need to resolve some of the current contradictions. For the social relations of agrarian labour to be fully equitable, there is a need to further redistribute land to former farm workers and clarify their tenure security, as well as for the state and NGOs to step up the provision of social services, enforcement of labour rights and implement agrarian labour quality and skills development programmes.

References

Adams, J. 1991. The rural labour market in Zimbabwe. *Development and Change*, 22, 297–320.

AIAS 2005/06. Inter-district household baseline and farm worker survey. Harare: AIAS.

Alexander, J. 2003. 'Squatters', veterans and the state in Zimbabwe. *In*: A. Hammar, B. Raftopolous, and S. Jensen, eds. *Zimbabwe's unfinished business: rethinking land, state and nation in the context of crisis*. Harare: Weaver Press.

Amanor-Wilks, D. 1995. *In search of hope for Zimbabwe's farm workers*. Harare: Dateline Southern Africa and Panos Institute.

Arrighi, G. 1970. Labour supplies in a historical perspective: a study of proletarianisation of African peasantry in Rhodesia. *Journal of Development Studies*, 6(3), 197–234.

Chambati, W. 2003. Land reform and agricultural labour: case study evidence from Mazowe and Chikomba districts. Harare: AIAS mimeo

Chambati, W. 2009. Land reform and changing agrarian labour processes in Zimbabwe. MM Thesis, University of Witwatersrand, South Africa.

Chambati, W. and S. Moyo 2004. Impact of land reform on former farm workers and farm labour process. Harare: African Institute for Agrarian Studies monograph.

Chambati, W. and G. Magaramombe. 2008. An abandoned question: farm workers. *In*: S. Moyo, K. Helliker and T. Murisa, eds. *Contested terrain: civil society and land reform in contemporary Zimbabwe*. Johannesburg: SS Publishing.

Chambati, W. and S. Moyo. forthcoming. Situation of former farm workers after land reform in Zimbabwe: empirical evidence from a national baseline survey. AIAS Monograph Series, Harare.

Clarke, D. 1977. *Agricultural and plantation workers in Rhodesia.* Gwelo: Mambo Press.

CSO 2000. *1999 indicator monitoring labour force survey.* Harare: Government Printers.

CSO 2001. *Agricultural production on large scale commercial farms.* Harare: Government Printers.

FCTZ 2002. Commercial and resettled farms pilot survey, Seke district. Harare: FCTZ mimeo.

Freund, B. 1984. Labour and labour history in Africa: a review of the literature. *African Studies Review,* 27(2), 1–58.

GoZ and IOM 2004. Zimbabwe third farm workers survey report: labour dimensions of the land reform programme. Harare: Unpublished Report.

GoZ 1998. *Poverty assessment survey main report.* Harare: Government Printers.

GoZ 2001. Land reform and resettlement: revised phase II. Harare: Ministry of Lands and Rural Resettlement.

GoZ 2007. Agricultural engineering, mechanisation and irrigation strategy framework: 2008 – 2058. Draft Strategy. Harare: Ministry of Agricultural Engineering, Mechanisation and Irrigation.

Hammar, A., B. Raftopolous and S. Jensen, eds. 2003. *Zimbabwe's unfinished business: rethinking land, state and nation in the context of crisis.* Harare: Weaver Press.

Hartnack, A. 2007. My life got lost: farm workers and displacement in Zimbabwe. *Journal of Contemporary African Studies,* 23(2), 173–93.

Helliker, K. 2008. Dancing on the same spot: NGOs. *In*: S. Moyo, K. Helliker and T. Murisa, eds. *Contested terrain: civil society and land reform in contemporary Zimbabwe.* Johannesburg: SS Publishing.

Hellum, A. and B. Derman. 2004. Land reform and human rights in contemporary Zimbabwe: balancing individual and social justice through an integrated human rights framework. *World Development,* 32(1), 1785–1805.

JAG/RAU 2008. Reckless tragedy: irreversible? A survey of human rights violations and losses suffered by commercial farmers and farm workers from January 2000 to 2008. Harare: Justice for Agriculture and Research Advocacy Unit Report.

Kanyenze, G. 2001. Zimbabwe's labour relations policies and their implications for farm workers. *In*: Amanor-Wilks, ed. *Zimbabwe's farm workers: policy dimensions.* Lusaka: Panos Southern Africa, pp. 86–114.

Kibble, S. and P. Vanlerberghe 2000. *Land, power and poverty: farm workers and the crisis in Zimbabwe.* London: CIIR.

Leavy, J. and H. White. 2003. Rural labour markets in sub-Saharan Africa. Institute of Development Studies, Sussex: Working Paper.

Loewenson, R. 1992. *Modern plantation agriculture: corporate wealth and labour squalor.* London and New Jersey: Zed Books.

Magaramombe, G. 2003. An overview of vulnerability within the newly resettled former commercial farming areas. Harare: Draft Report prepared for the United Nations Humanitarian Coordinator.

Magaramombe, G. 2010. Displaced in place: agrarian displacements, replacements and resettlement among farm workers in Mazowe district. *Journal of Southern African Studies,* 36(2), 361–75.

Masiiwa, M. and L. Chipungu. 2004. Land reform programme in Zimbabwe: disparity between policy design and implementation. *In*: M. Masiiwa, ed. *Post Independence land reform in Zimbabwe: controversies and impact on the economy.* Harare: Frederick Ebert Stiftung and University of Zimbabwe.

Moyo, S. 1995. *The land question in Zimbabwe.* Harare: SAPES Books.

Moyo, S. 2000. *Land reform under structural adjustment in Zimbabwe: land use change in the Mashonaland provinces.* Uppsala: Nordic Africa Institute.

Moyo, S. 2011a. Three decades of agrarian reform in Zimbabwe. *Journal of Peasant Studies,* 38(3), 493–531.

Moyo, S. 2011b. Land concentration and accumulation after redistributive reform in post settler Zimbabwe. *Review of African Political Economy,* 38(128), 257–76.

Moyo, S., B. Rutherford and D. Amanor-Wilks 2000. Land reform and changing social relations for farm workers in Zimbabwe. *Review of African Political Economy*, 84(18), 181–202.

Moyo, S. and P. Yeros. 2005. The resurgence of rural movements under neoliberalism. *In*: S. Moyo and P. Yeros, eds. *Reclaiming the land: the resurgence of rural movements in Africa, Asia and Latin America.* London and Cape Town: Zed Books, David Phillip, pp. 165–208.

Moyo, S. and P. Yeros 2007. The radicalised state: Zimbabwe's interrupted revolution. *Review of African Political Economy*, 34(111), 103–21.

Moyo, S., K. Helliker and T. Murisa, eds. 2008. *Contested terrain: civil society and land reform in contemporary Zimbabwe.* Johannesburg: SS Publishing.

Moyo, S., W. Chambati, T. Murisa, D. Siziba, C. Dangwa, K. Mujeyi and N. Nyoni. 2009. *Fast Track Land Reform baseline survey in Zimbabwe: trends and tendencies, 2005/06.* Harare: AIAS Monograph.

Musungwa, S. 2001. Beneficiary selection in the context of land reform programme, Draft Masters Dissertation, SARIPS. Harare.

Mutingwende, D. 2004. Zvimba district war veterans association land reform progress report. Paper presented at AIAS Provincial Land Reform Dialogue, Chinhoyi Public Service Training Centre, 23–24 August 2004.

Neocosmos, M. 1993. The agrarian question in southern Africa and 'accumulation from below': economics and politics in the struggle for democracy. . Uppsala: Nordic Africa Institute Research Report no. 93.

NRC 2003. *Displaced and forgotten.* Harare: Global Internal Displacement Project.

Oni, S. 1997. *The impact of ESAP on communal areas in Zimbabwe.* Harare: Friedrich Ebert and Stiftung.

Palmer, R. 1977. *Land and racial domination in Rhodesia.* London: Heinemann Educational.

PWC 2010. *Zimbabwe agricultural sector assessment study.* Harare: Price Waterhouse Coopers Unpublished Report.

Richardson, C. 2005. The loss of property rights and the collapse of Zimbabwe. *Cato Journal*, 25(2), 541–65.

Rubert, S. 1997. Tobacco farmers and wage labourers in colonial Zimbabwe, 1904–1945. *In*: A. Jeeves and J. Crush, eds. *White farms, black Labour: the state of agrarian change in southern Africa, 1910–1950.* Portsmouth, Oxford and Pietermaritzburg: Heinemann, James Currey, University of Natal Press.

Rutherford, B. 2001. *Working on the margins: black workers, white land beneficiaries in post colonial Zimbabwe.* Harare and London: Weaver Press, Zed Books.

Rutherford, B. 2004. Farm workers in Zimbabwe: an overview of the current moment. Draft paper for Britain-Zimbabwe Society Review.

Rutherford, B and L. Addison 2007. Zimbabwean farm workers in northern South Africa. *Review of African Political Economy*, 34(114), 619–35.

Sachikonye, L. 2003. *The situation of commercial farm workers after land reform in Zimbabwe.* Harare: Unpublished Report prepared for FCTZ.

Sadomba, W. 2008. *War veterans in Zimbabwe's land occupations: complexities of a liberation movement in an African post-colonial Society.* Thesis (PhD), Wagenigen University.

Scarnecchia, T. 2006. The 'fascist cycle in Zimbabwe'. *Journal of Southern African Studies*, 32(2), 221–37.

Scoones, I., N. Marongwe, B. Mavedzenge, J. Mahenehene, F. Murimbarimba and C. Sukume. 2010. *Zimbabwe's land reform: myths and realities.* New York and Harare: James Currey and Weaver Press.

Shivji, I. 2009. *Accumulation in an African periphery: a theoretical framework.* Dar es Salam: Mkuki na Nyota.

Sunga, I. 2003. *Emerging production systems and technological capabilities in resettlement areas: a case study of Chikomba District.* Harare: AIAS mimeo.

Tandon, Y. 2001. Trade unions and labour in the agricultural sector in Zimbabwe. *In*: B. Raftopolous and L. Sachikonye, eds. *Striking back: the labour movement and the post colonial state in Zimbabwe 1980 – 2000.* Harare: Weaver Press, pp. 221–49.

UNDP 2003. *Zimbabwe human development report 2003.* Harare: UNDP.

USAID 2010. *Zimbabwe agricultural sector market.* Harare: Unpublished report prepared by Weidemann Associates, Inc.

Utete Report 2003. Report of the presidential land review committee under the chairmanship of Dr Charles M.B. Utete, *Volumes 1 and 2: Main report to His Excellency The President of The Republic of Zimbabwe*. Harare: Presidential Land Review Committee (PLRC).

Vhurumuku, E., M. McGuire. and V. Hill. 1998. *Survey of farm worker characteristics and living conditions in Zimbabwe: verification report*. Harare: Agricultural Labour Bureau, Farm Community Trust of Zimbabwe and USAID FEWSNET.

Worby, E. 2003. The end of modernity in Zimbabwe? Passages from development to sovereignty. *In*: A. Hammar, B. Raftopolous, and S. Jensen , eds. *Zimbabwe's unfinished business: rethinking landstate and nation in the context of crisis*. Harare: Weaver Press.

World Bank 2006. *Agricultural growth and land reform in Zimbabwe: assessment and recovery options*. Harare: Green Book.

ZHRF and JAG 2007. Adding insult to injury: a preliminary report on human rights violations on commercial farms, 2000–2005. Harare.

ZIMSTAT 2011. *Poverty Datum Lines – June 2011*. Harare: ZIMSTAT.

Walter Chambati is a researcher at the African Institute for Agrarian Studies (AIAS) in Harare, Zimbabwe and Future Agricultures Consortium Research Fellow for 2011. He received a BSc. (Hon) in Agricultural Economics from University of Zimbabwe and a Masters in Public and Development Management from the University of Witwatersrand. His research interests are in rural labour issues and agricultural development in Africa. He is currently studying for a PhD at the School of Public and Development Management, University of Witwatersrand, focusing on agrarian labour changes after the land reform programme in Zimbabwe.

Who was allocated Fast Track land, and what did they do with it? Selection of A2 farmers in Goromonzi District, Zimbabwe and its impacts on agricultural production

Nelson Marongwe

Questions of who was allocated land under Zimbabwe's Fast Track land reform programme and how productive the beneficiaries have been are highly controversial. This article presents detailed empirical data on beneficiaries who were small and medium-sized commercial farms (the A2 model) in Goromonzi district, land allocation processes, and land use. Goromonzi District is one of the four districts that share a boundary with Harare, the capital city of Zimbabwe. A questionnaire survey targeting 65 A2 beneficiaries was implemented in 2003, while key informant interviews were done in 2006. Drawing on both primary and official data, the article shows that official criteria for selecting beneficiaries for A2 farms that emphasized the potential to use the land productively were ignored in practice. The institutions responsible for land allocation were captured by members of the ruling party and by representatives of the state security apparatus, and most beneficiaries were drawn from the governing or the local elite. Many lacked sufficient capital to invest meaningfully in commercial agriculture, did not have relevant farming experience, and were unable to put the bulk of their land into production for several years. As a result, in Goromonzi District the impact of Fast Track land reform on commercial agriculture has been negative.

Introduction

Who received land through Zimbabwe's Fast Track land reform after 2000, and what did the new owners do with their land? These two questions lie at the heart of the many controversies surrounding the Fast Track programme (Mamdani 2008, Moyo and Yeros 2005, Moyo 2011, Scoones *et al.* 2010). This article presents detailed empirical data on the identity of beneficiaries who were allocated so-called 'A2'[1] land (i.e. small and medium-sized commercial farms) in Goromonzi district,

Professor Ben Cousins deserves special mention. In the first instance, it was Ben who saw potential in my PhD proposal. I value the tremendous support that he gave me in proposal development and supervising the thesis itself. Together with Lionel Cliffe, Ben saw the potential for 'extracting' a publication from my thesis. I thank you for this. I also thank all the reviewers who, through their invaluable comments, contributed to the development of this paper.
[1]Under the Fast Track programme land was redistributed in terms of two main planning models: the A1 model involved small plots for household-based production, and the beneficiaries were to be the generality of the 'landless' population of Zimbabwe, while the A2

the processes through which their land was allocated, and their use of farm land in the first few years of occupation. Located in Mashonaland East Province, Goromonzi District is one of the four districts that share a boundary with Harare, the capital city of Zimbabwe. Drawing on both primary field data and official secondary data contained in government land reform audits, the article shows that official criteria for selecting beneficiaries for resettlement on A2 farms, drawn from policy prescriptions that emphasized the potential to use the land productively, were ignored in practice. In Goromonzi, the bulk of beneficiaries were drawn from the governing elite.[2] The institutions responsible for land allocation were captured by members of the ruling party, ZANU (PF) (Zimbabwe African National Union, Patriotic Front), and by representatives of the state security apparatus. Many of these beneficiaries lacked sufficient capital to invest meaningfully in commercial agriculture, did not have relevant farming experience, and were unable to put the bulk of their land into production for several years. As a result, in Goromonzi District the impact of Fast Track land reform on commercial agriculture has been largely negative.

The article draws on theories of the post-independence state in Africa to explain these processes and outcomes. Rather than representing the general interest, by the late 1990s the Zimbabwean state had been monopolized by the ruling party and become a vehicle for personal wealth accumulation by the governing elite, who practised a politics of patronage. The chaos inaugurated by land invasions created opportunities for the governing elite both to secure their control of the state in the face of a strong opposition challenge and also to acquire valuable assets for themselves. This exemplifies the 'instrumentalization of disorder' (Chabal and Jean-Pascal 2003) by powerful private interest groups occupying key positions in the post-independence state. The fact that beneficiary selection criteria for A2 farms drawn from official policy prescriptions were set aside in practice was thus no accident. The article argues that the conditions for economic and agricultural renewal in Zimbabwe require a state that governs in the interests of society in general, the 'instrumentalization of order', impersonal bureaucratic practices, and the allocation of land to those with the ability and resources to farm productively.

The larger study from which the data in this article is drawn focused on both A1 and A2 beneficiaries in Goromonzi District (Marongwe 2009[3]). Primary data collection included the administration of questionnaires in 2003 to a total of 233 randomly selected A1 and A2 farmers, and the collection of both quantitative and qualitative data on the socio-economic characteristics of beneficiaries, processes of land allocation, and selected aspects of agricultural production. A total of 65 respondents were A2 beneficiaries. Secondary data included official government documents, notably the Mashonaland East Province List of Screened A2

Applications (GoZ 2003) and the Mashonaland East Province Commercial Audit Reports for A2 farms (GoZ 2006). Follow-up field visits were undertaken in selected A2 farms in Goromonzi district in 2006. This article does not discuss the study's findings in relation to A1 beneficiaries, and focuses only on selected findings for the sample of 65 A2 farms.

Debating 'who got the land?' in Zimbabwe's Fast Track land reform

The question of who benefited from Fast Track land reform is highly controversial, and is at the heart of opposing interpretations of the programme. For the first half of the decade the debate was highly polarized, and pitted Sam Moyo and colleagues (e.g. Moyo 2001, Moyo and Yeros 2005) against a range of scholars that included Hammar and Raftopoulos (2003), Davies (2004), Sachikonye (2003, 2004), Cousins (2006) and many others. Moyo and Yeros (2005, 189) argued that the 2000–2002 land occupations represented a genuine, grassroots land movement whose social base was the 'rural semi-proletariat' and which united key elements of the urban and rural poor but also incorporated other class elements such as the black middle and capitalist class. It was thus a progressive 'cross-class nationalist alliance' (Moyo and Yeros 2005). The contradictory character of this alliance led over time to a shift in its internal balance of power, allowing 'the black elite [to] exercise its bureaucratic power not only to make room for the urban petty bourgeoisie on 7,260 small/middle capitalist farms but also for itself, appropriating 150, 000 hectares (0.5 percent of the acquired land) for the benefit of an estimated 178 elites' (Moyo and Yeros 2005, 193). Nevertheless, the fundamentally progressive character of Fast Track land reform by 2003 was evident in the extensive re-peasantization that it entailed, with over 93 percent of all beneficiaries being A1 farmers ('peasants') on over 40 percent of the redistributed land (ibid.).

Hammar and Raftopoulos (2003, 23) queried this kind of characterisation of farm occupations, calling attention to the dispossession of hundreds of thousands of farm workers, the allocation of plots along partisan lines that excluded opposition supporters, and extensive grabbing of prime farms by political, bureaucratic and business elites affiliated to the ruling party. The displacement of permanent farm workers is estimated at 150,000, while the total population affected is around 300,000, including temporary non-resident workers (Chambati, this collection). Raftopoulos (2003) argued that the nature of the occupations was such that they removed land redistribution from the arena of broader public accountability and helped to consolidate ZANU-PF's waning political support through violence and intimidation, reinforcing the use of land reform as a tool for political patronage (see also Alexander 2003 and Bernstein 2005). Sachikonye (2004, 13) remarked on 'the corruption that has permeated land allocation, especially amongst the elite'. Davies (2004) argued that land acquisition under Fast Track land reform had the character of 'personal wealth acquisition' rather than the primitive accumulation of capital. Byres (2004, 13) reiterated this argument and went on to note that Fast Track land reform 'has done very little or nothing to address the condition of the poor in rural Zimbabwe ... the major beneficiaries have been better-off Zimbabweans.'

More recently, the debate on who benefited from Fast Track has become less polarized, at least in relation to the A1 model. With the publication of new research findings, such as studies by Scoones et al. (2010), the AIAS (Moyo 2009), and Ruzivo Trust (Matondi 2011), there appears to be an emerging consensus among researchers that the majority of A1 beneficiaries were 'ordinary' Zimbabweans, and that such allocations occurred in the context of land occupations. Still contentious, however, is the degree to which allocations of A1 plots were biased towards ruling party supporters (Zamchiya 2011, for example).

In relation to A2 farm allocation processes, however, there is much less agreement among researchers. Some studies appear to simply accept the official view that A2 settler selection was guided by the processes prescribed by policy. Moyo *et al.* (2009, 18), for example, claim that 82 percent of beneficiaries in their sample declared that they had been formally allocated land by government. The authors assume that this is strong evidence of allocation in conformity with policy. This is flawed for two reasons. First, Fast Track processes were highly politicized, contested and complex, and relying on only one source of information on how land was allocated is inadequate. Second, formal land allocation in accordance with policy involved several key steps, including the completion of a form by the applicant, and then the assessment and scoring of all applications by bureaucrats in relation to key variables such as income, possession of property, cash flow, experience, etc. Yet the authors pay no attention to these procedures or their outcomes. Moyo *et al.* (2009, 22) provide no hard evidence for their claim that '... most A2 allocations were done by government bureaucrats ...'.

Other studies choose not to investigate A2 land allocation in any depth, perhaps on the assumption that the methods used to allocate A2 plots were generally straightforward and non-controversial and led to fair land allocations. For example, Scoones *et al.* (2010, 47) reveal that political connections played a key role in land allocation processes in some of their study sites in Masvingo Province. On Northdale Farm, where the A2 model was implemented, 'some of the invaders got plots, but most went to outsiders, including senior officials. Those who were well-connected could choose their plots at the offices in town using aerial photographs' (Scoones *et al.*). At Wondedzo farms, Scoones *et al.* note that 'As base commander, Comrade B decided to establish criteria for who would get self-contained plots in Wondedzo Extension and who would join the villages of Wondedzo Wares'. However, in relation to the settlement of sugar plots in Hippo Valley and Fair Range, the study simply states that '... in 2001 successful applicants were issued with offer letters' (Scoones *et al.* 2010). This implies that successful applicants were allocated land through prescribed processes and criteria. Yet, the government's A2 Audit Report for Masvingo (GoZ 2006, 48) shows that many politicians in the province forced their way into such farms, sometimes leading to direct confrontation with war veterans. In particular, there was heavy contestation over who would take over conservancies and other farms with wildlife on them.

The literature describes several by-products of the chaotic and highly politicized nature of land occupations: the development of intense conflicts over double farm allocations (PLRC 2003), boundary disputes (Marongwe 2002), illegal settlements (PLRC 2003) and the displacement of A1 settlers by A2 settlers (GoZ 2006: all provincial A2 Audit Reports). Analysts have noted the manner in which the formal planning procedures undertaken by designated institutions were ignored or undermined by the politically well connected. Gonese and Mukora (2003, 184), for example, remark that:

Empirical observation ... suggests a diversity of operational departures from policy provisions prompted by situational decisions and de-facto conditions.... Numerous decisions by respective Identification and Selection Committees have ... been ignored and/or reversed by political heavyweights whose actions not only compromise formal planning processes, but also seriously undermine the role and effectiveness of officials responsible for implementing programme experiences ...

In relation to A2 farms, intended for aspirant black medium-scale commercial producers, the character of the criteria for beneficiary selection implied an underlying logic: access to capital for farm investment, as well as the skills and expertise to manage a technologically advanced and productive commercial farm. This was seen as important given the specialized nature of many of the farming operations in place on land targeted for acquisition and settlement: timber in Manicaland Province; sugarcane in Masvingo Province; tobacco and horticulture in the Mashonaland provinces; wildlife in the Matebeleland provinces, Midlands and Masvingo provinces; and beef and dairy farming in all the provinces (Zimbabwe, Government 2006: Provincial Audit Reports). Improving agricultural productivity and growth have often been key rationales for redistributive land reform (Sen 1981, Bernstein 2005, 90–91), and these A2 selection criteria imply they were seen as important in Zimbabwe, too.

This emphasis on clear criteria for beneficiary selection and rational administrative procedures to apply them to applicants for land has a long history in Zimbabwe. Save for the early years of independence after 1980, when in some instances land occupations determined who received plots (Tshuma 1997, Ranger 1985), settler selection in Zimbabwe's pre-2000 land reform programme was, by and large, guided by policy prescriptions (Kinsey 1982, 97, Kinsey and Binswanger 1996, 115, Alexander 2003, 85). In the immediate post-1980 period, land allocation for smallholder farmers targeted the landless, the unemployed, the married or widowed with dependants, and those aged between 18 to 55 years (Government of Zimbabwe 1981, 10, Gonese and Mukora 2003, 180). Kinsey (2004, 1681) remarks that '... it is striking how the existing evidence points to a selection process for land reform beneficiaries that appears to have been remarkably equitable, and also efficient in targeting those who were government's and donors' priorities at the time'.

Over time the focus of land reform gradually shifted from welfare to improving agricultural productivity, culminating in the introduction of the Commercial Farm Settlement Scheme in the late 1990s. This was aimed at 'indigenising' the large-scale commercial farming sector, intending to produce a high calibre, productive group of black commercial farmers, and to create employment opportunities for college graduates who would otherwise be unemployed. Strict adherence to settler selection criteria was the fundamental tenet of this scheme, which then evolved into the A2 resettlement model after 2000.

Tensions between technical/rational approaches to land reform and the real world of politics are not unique to Zimbabwe. Deininger (2003, 149) suggests that the allocation of farms with highly developed and productive infrastructure has often proved difficult, creating the basis for conflicts among beneficiaries. More generally, given that the primary motivation for most land reform in developing countries has been political rather than economic, aimed at 'calming social unrest and allaying political pressures by peasant organizations rather than increasing productivity' (Deininger 2003, 146), political and social relationships and dynamics have often become the key determinants of land allocations (see also Khan 2004, Peters 2004). This means that the formal institutions tasked with land reform often become contested terrain. For this reason, the character of the state is key to understanding the political dynamics of state-sponsored land redistribution. The next section briefly reviews some of the literature on the post-independence African state and identifies some key concepts and analyses of relevance for the Zimbabwean case.

The state and land reform in Africa

The state can be defined as 'an organization, composed of numerous agencies led and coordinated by the state's leadership (executive authority) that has the ability or authority to make and implement the binding rules for all the people, as well as the parameters of rule-making for other social organizations in a given territory, using force if necessary to have its way' (Akude 2007, 2, quoting Migdal 1988). In Africa the state is the dominant player in the sphere of development. A heterogeneous institution, the African state is the locus of actions and contestations by differentiated classes representing competing economic interests and political ambitions. Broadly, the classes which tend to control the state are the national or governing elites, together with sections of the middle class. Hodder-Williams (1984, 95) argues that in Africa 'the lack of a developed indigenous private sector, of entrenched pressure groups and of secondary organizations results in the "monopolistic state"'. The monopolistic state is often merged with the ruling party. For Shivji (1991, 85), 'The Party becomes part of the state rather than civil society...Over time the party ceases even to be a ruling party, but a supreme political existence which holds the last word on the social good and the political truth'. This tallies with the assertion by Herbst (1990, 7) that 'in Africa and elsewhere in the Third World, some parties have evolved to such an extent that they must be considered part of the resource allocation process, even though there are variations across the continents'.

In an influential analysis, Fanon (1965) argues that in post-independence Africa political parties have tended to pursue the agenda of the governing elite or national bourgeoisie ahead of the interests of the populace and the peasantry. The historical path followed by liberation struggles gives rise to particular relationships between the peasantry/the masses and their leaders, who become the governing elite. This relationship then regulates the expansion and contraction of political space and governs the interaction between those at the grassroots and their leaders, and prolongs the domination of the peasantry. The authority of the national bourgeoisie over the masses is reaffirmed and constantly refurbished through the astute use of popular slogans. Referring to the top leader of the political party (who is often the president of the particular country), Fanon (1965, 135) notes that, 'Every time he speaks to the people he calls to mind his often heroic life, the struggles he has led in the name of the people and the victories in their name he has achieved, thereby intimating clearly to the masses that they ought to go on putting their confidence in him.' The state and the party continually impose themselves on the peasantry, creating the space for private advancement, inequality in the acquisition of wealth and its monopolisation (Fanon 1965). Subsequently, the party is transformed into an information service which becomes a key instrument for managing and controlling people. This description resonates strongly with the situation in post-independence Zimbabwe.

African states often fail to deliver much needed economic development. Weakened economically and increasingly malfunctioning, the institutions of the state lose their legitimacy in the eyes of the citizenry (Akude 2007, 2). Worried about losing power through democratic processes, the governing elite may then turn to populist rhetoric and 'tired nationalism' to ensure political survival (Dzingirayi 2010). Raftopoulos (2003) suggests that this was precisely the dynamic at work when the Zimbabwe government steered the 2000 land occupations, following the rejection of the Draft Constitution in early 2000 and in the face of impending national elections later that year.

Akude (2007) argues that national elites in Africa often lack the required incentives for the pursuit of economic development in a manner that strengthens democratic state institutions. In the Western world, there is a clear separation of political power and economic power (or private wealth), which creates the need for symbiotic but arms-length relations between those who possess political power and those providing the economic resources required to fund public administration. This then drives the logic of 'good governance' (Akude 2007). However, conditions are materially different in most of Africa. With states largely funded by foreign firms and donors, assuming powerful positions within the state becomes the key strategy for elites wishing to pursue personal aggrandisement. Put differently, the ruling class 'does not depend on the economic activities of its citizens to maintain its political dominance' (Akude 2007, 8).

The character of the African state and the subsequent manipulation of political space by the dominant classes has embedded the state deeply in an 'economy of affection', where social relations of kinship, ethnicity and patronage dominate (Chabal and Daloz 2003, Hyden 1983). Hyden (1983) argues that the objectives of the economy of affection are geared to survival in risky environments and to maintenance of 'primordial' social relations of kin, clan, ethnicity, etc. These allow the accumulation of prestige and power, and of the resources needed to build, maintain and expand followings through patronage rather than the accumulation of economic assets required for productive investment and capitalist development. The patronage politics informed by the economy of affection and 'social' forms of accumulation permeate the African state and its functioning, and state-based actors mobilise their followings on a political, ethnic or regional basis. According to Hyden, this means that they cannot provide the impersonal types of instrumentalism needed to undertake a project of national economic development. Instead, 'personalised relations' underpin political dynamics and undermine legal and bureaucratic rationality and institutions.

The failings of the African state identified in this brief literature review help explain why the allocation of A2 plots in Zimbabwe did not take place in line with policy prescriptions. As illustrated by detailed empirical evidence from Goromonzi District, land allocation was deeply influenced by social and political relations of patronage, with negative impacts on agricultural output. This was facilitated by the hijacking of the institutions through which the allocations were made, by both ruling party members and representatives of state security apparatuses.

Who selected the beneficiaries for A2 farms? Evidence from government audit reports and other sources

As indicated above, beneficiary selection for land on A2 resettlement schemes was designed to take place through an elaborate administrative process. After completing forms, applicants were required to develop a cash flow statement and a project proposal for their farming enterprises. The interest that this generated among those in (mostly urban-based) employment, including both low and high-income groups, was demonstrated by long-winding queues of people waiting to submit their application forms. Images of these queues were shown repeatedly in government print and electronic media, seeking to demonstrate that Fast Track land reform had popular support.

The main criteria used to assess the applications were income, gender, property, cash flow, experience, qualification and training, and recommendations were to be based on the marks scored. A Provincial Assessment Committee was constituted to assess the applications, and comprised senior officials of the provincial offices of different government departments, such as those responsible for local government, physical planning, lands, agricultural research and extension, veterinary services and district development, plus one war veteran (Marongwe 2009, 273). In Mashonaland East Province alone, a total of 14,246 applications were received and of these 48 percent were for small-scale farms, 24 percent for A1 plots, 15 percent for peri-urban plots, 11 percent for medium-scale farms, less than 2 percent for large-scale farms. This compares to the national total of just over 15,000 A2 land parcels that had been allocated by 2003[4] (GOZ 2003).

Land identification committees were established to identify farms for allocation to successful applicants, but these were not manned by technocrats from the relevant government departments (e.g. the Department of Lands, or AREX[5]). Instead, their membership was dominated by representatives of security ministries and Zimbabwe African National Union-Patriotic Front (ZANU-PF) as well as chiefs and war veterans, both aligned to or controlled by the ruling party (Marongwe 2009, 279). The committee on peri-urban settlement in Harare had two ZANU-PF party members of Harare Province, two members from the Central Intelligence Office, one member from the Zimbabwe Republic Police (ZRP) and one member from the War Veterans Association (Marongwe 2009, 280). At the district level, membership of the district resettlement committee followed a similar format.

The efforts of the technical teams that had assessed the A2 application forms came to naught as the process appeared to have little impact on who was allocated land in practice. The ruling party's distrust of civil servants perhaps explains why the parallel process of land allocation by land committees was put in place (Alexander 2006). There is no reference in any government records to the role played by the technical process of assessing the A2 application forms. There is also clear evidence of bias against opposition party supporters, prompted by public statements by political leaders at national and provincial levels that they would not be allocated land, and backed up the statements and actions of war veterans as leaders of land occupations.

The Government's own land reform audit reports (GoZ 2006, all eight Provincial A2 Audit Reports) confirm the bypassing of policy prescriptions governing beneficiary selection, creating problematic situations in land allocation to both A1 and A2 beneficiaries. In Mashonaland East Province, the Ministry of Lands and Agriculture also allocated land from its Head Office, bypassing provincial leadership (GoZ 2006, 14, Mashonaland East Provincial Audit Report). In Manicaland, the District Land Identification Committee's functions were usurped by the Committee

---

[4]By 2010 around An estimated 16 386 A2 farms were in place by 2009. A2 farms were in existence.

[5]Department of Agricultural Research and Extension.

of Seven, war veterans, traditional leaders, councillors and political leaders who ended up allocating land to A1 and A2 beneficiaries un-procedurally and under unclear circumstances (Mutare Rural District Council Minutes, 15 January 2004). Some allocations were done directly from the Ministry of Lands in Harare without prior recommendation from the Provincial Land Identification Committee. In Mashonaland West, prominent politicians took charge of land allocations, leading to administrative problems and 'unofficial' settlements, especially in Hurungwe and Kadoma districts (GoZ 2006 Mashonaland West A2 Audit Report). In Masvingo, especially in Chiredzi, Masvingo and Mwenezi Districts, war veterans, politicians and traditional leaders ended up controlling the functions of the District Land Identification Committees. Base commanders in Gutu and Masvingo controlled land allocation and not the Committee of Seven (GoZ 2006, 46, Masvingo A2 Audit Report). In Matebeleland South, the Provincial Land Identification Committee was over-shadowed by the District Land Identification Committees such that farm allocations took place at the district level, mainly because of the work of powerful politicians (Parliament of Zimbabwe 2002, 3, GoZ 2006, 49 Matebeleland South A2 Audit Report). Some errant councillors, war veterans and government officials allocated land in the Gweru, Mvuma and Shurugwi Districts of the Midlands Province (GoZ 2006, Midlands A2 Audit Report).

The Portfolio Committee on Lands, Agriculture and Water Development, Rural Resources and Resettlement (Parliament of Zimbabwe 2003, 6) noted that 'In some instances, it is very clear that senior politicians do not respect these institutions resulting in some of these politicians appropriating the functions of the land identification committees by deciding what farm must be settled ... In some cases they also decided who must be settled... The appropriation of and bypass of legitimate institutions ... results in double and competing allocations'. In Manica-land, minutes of the Mutare Rural District Council revealed that 'parallel allocation of land was seen being done by traditional leaders, ZANU-PF leadership and war veterans, thereby promoting squatting and distorting the planning process' (Mutare Rural District Council Minutes, 15 January 2004). Effectively, the 'disorder' associated with Fast Track land reform had multiple effects – it undermined formal planning and allocation, it opened spaces for those who wanted land (occupations), and it created a veil around which elites could grab land, deploying nationalist rhetoric to support this.

There were also cases where political and social connections helped shield some white farmers from having their farms acquired. The owner of an estate in Mutare had his farm spared, and is described in a key government document as 'very forthcoming and straightforward, contributed to ZANU-PF party fund raising campaigns and also assists A1 and A2 farmers' (Government of Zimbabwe 2007, 2). In Mutasa District, the owner of Hamu[6] farm was also spared because he is '...cooperative and supported by local communities. Victory of ZANU-PF party in the last elections attributed to him' (GoZ 2007). In Shamva District, the owner of the 571-hectare Slowman farm was saved by both economic and political considerations. In addition to employing over 100 skilled workers and 400 unskilled workers, the farmer exported citrus and pork products. The farmer also supported the surrounding community with tillage, planting and harvesting services. The farmer provided transport to the community, assisted in the Maguta inputs programme, and

---

[6]Pseudonyms are used.

supported the ruling party in cash and kind, especially for national events like the Independence Celebrations (GoZ 2007).

Further evidence of the use of political influence in the selection of beneficiaries is provided by court proceedings. An example is the case of Minister of Lands, Agriculture and Rural Resettlement vs Karori (Pvt) Ltd (2002) in a case brought before the Administrative Court of Zimbabwe over the acquisition of Lot 5 Lawrencedale farm. The judge confirmed that irregularities and deviations from procedure took place in connection with the process of acquiring and allocating the farm.

These examples suggest that land allocation processes under Fast Track land reform were politically driven and open to manipulation by powerful interest groups as well as contestation between such groups. The technical/rational process prescribed by policy prescriptions was essentially set aside. What does the evidence from Goromonzi show?

Beneficiary selection for A2 farms in Goromonzi district

Less than 5 percent of A2 beneficiaries in Goromonzi had their names on the List of Screened Applicants for Mashonaland East, implying that 95 percent of beneficiaries did not make applications through the formal channels and were therefore not assessed in terms of the official criteria (author calculations, Mashonaland East List of Screened Applicants Excel Sheet, Ministry of Lands, Agriculture and Rural Resettlement). As a result of the high numbers of applicants (for instance in Mashonaland Central, close to 10 000 people applied for land), comparable analysis of data from other provinces was not possible. As argued elsewhere in this paper, Goromonzi District was the hub of commercial farming and its location less than 40 km from Harare could have shaped some of the institutional and process issues of settler selection.

Table 1 shows findings from the questionnaire survey of the sample of 65 A2 beneficiaries conducted in 2003 (Marongwe 2009, 336). Civil servants constituted the largest category of beneficiaries at 26 percent of the total. The Zimbabwean National Army followed closely with 18 percent, followed by businessmen at 14 percent. Top government officials, a category that included cabinet ministers, deputy ministers, and diplomats or government representatives in foreign ministries and other

Table 1. A2 beneficiaries in Goromonzi by occupation/employer, in Marongwe's 2009 study.

Occupation/employer	n	Percentage
Government ministers, diplomats and other leading members of ZANU-PF	6	10%
Zimbabwe National Army	11	18%
Zimbabwe Republic Police and President's Office	4	6%
Business people	8	14%
Private sector	5	8.5%
Civil servants	16	26.5%
War veterans	3	5%
Unemployed/ordinary farmers	7	12%
Total	60	100%

Source: Marongwe (2009, 336) reporting survey data 2003.

prominent figures, constituted 10 percent of the beneficiaries. The unemployed constituted about 12 percent. Those in permanent employment made up the bulk of those who were allocated land.

Most A2 beneficiaries in Goromonzi can be classified as either members of the governing elite or a local elite based in the district. The governing elite is based either at the national or provincial levels and most of them are fully employed and occupy top positions in Government, the private sector or within ZANU-PF, the then ruling party. In this category are cabinet ministers, former ministers, the provincial governors of Harare and Mashonaland East Provinces, ZANU PF provincial party chairman, top officers in the Zimbabwe National Army, and successful businessmen. These beneficiaries were strategically well-placed to influence the outcome of land reform. This category of beneficiary was generally allocated either whole farms or farm units comprised of a consolidation of new subdivisions generated by land reform.

Unpacking the category 'business people', it is apparent that members of this category were based either in Harare or in Goromonzi District. More than half were running supermarket chains. Another operated tailoring shops and hair salons in Harare. A female farmer grew up in Mutoko District in Mashonaland East and had experience in growing vegetables. Another businessman operated a transport enterprise, specializing in lorry transport. One businessman was a property developer and had properties in Harare, Ruwa and Beatrice. Some had attributes (e.g. expertise in growing vegetables, or managing a transport business) which suggested they might succeed in farming ventures. The study could not establish whether or not these business people were ZANU-PF members.

The Government's 2006 land audit used similar categories in reporting the occupation or employment status of A2 beneficiaries in Goromonzi (see Table 2). In Marongwe (2009) business people constitute 14 percent of the beneficiaries compared to 12.3 percent in the Government audit report. There are notable differences in the proportion of other categories. These are 24 percent[7] and 8.5 percent[8] for security services, 26.5 percent and 16.6 percent for civil servants, 5 percent and 17.3 percent for war veterans and 10 percent and 1.2 percent for government ministers respectively. The large sample size in the Government audit report (n = 431) could be the main reason for the variation as Marongwe (2009) could have missed some important dynamics due to the small sample size. Further, interpretation of categories could be different. For instance, Marongwe (2009) puts government ministers, diplomats and top ZANU-PF officials in one category, while the Government audit report uses the category 'ministers' only.

A detailed case study of Dannah B farm carried out in 2006 is a revealing process of both the identities of beneficiaries and of allocation processes. Seven people were allocated A2 plots of different sizes:

Joseph[9] is aged 47 years and was allocated 60 hectares in July 2001. Joseph was a clerk in the Ministry of Local Government, Public Works and National Housing at the time

---

[7]Refer to Table 1: 24 percent comprises 18 percent from the Zimbabwe National Army and 6 percent from the Zimbabwe Republic Police and the President's Office
[8]Refer to Table 2: 8.5 percent comprises 3.3 percent from the Zimbabwe Republic Police, 1 percent from the Air Force of Zimbabwe and 4.2 percent from the Zimbabwe National Army.
[9]Pseudonyms are used here.

Table 2.   A2 beneficiaries in Goromonzi District by occupation/employer, in government's 2006 land audit (n=431).

Category	Number	Percentage
Business People	53	12.3
Civil Servants	72	16.6
War veterans	75	17.3
Ministers	5	1.2
Ordinary members	148	34.3
Politicians	2	0.5
War Collaborators	6	1.4
ZEPDRA	20	4.7
Zimbabwe Republic Police	14	3.3
Zimbabwe National Army	18	4.2
Air Force of Zimbabwe	4	1.0
Others	14	3.2
Total	431	100%

Source: Government of Zimbabwe, (2006, 16).

of allocation, and was based in Harare. He retired from service in 2004, Joseph's wife works in the President's Office and she is the one who applied for the land. The application form was submitted to the responsible Provincial Administrator's office via the President's Office and in September 2001, the couple was allocated the plot.

Gift is aged 57 years and was allocated 120 hectares at the farm. He is a war veteran who participated in the occupation of many other farms in Goromonzi. Gift's wife is a teacher. He sent his A2 application form to the District Administrator's office and was allocated the plot in September 2003.

George is aged 52 years and was allocated a 140-hectare plot. He is employed under the Ministry of Local Government, Public Works and National Housing while his wife is a housewife. George's application form was submitted through the ministry and he settled on the farm in September 2002.

Jeremiah is 47 years of age and was allocated 100 hectares. He is a retired soldier who had risen to the rank of Major and is a war veteran. Jeremiah applied for land through the army and was allocated his plot in 2002. Initially he had been given an A1 plot in Marondera but surrendered it when he was allocated the A2 plot.

Alex is aged 42 and was allocated 200 hectares. He holds a degree in Sociology and is employed by the President's Office. His wife is a teacher. Alex submitted his application through his employer and in September 2002 he had taken occupancy at the plot.

Kombo is 53 years old and was allocated 160 hectares. He has two wives, both of whom are housewives. Kombo was employed by the Ministry of Youth and Employment Creation until he retired in 2003. He applied for land through the Ministry and was allocated the plot in 2002. Previously he was farming at a family farm in Chivhu, but he had moved only 50 of his 200 herd of cattle to his A2 plot.

Amon is aged 55 years and was allocated 120 hectares. Amon is a retired member of the Presidential Guard and is also a war veteran. He has two wives who run businesses in Harare. Amon sent his application through the District Administrator's Office and was allocated the plot in 2001.

Of the seven beneficiaries on Dannah B farm, none were 'ordinary' people, meaning that they occupied positions of authority as war veterans, civil servants or employees of the President's office. When asked to explain the plot allocation process, all eight gave short responses to the effect that they simply applied and were allocated land. The clear impression was given that they did not want to entertain any further questions on the subject. None had their names on the official Mashonaland East List of screened applicants. None were knowledgeable about the 'scores' they had attained during the selection process and how this matched their farm allocations. It is clear that the size of these A2 plots was relatively modest, and that none of these beneficiaries could be seen as members of the governing elite.

Commercial farms allocated to members of the local elite generally conformed to maximum farm sizes by agro-ecological region as prescribed by Statutory Instrument 419 of 1999. With farms in Goromonzi district located in Natural Region 2a, land allocations were not to exceed 350 hectares. Many received subdivisions much smaller than this. The local elite category includes civil servants based at either the provincial or district level, with a few based at the national level. Many were employed by the Ministry of Lands and Agriculture, the ministry that was officially in charge of land allocation. Others were locally based civil servants and these included teachers and nurses. The District Administrator of Goromonzi and officials from the Goromonzi Rural District Council were also among the beneficiaries, as were local businessmen.

In contrast, members of the national elite were often allocated whole farms, and often these were much larger than the prescribed maximum farm sizes. Table 3 was constructed using data from both the author's 2003 survey and government land audit reports, and illustrates high profile land allocations in Goromonzi. There were at least six cases where whole farms were allocated to members of the governing elite: the Governor and Resident Minister of Mashonaland East, the ZANU-PF provincial party Chairman and his predecessor, one cabinet minister, the son of a former cabinet minister who was declared a national hero when he subsequently died, a senior army officer and a senior official in the Prisons Services Department and former Brigadier in the Army. A female cabinet minister was given one subdivision plot of 122 hectares, which was the smallest farm allocated to any cabinet minister in Goromonzi district.

If the maximum farm size for Goromonzi district had been maintained at the upper limit of 350 hectares, a number of additional farms could have been demarcated. For the top 10 allocations to governing elites, it can be seen that an additional 17 farms could have been created (see Table 4). Only 10 instead of a potential 27 farms were created, or 37 percent of the potential number of beneficiaries. The number of those prejudiced as a result would be even higher if the farms had been sub-divided into small-scale commercial farms rather than large-scale farms. These estimates illustrate how the governing elite, by ignoring policy prescriptions and selectively implementing maximum farm-size regulations, deprived others of the opportunity to benefit from Fast Track land reform.

The Goromonzi data on beneficiary selection compared to other findings

Other studies (e.g. Scoones *et al.* 2010, Moyo *et al.* 2009, Zamchiya 2011, this collection) use similar categories to those in Tables 1 and 2 above for analyzing the social and political identity of A2 beneficiaries. However, major differences are evident in relation to their relative proportions of the total number of beneficiaries. Scoones *et al.* (2010, 53) report that in their 57 A2 study sites in Masvingo Province, 56 percent of beneficiaries were 'ordinary' people from both urban and rural areas,

Table 3.   Land allocations to members of the governing elite in Goromonzi District: the top 20 cases.[1]

	Occupation/status of beneficiary	Size of Farm (ha)	Comment
1	Former ZANU-PF provincial chairman & former board member in the private sector	1606	Wholesale farm allocation
2	Senior prison services official and a retired top army official	1028	Consolidation of 14 subdivisions
3	Senior army official	1020	Two farms consolidated to make one
4	Governor of Mashonaland East Province	941	Wholesale farm allocation
5	Chairman of ZANU-PF Mashonaland East Province & chairman of many other private boards in the private sector	661	Wholesale farm allocation
6	Son of a late cabinet minister who was declared a national hero	610	Wholesale farm allocation
7	Senior official in the Central Intelligence Organization and former ambassador	432	Wholesale farm allocation
8	Cabinet minister	400	Consolidation of five plots
9	Cabinet minister	366	Consolidation of four subdivisions
10	Governor and former cabinet minister	357	Consolidation of three subdivisions
11	Member of parliament	351	Consolidation of five subdivisions
12	Senior official in the Central Intelligence Organization	327	One subdivision
13	Deputy director in the Office of the President	211	Consolidation of three subdivisions
14	Senior ZRP official	290	Wholesale farm allocation
15	Former chairperson of Education Committee	144	Consolidation of two subdivisions
16	Cabinet minister	122	One subdivision
17	Ex- Member of Parliament, Chairperson of ZANU-PF Mashonaland East Women's League	120	Single plot
18	District administrator of Goromonzi	75	One subdivision
19	Senior police officer	75	One subdivision
20	Deputy secretary in a government ministry	45	One subdivision

Note: [1]This is based on sizes of farms allocated to national elites.
Source: Marongwe (2009, 328), using survey data cross-referenced with government and audit reports (GoZ 2003, GoZ 2006).

26.3 percent were civil servants, 10.5 percent were business people, 5.3 percent were former farm-workers, and 1.8 percent were employees in the security services. War veterans constituted a separate category that cut across these occupational categories, and comprised 8.8 percent of A2 beneficiaries, with 60 percent of these being 'ordinary' people, 20 percent being from the security services and another 20 percent from the ranks of business people (Scoones et al. 2010).

When compared with Table 1, the proportion of civil servants in the two studies is almost the same, but in relation to all other categories the differences are significant. Particularly stark differences are evident in relation to 'ordinary' people (12 percent vs 56 percent) and members of security services (24 percent[10] vs 1.8 percent). The proximity of Goromonzi to Harare and the attractions of the district as a prime farming area help to explain the larger proportion of beneficiaries who were powerful security services employees in Goromonzi than elsewhere (see Table 5 for a comparative analysis of findings from Marongwe 2009, Moyo *et al*. 2009 and Scoones *et al*. 2010).

Table 4. How whole farm allocations to national elites prejudiced other potential A2 beneficiaries in Goromonzi: the top 10 cases.

Case	Size of farm allocated to national elites	Potential no. of plots if policy was adhered to	No. of farmers prejudiced
1	1606	5	4
2	1080	3	2
3	1028	3	2
4	1020	3	2
5	941	2	1
6	863	3	2
7	782	2	1
8	698	2	1
9	685	2	1
10	661	2	1
Total		27	17

Source: Author calculations using a maximum farm size of 350 ha which is applicable for Goromonzi.

Table 5. Comparative analysis of the three studies (Marongwe 2009, Moyo *et al*. 2009 and Scoones *et al*. 2010).

Occupation/employer	Marongwe 2009	Moyo *et al.* 2009[2]	Scoones *et al.* 2010[3]
Government ministers, diplomats and other leading members of ZANU-PF	10%	**	**
Zimbabwe National Army	18 %	8.9%[4]	1.8 %[5]
Zimbabwe Republic Police and President's Office	6%	**	**
Business people	14%	2.3%	10.5%
Private sector	8.5%	17.1%	**
Civil servants	26.5%	8%	26.3%
War veterans	5%	**	8.8%
Unemployed/ordinary farmers	12%	**	56%[6]

**Indicates that the study does not use this categorization of beneficiaries.
Notes: [2]Based on Table 2.11, page 28. The data are not specific to Goromonzi.
[3]Based on Table 2.6, page 53 and Table 2.7 for figures on war veterans. Data do not sum to 100 percent as the study does not use war veterans as a separate category.
[4]The study uses the category 'civil service' (ununiformed)
[5]The study uses the category 'security services.'
[6]This comprises 12.2 percent from 'ordinary in rural areas' and '43.8 percent from ordinary from urban areas.'
Source: summarized from Marongwe (2009, 336) (Moyo *et al*. 2009, 28) and Scoones *et al*. 2010.

---

[10]See footnote 7.

Moyo *et al.* (2009) make the case that members of the national elite, or 'political cronies', constitute only a small proportion of the total number of land reform beneficiaries. Missing in this analysis is the long chain of beneficiaries who accessed land on the basis of their social and political relations with members of the governing or local elite as relatives or friends. In addition, multiple-farm ownership by members of the governing elites is often camouflaged within such relations. In addition, Moyo does not discuss the proportion of members of the governing elite who own farms. Is it desirable for all cabinet ministers, deputy ministers, permanent secretaries of ministries and other important government officials to own farms? This key question is rarely asked.

Agricultural production by A2 farmers in Goromonzi

It is clear that in Goromonzi processes of land allocation were highly politicized and policy prescriptions on criteria for selection of A2 beneficiaries were largely by-passed. What were the consequences for levels of agricultural production? In addition to describing levels of land utilization and patterns of production, this section discusses a number of factors that constrained crop production on A2 farms, such as insecure tenure, access to inputs and under-utilization or vandalization of farm infrastructure and equipment.

Land utilization

Data from the questionnaire survey revealed that about 81.5 percent of beneficiaries interviewed had cultivated some crops in 2002/2003. Table 6 shows the proportion of the total arable land on the farms that was being cultivated in the 2002/03 cropping season. More than 50 percent of the farmers had cultivated between 1 and 40 percent of their plot size. The largest quota of 23 percent of the total farmers had cultivated between 31 and 40 percent of the arable plots. The highest percentages of 100 percent, 95 percent, 94 percent and 81 percent were experienced on 14ha, 84ha, 17ha and 84 hectare plots respectively. The largest amounts of land being cultivated by individual beneficiaries were located on the larger farms, but the proportion cultivated rarely exceeded 35 percent of the farm size, and the farms with the largest individual areas under crops were cultivating around 30 percent of total arable land.

Table 6. Proportion of cropped area for A2 beneficiaries in the Goromonzi study sample (n = 53).

% of Cropped Area	No. of Farms	% of Total
1–10%	2	3%
11–20%	9	17%
21–30%	7	13%
31–40%	12	23%
41–50%	9	17%
51–60%	6	11%
61–70%	2	4%
71–90%	3	6%
91–99%	2	4%
100%	1	2%
Total	53	100%

Source: Field Data 2003.

Data in the 2003 and 2006 government audit reports indicate that under-utilization of land in Goromonzi, as evidenced in the existence of vacant plots, was a major problem: about 105 beneficiaries (out of 438) had not taken occupation of their plots by January 2003 and there was no agricultural activity taking place. Over 12,000 hectares of land were lying idle. By 2006, there were some 75 vacant plots, but the hectarage involved was even higher at 35,227 hectares. It could be that the vacant plots were significantly bigger than was previously the case.

Comparable figures from Moyo *et al.* (2009, 53) show that 18.3 percent of A2 farmers in the surveyed districts (n = 421) had not cultivated any land in the 2004/05 season (implying under-utilization) while 21.6 percent had cultivated between one and 20 percent of their total plot sizes (not total arable land). These studies taken together suggest that as recently as 2006 the A2 model had failed to address the problem of chronic under-utilization of land inherited from the large-scale commercial farming sector (Riddell 1978, Weiner *et al.* 1985).[11]

Production patterns

In the pre-2000 period, large-scale farmers in Goromonzi practised mixed farming. Some were involved in intensive and specialized forms of agricultural production systems. The main agricultural products were maize grain, seed maize, tobacco, wheat, soya and sugar beans, horticulture products, sunflower, ground-nuts, sorghum, livestock, dairy and wildlife. Data from 46 randomly selected large-scale commercial farms show that 45 out of the 46 farms were involved in the production of maize, making the crop the most commonly grown by the farmers (Table 7 refers). As discussed above, this trend where maize was dominant is similar to the current situation under Fast Track land reform. What could be different is the proportion of land cultivated and the use of the maize harvested (many of the white farmers grew maize for use as stock feed). Beef production was the second most common activity with about 89 percent of the farmers being involved in this form of land use. Tobacco

Table 7. Production systems on 46 large-scale commercial farms in Goromonzi in the pre-2000 period.

Type of Crop/Production System	No. of farmers involved (n = 46)	% of Total
Maize	45	97.8
Tobacco	27	58.7
Soya Beans	22	47.8
Livestock (Beef)	41	89.1
Dairy	2	4.3
Horticulture	20	43.5
Wildlife	1	2.1
Seed Maize	14	30.4
Wheat	14	30.4
Vegetables (Onions, Tomatoes, Butternuts)	9	19.6
Sunflower	5	10.9
Ground-nuts	2	4.3
Sorghum	6	13.0

Source: Marongwe (2009, 386).

---

[11]More recent data on land utilization is not readily available.

occupied the third place followed by horticulture. A key feature is the diversified nature of land uses at the farm level. From the data of the 46 farms, a farmer was engaged in the production of an average five types of crops, which is significantly different from the findings of this study where on average a beneficiary cultivated only two crops. Specialized production systems like maize seed, wheat seed, horticulture and wildlife were a common feature in the district. Several of the beneficiaries were contracted by Chibuku breweries to produce red sorghum. Some of the farms that ventured into beef production were also involved in sheep and goat production.

Data from the questionnaire survey showed that in terms of cropping patterns, there was a strong bias toward maize production (for grain) (Marongwe 2009, 379). The simplest evidence is that out of 65 farmers, only three of these had not cultivated maize in the 2002/2003 agricultural season. Of the estimated 4178 hectares under cultivation, about 77 percent was under maize, the balance being largely between soya beans or vegetables. More than six percent of the farmers had cultivated at least two types of crops, with maize being one of them.

One significant impact of Fast Track land reform has thus been the reduction in the diversity of crops produced. Data from the Mashonaland East Provincial Audit Report shows the minimal involvement of land reform beneficiaries in tobacco production. The reasons for this might include the fact that farmers were not assisted by the government with inputs as happened in relation to livestock and other crops such as maize and wheat. Further, the technical skills required for tobacco production are demanding and few of the beneficiaries had such skills or relevant farming experience. However, it is evident from recent data that nationally tobacco production is on the recovery path. For instance, tobacco output in 2011 was 129.9 million kg, representing an 11 percent increase over the 2010 production figures[12] Another important contrast is that farmers in the pre-2000 period grew maize mostly for use as stock feed.

Unstable and insecure land tenure and its impacts on production

Lack of certainty and insecurity of land tenure on the redistributed farms in Goromonzi had negative impacts on production for several years. One reason was the lack of finality in the land acquisition process. In addition to unexplained withdrawal of offer letters, some farms were de-listed or dropped from the land acquisition list years after being occupied. For example, eight A2 farmers had been allocated plots, ranging in size from 18.4 to 32.8 hectares each, on Mashonganyika Farm. However, the farm was later de-listed and in 2003 the beneficiaries were awaiting relocation; none had planted any crops during the 2002/2003 agricultural season (Government of Zimbabwe 2003). This was a recipe for conflicts, which in the end undermined agricultural production (Marongwe 2009, 389).

In other examples, the relocation of 60 A1 settlers from Colga Farm following its designation as an A2 farm, and the pending relocation of about 30 households at Oribi Farm, meant that such beneficiaries did not engage in meaningful agricultural activities. On Clovadale B Farm, 10 beneficiaries had been allocated plots ranging in size from 15 to 47 hectares. Only two of the 10 beneficiaries had taken up their plots by the 2002/2003 agricultural season. The farm was however de-listed and the farmers were relocated elsewhere. The 2006 audit report for Mashonaland East showed that 118 A1 beneficiaries were to be relocated from Balkiza Farm in

---

[12]http://allafrica.com/stories/201108151221.html; downloaded 25 August 2011.

Goromonzi following its reallocation to the Zimbabwe National Army (Government of Zimbabwe, 2006, 52; Mashonaland East A2 Audit report). Further, Chabwino and Mukwene farms were re-designated as A2 farms, forcing the relocation of A1 beneficiaries (Government of Zimbabwe 2006, 54). An A2 beneficiary at Alymersfield Farm and employed in the President's Office was seeking the eviction of two other A2 beneficiaries (Government of Zimbabwe 2006, 48).

Access to farming inputs

Access to farming inputs was a huge challenge for land reform beneficiaries in the period of the study, as acknowledged in most other studies. There were essentially two methods of sourcing farming inputs for A2 beneficiaries in Goromonzi. One was to apply for a loan from the Grain Marketing Board (GMB) and then purchase the inputs from the same institution. The other was to use their own resources to purchase farming inputs from commercial suppliers. Almost 70 percent of the A2 beneficiaries in the study sample utilized the GMB input loan scheme suggesting their dependence on state resources – in contradiction of the policy that A2 beneficiaries were not supposed to receive state support. Heavy reliance on the government input scheme was either an indication of lack of own resources, or perhaps that the facility was cheaper. About 23 percent of the beneficiaries used their own resources.

Comparable data from Moyo *et al.* (2009, 69) show access to agricultural inputs was problematic for many land reform beneficiaries: '... the majority of the new farmers are resource-constrained and thus cannot afford to meet their input requirements from the market even when inputs are available'. Only 58.7 percent of A2 beneficiaries in the AIAS study sites had used inorganic fertilizer in maize production, with the percentage being much lower for the other crops.

Under-utilization and vandalization of production infrastructure

A common feature under Fast Track land reform has been the under-utilization, and in some cases vandalization, of productive infrastructure such as irrigation equipment, tobacco barns and greenhouses. The range of irrigation equipment included centre pivots, drip irrigation, sprinkler irrigation and canals for flood irrigation. Government reports show that in 2002/03 farming season at least 117 beneficiaries in Mashonaland East were not utilizing various forms of irrigation infrastructure (summarized from Goromonzi District Commercial A2 Audit Report, Excel Sheet, Department of Lands and Rural Resettlement). Further, 14 and 99 beneficiaries were not utilizing greenhouses and tobacco barns respectively. A total of 230 beneficiaries were not utilizing various types of production infrastructure that they found on the farms in Mashonaland East, with Goromonzi District being the most affected (Goromonzi District Commercial A2 Audit Report). Some beneficiaries clearly did not have the capacity to use the equipment (GoZ 2006, 14, 27).

Vandalization of productive infrastructure was perpetrated by a variety of actors. In some cases departing white farmers, frustrated by their failure to receive compensation, destroyed their farming equipment and other productive infrastructure. Key informant interviews in Goromonzi identified three such cases in A2 schemes. For example at Banana Grove Farm where 13 A2 farmers had been allocated plots, it was reported that the former owner vandalized the engines used for pumping water before he left. Thus although the underground piping of the irrigation system was still in place, the lateral pipes and sprinklers were no longer available. A similar situation was reported at Vuta Farm where 25 A2 settlers had been settled. The engines for pumping water were not working while the lateral pipes and sprinklers had gone missing. At Chinyika Farm where 13 A2 beneficiaries were put on the land, the underground main supply was still in place but the hydrants had been tampered with and hence were no longer functioning.

Violence, vandalization of property and the 'grabbing' of crops in the fields by beneficiaries also occurred. The confiscation of farm equipment such as combine harvesters, tractors, irrigation pipes, water pumps, planters, and water tanks from the displaced white farmers resulted in contestations among the beneficiaries. A government audit report found that 'many beneficiaries are holding onto large quantities which they took without the authority of the state. Some of the equipment has been sold and are being sublet for personal gains. Some beneficiaries are stripping the equipment for spare parts' (*sic*). (Government of Zimbabwe 2006, 46, Mashonaland East A2 land audit report).

Goromonzi District is awash with examples of the abuse of power in the allocation of farm equipment. At Alymersfield Farm, a beneficiary who was an employee of the President's Office 'was refusing to share infrastructure which they all found on the farm which included irrigation pipes, water pumps, a milling plant, a T35 truck, 2 motor-bikes, a combine harvester, an electricity generator, grading shades etc. As a result only 27 ha out of a possible 90 ha are under wheat' (GoZ 2006, 48). The beneficiary was seeking the eviction of two other beneficiaries so that he would be left with the whole farm. At Entre Rios Farm, a beneficiary who was a brigadier in the army confiscated 11 cattle that had strayed into his plot. Further, the beneficiary had closed and confiscated irrigation pipes belonging to his female neighbour. Such practices have inevitably undermined agricultural production.

Key findings and their wider implications

The A2 resettlement model was intended as a vehicle for the allocation of medium-scale commercial farms owned by indigenous black Zimbabweans, and government appeared at first to emphasize the importance of selecting beneficiaries who would be able to make good use of the formerly highly productive redistributed farms. Despite the establishment of an elaborate process to govern the assessment of A2 applicants and allocate land to them, empirical data from Goromonzi suggest that this was merely cosmetic. The impression was created of a well-publicized, transparent and technically-driven process of land allocation in accordance with agreed criteria, but this served only to camouflage the power relations and social connections that in practice determined who was allocated land. In Goromonzi it was members of the governing elite and the local elite who received most A2 farms.

One consequence of these politicized and manipulated processes of land allocation was that many of the new farm owners in Goromonzi district lacked the resources or the skills required for commercial farming. The previous owner-operated system of farming, often with a resident farm manager in place, was replaced by a new farming model which has not proved successful to date. This is centred on beneficiaries who for the most part are not resident on their farms, with a significant proportion being civil servants with scant experience of farming. In general, few of these beneficiaries have complied with the policy requirement that they hire competent farm managers (Marongwe 2009, 432). Agricultural production on these farms faces monumental challenges in relation to low levels of productivity, inadequate infrastructure, under-utilization of land, and lack of extension support. Many of these problems are rooted in the disregard of policy prescriptions on criteria for land allocation. In fact, it can be argued that the capture by the elites replicated in many ways the problems of the past, including multiple ownership and absenteeism.

Can these findings be generalized across Zimbabwe? Most scholars acknowledge that Fast Track land reform has varied greatly in its processes and outcomes across different localities (e.g. Scoones *et al.* 2010, xii, 43). Generalizations on the basis of one case are indeed difficult to make, and the temptation is to see every case study as unique. It is also true that the proximity of Goromonzi to the capital, Harare, home to most of the governing elite, as well as the fact that it contains highly productive farmland, made it particularly attractive to the elite. The extent of farm ownership by the governing elite might well be higher here than elsewhere.

On the other hand, the broader study (Marongwe 2009) demonstrates that the use of secondary data from government audit reports, court records and minutes of meetings provides grounds for generalization around specific issues such as, *inter alia*, the significance of political and social relations in land allocation, disorder and conflicts following land occupations, under-utilisation of land and skewed access to productive infrastructure. It seems highly likely that A2 land allocations in many places were subject to power dynamics and patronage, and that policy prescriptions were ignored in the majority of cases.

Given the commercial orientation of A2 schemes, such departures from official policy were both inappropriate and highly undesirable and have generated many of the problems bedevilling Fast Track land reform in the high potential agro-ecological regions. Studies elsewhere have identified the key characteristics that beneficiaries need to be evaluated on for successful land reform as including age, education, supply of family labour, farming experience, capital assets, non-farming skills, poverty status, marital status and health status (van Rooyen and Njobe-Mbuli 1996, 467). This approach has been criticized for being apolitical and technocratic in character, but in my view it remains the best available option for the promotion of commercial farming in land reform contexts. As efforts continue to correct some of the anomalies associated with Fast Track land reform and resuscitate agricultural production, a complete return to policy prescriptions in relation to, *inter alia* competency, possession of resources, farming experience, commitment to farming and farm-size regulations remains overdue.

Land occupations allowed power to be grabbed by war veterans, security personnel and political leaders, who used it to influence land allocation. The flip side was the weakening of formal institutions, including those of government departments involved in land reform, land allocation institutions at district level and below, and the courts. The dramatic decline in state capacity after 2000 was married to what might be termed the deliberate creation of a 'culture of chaos'. Given that there are powerful interest groups in Zimbabwe who have benefited, and continue to benefit, from the disorder created by Fast Track land reform, a return to the rigours of rational planning frameworks and the use of technical prescriptions will face resistance. As argued by Chabal and Daloz (2003) disorder can be a resource useful for different political factions contesting for power in society. In this regard, only a strong, determined *and democratic* state, supported by institutions with similar attributes, will be able to correct the anomalies that were created under Fast Track land reform.

References

Akude, J. E. 2007. The failure and collapse of the African state: on the example of Nigeria. *FRIDE, Comment*, September, 1–12.

Ayoob, M. 1996. State making, state breaking and state failure: Explaining the roots of Third Wworld insecurity. *In*: Luc van de Goor, *et al. Between development and destruction: an inquiry into the causes of conflict in post-colonial societies*. New York: St Martins Press.

Alexander, J. 2003. Squatters, veterans and the state in Zimbabwe. *In*: Hammar, A.; B. Raftopoulos and S. Jensen, eds. *Zimbabwe's unfinished business: rethinking land, state and nation in the context of crisis*. Harare: Weaver Press, pp. 83–117.

Alexander, J. 2006. *The unsettled land: state making and the politics of land in Zimbabwe 1893–2003*. Oxford, Harare and Athens: James Currey, Weaver Press and Ohio University Press.

Barker, R.L. ed. 1999. *The social work dictionary*, 4th edition. Washington, DC: NASW Press..

Bendix, R. 1960. *Max Weber: an intellectual portrait*. London: Heinemann, Prentice Hall.

Bernstein, H. 2005. Rural land and land conflicts in sub-Saharan Africa. *In*: Moyo, S. and Yeros, P. eds.*Reclaiming the land: the resurgence of rural movements in Africa, Asia and Latin America*. London and New York: Zed Books, pp. 67–101.

Bernstein, H. 2003. Land reform in Southern Africa in world-historical perspective. *Review of Political Economy*, 96, 21–46.

Byres, T. J. 2004. Introduction: contextualizing and interrogating the GKI case for redistributive land reform. *Journal of Agrarian Change,*. 4 (1 and 2), January and April, 1–16.

Chabal, P. and Jean-Pascal Daloz. 2003. Africa works, disorder as political instrument. *Political Science Journal*, 102 (408), 429–446.

Chaumba, J., Scoones, I. and Wolmer, W. 2003a. From jambanja to planning: the reassertion of technocracy in land reform in south-eastern Zimbabwe,? *Journal of Modern African Studies*, 41 (4), 533–554.

Chaumba, J., I, Scoones, I., and Wolmer, W. 2003b. New politics, new livelihoods: agrarian change in Zimbabwe *Review of African Political Economy*, No. 98.

Chimedza, R. 1994. Rural financial markets. *In*: M. Rukuni and C.K. Eicher, ed. *Zimbabwe's agricultural revolution*. Harare: University of Zimbabwe Publications.

Christodoulou, D. 1990. *The unpromised land: agrarian reform and conflict world-wide*. London and New Jersey: Zed Books.

Cousins, B. 2003. The Zimbabwean crisis in its wider context. *In*: A. Hammer; B Raftopoulos and S. Jensen, eds. *Zimbabwe's unfinished business: rethinking land, state and nation in the context of crisis*. Harare: Weaver Press, pp. 263–316.

Cousins, B. 2006. Review essay: debating politics of land occupations. Cape Town: Programme for Land and Agrarian Studies, School of Government, University of Western Cape, South Africa.

Davies, R. 2004. Memories of underdevelopment: a personal interpretation of Zimbabwe's economic decline. *In*: B. Raftopoulos and T. Savage, *Zimbabwe: injustice and political reconciliation*. Harare: Weaver Press, pp. 19–42.

Deininger, K. H. 2003. *Land policies for growth and poverty reduction*. A World Bank Policy Research Report, Washington, DC: World Bank and Oxford University Press.

Dzingirayi, V. 2010. *The impact of political crisis on natural resources: a case study of Zimbabwe*. Centre for Applied Social Sciences Working Paper, University of Zimbabwe, Harare.

Engberg-Pedersen, L.and N. Webster. 2002. Introduction to political space. *In*: idem eds. *In The name of the poor: contesting political space for poverty reduction*. London and New York: Zed Books, pp. 1–29.

Fanon, F. 1965. The national bourgeoisie. *In*: C. Allen and G. Williams, eds. 1982. *Sociology of developing societies: sub-Saharan Africa*. London: The Macmillan Press, pp. 166–168.

Ghimire, K. B. 2001. *Regional perspectives on land reforms: considering the role of civil society organizations. In*: K.B. Ghimire ed. *Whose land? Civil society perspectives on land reform and rural poverty reduction: Regional experiences from Africa, Asia and Latin America*. Geneva: UNIRISD, pp. 13–5.

Gonese, F. and C.M. Mukora. 2003. Beneficiary selection, infrastructure provision and beneficiary support. *In*: M. Roth M. and F. Gonese, *Delivering land and securing rural livelihoods: post independence land reform and resettlement in Zimbabwe.* Harare: CASS and the Land Tenure Centre, pp. 173–198.

Goredema, C. 2004. Whither judicial independence in Zimbabwe. *In*: B. Raftopoulos and T. Savage. *Zimbabwe: injustice and political reconciliation.* Harare: Weaver Press, pp. 99–118.

Government of Zimbabwe (1980, 1981, 1983 and 1985): Editions of intensive resettlement: policies and procedures. Government of Zimbabwe, Harare.

Government of Zimbabwe 2003. Goromonzi District A2 audit report. Ministry of Lands, Land Reform and Resettlement, Harare.

Government of Zimbabwe 2003. Mashonaland East list of screened A2 Applicants. Harare.

Government of Zimbabwe (2005). Preliminary A2 land audit report, Ministry of Lands, Land Reform and Resettlement and The Informatics Institute, Harare: SIRDC.

Government of Zimbabwe 2006. A2 land audit reports for: Manicaland, Mashonaland Central, Mashonaland East, Mashonaland West, Matabeleland North, Matabeleland South, Masvingo, Midlands. Ministry of Lands, Land Reform and Resettlement and The Informatics Institute, Hararc: SIRDC.

Government of Zimbabwe 2006. National A2 land audit Report. Ministry of Lands, Land Reform and Resettlement and The Informatics Institute, Harare: SIRDC.

Government of Zimbabwe 2007. Summary of number of white farmers to remain and number of white farmers before the Land Reform Programme. Ministry of Lands, Land Reform and Resettlement, Harare.

Hammar, A. and B. Raftopoulos. 2003. Zimbabwe's unfinished business: rethinking land, state and nation. *In*: A. Hammar, B. Raftopoulos and S. Jensen, eds. *Zimbabwe's unfinished business: rethinking land, state and nation in the context of crisis.* Harare: Weaver Press, pp. 1–47.

Herbst, J. 1990. *State politics in Zimbabwe.* Harare: University of Zimbabwe Publications.

Hodder-Williams, R. 1984. *An introduction to the politics of tropical Africa.* London: Allen and Unwin.

Hyden, J. 1983. *No shortcuts to progress: African development management in perspective.* University of California Press.

Khan, M. H. 2004. Power, property rights and the issue of Land Reform: A general case illustrated with reference to Bangladesh. *Journal of Agrarian Change,* 4. (1 and 2), 73–106.

Kinsey, B. H. 1982. 'For ever gained': resettlement and land policy in the context of national development in Zimbabwe. *Africa,* 52 (3).

Kinsey, B. 2004. Zimbabwe's land reform programme: under investment in post-conflict transformation. *World Development,* 32 (10), 1669–1696.

Kinsey, B. and H. Binswanger. 1996. Characteristics and performances of settlement programmes: a review. *In*: J. Van Zyl, ed. *Agricultural land reform in South Africa: policies, markets and mechanisms.* Cape Town: Oxford University Press.

Mamdani, M. 1990. The social basis of constitutionalism in Africa. *Journal of Modern African Studies,* 28 (2), pp. 359–374.

Mamdani, Mahmood. 2008. Lessons of Zimbabwe. *London Review of Books* 30 (23), 17–21.

Marongwe, N. –2002. *Conflicts over land and other natural resources in Zimbabwe.* Harare: ZERO Publications.

Marongwe, N. 2009. Interrogating Zimbabwe's Fast Track Land Reform and resettlement programme: a focus on beneficiary selection. Unpublished PhD Thesis, a thesis submitted in partial fulfilment of the requirements for the degree of Doctor Philosophiae in the Department, Institute for Poverty, Land and Agrarian Studies (PLAAS), University of the Western Cape (UWC).

Matondi, P. B. 2011. Inside the political economy of the Fast Track land reform in Zimbabwe and its local understanding. Ruzivo Trust: Harare, Mimeo.

Migdal, J. 1988. *Strong societies and weak states.* Princeton: Princeton University Press.

Mosca, G. 1939 *The ruling class.* New York: McGraw-Hill.

Moyo, S. 2001. The land occupation movement and democratization in Zimbabwe: contradictions of neo-liberalism. *Millennium, Journal of International Studies,* 30 (2), 311–330.

Moyo, Sam. 2011. Three decades of agrarian reform in Zimbabwe. *Journal of Peasant Studies,* 38 (3), 491–532.

Moyo, S. and Yeros, P. 2005. The resurgence of rural movements under neo-liberalism. *In*: idem, eds. *Reclaiming the land: the resurgence of rural movements in Africa, Asia and Latin America*. London and New York: Zed Books, pp. 8–64.

Moyo, S. *et al.* 2009. Fast Track Land Reform baseline survey in Zimbabwe: trends and tendencies. Harare: Africa Institute for Agrarian Studies.

Parliament of Zimbabwe. 2003. *Report by the portfolio committee on lands, agriculture, water development, rural resources and resettlement*. Harare: Parliament of Zimbabwe,

Peters, P. E. 2004. Inequality and social conflict over land in Africa. *Journal of Agrarian Change*, 4 (3), July; 269–314.

PLRC 2003. *The presidential land review committee on the implementation of the Fast Track Land Reform Program, 2000–2002*. Harare: Government Publications.

Raftopoulos, B. 2003. The state in crisis: authoritarian nationalism, selective citizenship and distorting of democracy in Zimbabwe. *In*: A. Hammar, B. Raftopoulos and S. Jensen, eds., *Zimbabwe's unfinished business: rethinking land, state and nation in the context of crisis*. Harare: Weaver Press, pp. 217–241.

Ranger, T. 1985. *Peasant consciousness and guerrilla war in Zimbabwe: a comparative study*. London: James Currey.

Riddell, R. C. 1978. *The land problem in Rhodesia: alternatives for the future*. Gweru: Mambo Press.

Sachikonye, L. M. 2003. The situation of commercial farm workers after land reform in Zimbabwe., A report prepared for the Farm Community Trust of Zimbabwe. London: CIIR.

Sachikonye, L. M. 2004. The promised land: from expropriation to reconciliation and jambanja. *In*: Raftopoulos, B. and T. Savage eds., *Zimbabwe injustice and political reconciliatio*. Harare: Weaver Press, pp. 1–18.

Saturnino Jr, M. Borras 2007. *Pro-poor land reform: A Critique*. University of Ottawa Press.

Scoones, I. *et al.* 2010. *Zimbabwe's land reform: myths and realities*. Harare: Weaver Press.

Sen, A. 1981. *Poverty and famines: an essay on entitlement and deprivation*. Oxford: Clarendon Press.

Scott, J. C. 1998. *Seeing like a state: how certain schemes to improve the human condition have failed*. New Haven, CT and London: Yale University Press.

Shivji, I. 1991. The democracy debate in Africa: Tanzania. *Review of African Political Economy*, No. 50, 79–91.

Tshuma, L. 1997. *A matter of (in) justice, law, state and the agrarian question in Zimbabwe*. Harare: SAPES Books.

van Rooyen, J. and B. Njobe- Mbuli. 1996. Access to land: selecting the beneficiaries. *In*: J. Van Zyl, J. Kistern and H. P. Binswanger, eds. *Agricultural land reform in South Africa: policies, markets and mechanisms*. Cape Town: Oxford University Press, pp. 461–495.

Weiner, D. Moyo, S. B. Munslow and P. O'Keefe. 1985. Land-use and agricultural productivity in Zimbabwe. *Journal of Modern African Studies*, 23 (2), pp. 251–285.

Zamchiya, P. 2011. A synopsis of land and agrarian change in Chipinge District, Zimbabwe. This collection.

Nelson Marongwe is a Rural and Urban Planner by profession, with over 15 years of experience working in government, the NGO sector and individual research consultancy. After graduating from the University of Zimbabwe in 1991, he was employed in government as a Town Planning Officer, working under the Department of Physical Planning, Ministry of Local Government, Rural and Urban Development. A two year stint in Government was enough for him. He enrolled for a Masters Degree in Environmental Policy and Planning in 1994, and graduated in 1995. For the period 1996 to 2003, he was employed as a research fellow with the NGO ZERO-Regional Environment Organization. He left full-time employment in 2003 to work as a consultant. His key areas of specialization are land reform, land administration, rural livelihoods, social protection, evaluation of development projects and social services provision. His PhD Thesis (2009) was entitled Interrogating Zimbabwe's Fast Track Resettlement Programme: a focus on Beneficiary Selection, under the supervision and mentorship of Professor Ben Cousins, University of the Western Cape, School of Government, Institute for Poverty, Land and Agrarian Studies, South Africa. In 2010 he worked as a Technical Advisor with the USAID/ARD's Sudan Property Rights Program.

A synopsis of land and agrarian change in Chipinge district, Zimbabwe[1]

Phillan Zamchiya

This paper extends the analysis of Zimbabwe's Fast Track land reform to the district of Chipinge in Manicaland province in south eastern Zimbabwe, where particular agro-ecological, political and social dynamics are important. In the three A1 resettlement schemes studied, political loyalty and patronage largely explain how the new beneficiaries acquired land. Most scholarly work, media and advocacy reports acknowledge the role of political patronage in the acquisition of A2 farms but they underplay this on A1 resettlement schemes. Based on empirical data, I argue that some A1 land reform beneficiaries are clients of patronage networks. Even though the new A1 farmers have other legitimate claims to land they are being subordinated to a partisan state and authoritarian ruling party that is willing to exclude other 'ordinary' people with 'wrong' or weak political ties in a highly politicised landscape. Thus, the paper argues that in these cases the Zimbabwe African National Union-Patriotic Front (ZANU-PF)'s governing elite manipulated autochthonous, historical, political, social reproduction and livelihoods grievances among different groups of people to set in motion a party politicised Fast Track Land Reform (Fast Track) project meant to reassert its political hegemony.

Introduction

Since 2000, Fast Track has radically transformed the agrarian structure of Chipinge district from one dominated by white owned large-scale farms to one dominated by a large group of smallholder producers. Who are these new farmers? How did they get land and why did they seek the resource? The gist of my argument is that the ZANU-PF (Zimbabwe African National Union- Patriotic Front) ruling elite manipulated people's diverse claims to land by ensuring that political loyalty and patronage took centre stage in beneficiary selection across my three A1 study sites. This set a political landscape where land reform beneficiaries have to continuously reassert their legitimacy through political loyalty to guarantee their tenure. To a certain degree, this is more widely acknowledged on A2 schemes (see Marongwe 2010; Moyo *et al.*

---

[1] I am so grateful to Jocelyn Alexander. I thank Ben Cousins for his ever critical but constructive comments on my work. I first presented this paper at the Institute of Poverty, Land and Agrarian Studies (PLAAS) in Cape Town, South Africa on 17 February, 2011. I appreciate helpful (re) direction from the PLAAS team and those attending. Many thanks for feedback from University of Oxford colleagues where I did a second presentation on 17 June 2011. I would also like to thank the anonymous reviewers for their helpful comments. However, I take full responsibility for this final article.

2009; Scoones *et al.* 2010) where land was mainly allocated through state administrative procedures highly open to patronage. My study focuses on three A1 schemes where land was acquired through a highly politicised land invasion and state administrative procedures open to manipulation in ways that resulted in patronage-based land acquisition more typical of A2 schemes.

In Zimbabwe's Fast Track policy framework, the A2 model refers to medium and large scale farms and the A1 model can be of two small-scale types: either village or self-contained with farm sizes ranging between 12 and 70 hectares (GoZ 2001) and for details about these schemes see the editorial overview. However, politics dominated farm allocations on the ground and what happened is far from the government designed framework. For example, land sizes range between 3 and 15 hectares on my study sites namely Glen View Lot 1, Wedgehill and Wolfscrag farms. Glen View Lot 1 is villagised and Wedgehill farm consists of both villagised and self-contained farms separated by the main road. On Wolfscrag farm land occupiers pegged self-contained style farms but there was communal grazing land within the resettlement scheme. The beneficiaries' A1 offer letters indicate the different land sizes but do not mention the two variants of villagised and self-contained.

In order to substantiate my argument I present this article in five interrelated parts. First, I give an overview of agrarian change in the district. Second, I explore the process of how the land was acquired through formal allocation, occupation and formal allocation after re-placement of original invaders. Third, I focus on beneficiaries and those excluded to determine who got land, how and why. Fourth, I examine the motivations of new farmers in seeking and acquiring land and interrogate how they became subordinated to partisan politics. The fifth section summarises the findings in line with my argument.

Agrarian change in Chipinge district

Chipinge district is located in south eastern Zimbabwe, in the province of Manicaland, with a significant population of 283,671 people (CSO 2002) covering a total of 539,303 hectares. Chipinge is geo-physically divided into five broad natural regions depending on annual patterns of rainfall. Region 1 receives over 1000 mm; region 2 about 750–1000mm; region 3 between 500–700mm; region 4 receives 450–600 mm and region 5 less than 500mm (GoZ 1986). It can be broadly divided into the high veld covering region 1 and 2 and the low veld covering regions 3 to 5. My three resettlement schemes are situated in the high veld where most former white farmers and high value farms were concentrated. According to Government recommendations agriculture varies from region to region. Region 1 is designated for specialised and diversified farming; region 2 for intensive farming; region 3 for semi-intensive farming and region 4 and 5 for semi-extensive and extensive farming respectively. Chipinge is the only district that covers all the five natural regions in the country so it is representative of the country's geography and it provides an opportunity to capture regional variation and the underlying dynamics between households resettled in the high veld and those left out in the low veld regions.

In the past decade, Chipinge experienced sweeping land reform. In June 2000, the government gazetted 13 white-owned commercial farms in the district for seizure (GoZ 2000). As is typical of Fast Track, the land invasions were characterised by violence and disorder in such a way that land occupations on the ground did not reflect the gazetted list. The Commercial Farmers Union (CFU), which represents large-scale mostly white farmers, reported invasions on Watershed Farm, Reitvlei Farm and Skyline Estate in May 2000 before the gazette (CFU 2000). Apart from their race, we know very little about the beneficiaries on the acquired farms.

Prior to Fast Track, Chipinge district was dominated by individual and family owned large-scale commercial farms and corporate estates owned by companies covering 179,479 hectares of land with 141,230 hectares in the high veld which is a high potential agro-ecological zone. Given the history of colonial displacement most of the black households subsisted in poor communal areas of Musikavanhu, Mutema, Ndowoyo, etc. in 1980. Following the first phase of resettlement in the 1980s, blacks acquired 19,699 hectares in the whole district and within this reconfiguration some black households were resettled on 17, 903 hectares alongside white farmers in the high veld. Table 1 shows the land classification and distribution in the district prior to the year 2000.

Between 2000 and 2011, 107,741 hectares belonging to former white commercial farmers in the district were redistributed mostly to small scale agricultural producers and medium to large-scale black commercial farmers. Within this change, black farmers now occupy 96, 944 hectares of land in the high veld a significant increase from 17, 903 hectares before 2000. Table 2 shows the new agrarian structure in Chipinge district.

Corporate estates such as Tanganda, Southdowns and Makande retained large hectares of land maintaining a similar agrarian pattern but the same cannot be said of the former white-owned large scale commercial farms. Moyo (2011,259) aptly argues that the 'Zimbabwean state made concessions to some sections of *domestic and international* capital at the height of...confrontation' (my emphasis added). In Chipinge, some people led by war veterans invaded Makande and Southdowns estates which produced bananas, coffee and tea respectively but they were driven out violently by the anti-riot police in late 2000. One respondent working in the government district department of social welfare recalled, 'the war veterans wanted to resist but they were beaten by the riot police and no one ever returned to that place' (Government worker, Chipinge, 2010). This was after Zimbabwe's Vice

Table 1. Chipinge district: pre-2000 agrarian structure.

Region	LSCFA[1]	SSCFA[2]	Communal Land	Old Resettlement	Forest Land	Safari Area	Others	Total
1	138,621	4822	7962	9837	2598	–	1790	148,447
2	2609	815	11,787	8066	–	–	–	22,339
3	6286	2578	27,125	1796	–	1250	–	46,194
4	7626	1015	112,863	–	–	6641	–	130,172
5	24,337	2942	134,720	–	–	18,209	2843	192,151
Total	179,479	12,172	294,457	19,699	2598	26,100	4633	539,303

Source: data compiled by author from documents and interviews granted by Chipinge district departments of Agricultural technical and extension services (Agritex)[3]; lands and rural resettlement[4]; local government and urban development[5]; and police internal security and intelligence (PISI)[6].\
Notes: [1]Large Scale Commercial Farming Area; [2]Small Scale Commercial Farming Area; [3]Agritex is a department within the Ministry of Agriculture, Mechanisation and Irrigation Development. Its main mission is to provide administrative, technical and advisory support to farmers; [4]Its mission is to acquire, redistribute and manage land. designated for agriculture; [5]The department is responsible for local government affairs including land planning; [6]PISI is responsible for matters of national security, since land has become so politicised, they gather information on developments in the sector.

Table 2. Chipinge district: post-2000 agrarian structure as of 2010.

Region	LSCFA	SSCFA	Communal Land	Resettlement A1 and A2/ Informal	Old Resettlement	Forest Land	Safari Area	Others	Total
1	42,615	4822	7962	96,006	9837	2598		1790	148,447
2	1671	815	11,787	938	8066				22,339
3	5432	2578	27,125	854	1796		1250		46,194
4	6845	1015	112,863	781			6641		130,172
5	15,175	2942	134,720	9162			18,209	2843	192,151
Totals	71,738	12,172	294,457	107,741	19,699	2598	26,100	4633	539,303

Source: data compiled by author from documents and interviews granted by Chipinge district departments of Agricultural technical and extension services; lands and rural resettlement; local government and urban development; and police internal security and intelligence (PISI).

President, Mrs Joice Mujuru, visited Chipinge town in 2000 and she told government officials that agro estates were not going to be occupied in Chipinge because they provided fair employment, schools for the local children and also subsidised electricity costs for the town residents. This resonated with the broader local concerns even though some war veterans concluded that the Vice President had shares in the estates. Evictions of land invaders led by war veterans echo Sadomba's (2011,150) argument that the relationship between ZANU-PF and war veterans 'involved complicity as well as contradictions, alliances as well as antagonisms, authority as well as subordination'. However, the relationship is uneven because the ZANU-PF ruling elite has access to the coercive state apparatus and strategically used such to silence war veterans as shown above.

In a recent development, some communal farmers from Chisumbanje in the low veld of Chipinge district are bound to permanently lose their agricultural land to Macdom Pvt (Ltd) and Ratings Investment, a corporate company involved in bio-fuel production owned by Billy Rautenbach alleged to have links with the ZANU-PF ruling elite. According to an interview with a Platform for Youth Development (PYD) member, a local Non- Governmental Organisation (NGO) doing develop-ment work in Chipinge, about 40,000 hectares of land has been identified by the company for sugar cane production affecting over 80 villages. The company is expected to start producing bio-fuel in mid-2012. The ongoing contestations between the villagers and the company including effective lobby by PYD prompted the Prime Minister, Morgan Richard Tsvangirai, to visit the villagers on 3 July 2011 where he promised to take the issue to cabinet (PYD member, pers.comm, 2011). This development requires further scholarly investigation beyond the focus and scope of this paper.

By 2011, a total of 4,881 households had been resettled on A1 farms in the high veld where most of the land reform occurred. Table 3 shows the distribution of new A1 farmers in the high veld. On the other hand, a significant number of black farmers, that is 922, acquired A2 farms in the high veld and their distribution is shown in Table 4.

However, there are other land occupiers that have not been officially recognised by the government and are called squatters such as those who stay at Mahlasela farm

Table 3. Distribution of A1 small scale farmers across natural regions 1 and 2 in Chipinge district as of 2010.

Ward	Natural region	Number of farmers	Acquired land hectares
6	1&2	728	5915
7	1	172	3305
8	1&2	633	3565
10	1	698	5373
11	1	723	3973
12	1	783	3874
13	1	614	2596
17	2	130	3670
19	1	400	3861
Total		**4 881**	**36,132**

Source: data compiled by author from diaries, documents and interviews granted by Chipinge district departments of Agricultural technical and extension services; lands and rural resettlement; local government and urban development; and police internal security and intelligence (PISI).

Table 4. Distribution of A2 farmers across natural regions 1 and 2 in Chipinge district as of 2010.

Ward	Natural region	Number of farmers
6	1&2	2
7	1	77
8	1&2	33
11	1	52
12	1	723
13	1	35
Total		**922**

Source: data compiled by author from diaries, documents and interviews granted by Chipinge district departments of Agricultural technical and extension services; lands and rural resettlement; local government and urban development; and police internal security and intelligence (PISI).

about 5 kilometres (km) west of Chipinge town. At the beginning of 2011 there were only 7 white farmers left in the district out of 132 farmers before Fast Track. The remaining white farmers remain targets of threats in a wave of erratic land occupations and coordinated elite capture. For example, various media reports indicated that Noah Taguta, a ZANU-PF supporter, mobilised youth militias to invade Spillemeer dairy farm owned by François Kotze on 1 June 2011 (see Saxon 2011).

There is a new form of inequality in the ownership of land as a means of production defined by class. The nature of unequal ownership no longer reflects a dominant white race. A new dual agrarian structure has emerged but the main difference with the pre-Fast Track distribution is the scale and degree of racialisation. In a review of land reform initiatives undertaken in a number of African countries, 'it is argued that control of land has been retained by existing powerful social groups' (Sender and Johnston. 2004, 156). Out of 96,944 hectares of land redistributed in the high veld, A1 farmers, a total of 4881 households, occupy 36,132 hectares of land whereas 922 A2 farmers own the bulk of the remainder with a few hectares occupied by informal settlers. I could not establish the exact number of A2 hectares given the political sensitivity of the A2 cases and my bias toward A1 farms. Whereas Fast Track in Chipinge district radically reconfigured the ownership of land in terms of race, it has not adequately addressed issues of political and social differentiation. On the other hand, corporate estates with the support of ZANU-PF ruling elite maintain huge tracts of land in the high veld and are pushing to displace communal farmers in the low veld.

Following Fast Track, agriculture was transformed in Chipinge district. The suitable farming activities for the high veld as stipulated by the GoZ (1986) were tea, coffee, dairy, sheep, forestry, tobacco, dairy, beef cattle and fruits. However, maize, vegetables and beans now dominate. Even goat rearing which was recommended for natural region 5 in the low veld areas of Chisumbanje, Chibuwe, Mutema and Manesa, etc. is now a common practice in the high veld. There is a new trajectory from cash crops to food crops due to three primary reasons; need for household food consumption; inadequate knowledge on production of cash crops; lack of production technology destroyed during invasions as I detail later. Scoones *et al*'s (2010) findings in Masvingo typify this trajectory where former cattle ranches have been turned into crop fields by new farmers. With the advent of Fast Track there is a shift from the policy considerations and agricultural recommendations of the past and the impact on the livelihoods of beneficiaries and wider economy is still to be explored in Chipinge.

In terms of Fast Track dynamics, Chipinge is under researched so this will be one of the first detailed studies following the event and there are four justifications to this focus. First, Chipinge district has a particular political history which brings different dynamics to Fast Track. Since Zimbabwe's independence it has been a stronghold of an opposition party called Zimbabwe African National Union (Ndonga) led by Reverend Ndabaningi Sithole, a nationalist and founder President of ZANU in 1963. Reverend Sithole was deposed from the leadership in internal party squabbles for allegedly denouncing the struggle in the mid-1970s and was replaced by Robert Mugabe. ZANU developed into ZANU-PF but Ndabaningi continued to lead a splinter ZANU (Ndonga). In post independent Zimbabwe, Reverend Sithole perceived his life to be in danger from Mugabe's state security agents and he went into exile from 1983 to 1992 (Sithole 1993). Many people from Chipinge, Ndabaningi's home area, felt that his persecution was ethnically motivated and were inclined to support Ndonga politically (see Sithole 1993). During Ndabaningi's absence, ZANU Ndonga won the Chipinge constituency in the 1985 parliamentary elections.[2] In 1990, ZANU- PF wanted to promote a one party state but managed to win 117 out of 120 contested parliamentary seats. Two of the three parliamentary seats it lost were in Chipinge district namely; Chipinge North and Chipinge South. It is such kind of political resistance that did not go down well with ZANU-PF and resulted in so much political surveillance in the district.[3] Upon Reverend Ndabaningi's return from exile he contested on a Ndonga party ticket and won Chipinge South in 1995. However, he was charged with treason for trying to topple Mugabe's government and was convicted and barred from representing his constituency. Even though the opposition retained Chipinge South in the 24–25 June 2000 parliamentary election, for the first time ZANU-PF won Chipinge North, at the height of Fast Track. ZANU-PF's new constituency geographically covered the high veld where Fast Track occurred. In the 29 March 2008 general election ZANU-PF managed to win one parliamentary seat in the district, Chipinge central, which covers most of the Fast Track resettlement schemes and lost all the other three seats namely Musikavanhu, Chipinge East and Chipinge West in the low veld to the opposition. In as much as this gives us a general trajectory, voters have other social, political and economic needs that influence how they vote.

Second, the study will contribute to the growing body of work and provide a comparative basis, in some aspects, that helps to illuminate the national picture of Fast Track. There are field based seminal studies, generating ongoing academic debates that try to unpack beneficiary selection in detail, but are based on Masvingo province, Goromonzi, Mazowe, Shamva and Mangwe districts[4]. Moyo *et al.* (2009) have done a broad baseline survey which covers six districts. This paper therefore adds to the debate using Chipinge district as a detailed case study.

---

[2]Constituency boundaries keep on changing within districts in Zimbabwe. In Chipinge, there was 1 constituency in 1985, increasing to 2 in 1990 and 4 for the lower house parliamentary seats in 2008.

[3]During fieldwork I was assigned a security official Mr W.M. to monitor my activities and I had to report to him regularly as part of state surveillance

[4]The Mazowe, Shamva and Mangwe surveys will be published into a book in 2012

Third, substantial literature looks at how and why new farmers got land (see Fontein 2006, Marongwe 2008, Matondi 2011, Scoones *et al*. 2010) but detailed attention remains thin in Manicaland province and Chipinge district in particular. We are told a total of 13,000 households were allocated A1 farms by 2003 on 273,176. 87 hectares in the province (Utete 2003) but the question is how and why. Fourth, a slightly different methodology that pays attention to how some beneficiaries came from the communal areas whereas others did not show that there was politically-motivated exclusion. However, a more detailed study linking communal areas and resettlement schemes, à la Kinsey (1999), can bring out the full implications on those excluded.

Before I delve further into my empirical findings, I give a historic overview of beneficiary selection under Zimbabwe's land reform from 1980 then focus on recent debates on Fast Track settler selection. For a detailed overview of the pre-2000 resettlement programme see Dekker and Kinsey's (2011) essay in this collection.

1980-1989: From pro-poor to pro-elite

Let me just emphasise that in early 1980, the land reform programme was meant to benefit the poor strata of the communities that is the 'need[y], landless, [formally] unemployed and refugees' Alexander (1994,333). Beneficiaries had to be aged between 25 and 55 years and fit in the following three categories:

1) Refugees or those who had been displaced during the war. This group included urban refugees and former inhabitants of closed villages (keeps).
2) Landless villagers in communal areas.
3) Smallholder farmers without adequate land to sustain their livelihoods.

However, as early as in 1982, the government feared that the types of farmers belonging to the three categories above would not increase agricultural productivity because they were poor so it incorporated the better off communal farmers (Alexander 1994, 333). Nevertheless the government failed to buy much of the land which was transacted in foreign currency on a willing seller willing buyer basis. Much of the land was acquired by the black ruling elite such that by the mid-1980s estimations put the number of black elite commercial farmers at 300 (Alexander 1994, 337) including ten government Ministers who doubled as members of the CFU (Palmer 1990,175). By 1989 the government managed to resettle 52,000 households which was a huge land reform programme despite failing to meet its target of resettling 162,000 households at the end of the decade.

In the early 1990s, the government adopted the Economic Structural Adjustment Programmes from the Bretton Woods institutions which meant a reduction in state expenditure, market deregulation and need to restore macro-economic stability. As observed by Moyo (2007, 32) the macro-economic policies influenced the government to focus on 'creating efficient and capable agrarian capitalists' as opposed to pro-poor redistributive land reform. In the mid 1990s, it was becoming apparent that land reform was not meeting the expectations of the peasants, poverty was increasing in the rural areas and the situation was exacerbated by the deteriorating Zimbabwe economy. As ZANU-PF popularity waned the government resorted to the emotive land issue as an attempt to regain the confidence of the masses (Zamchiya 2007). The government designated 1471 farms for compulsory acquisition in 1997 which was followed by intermittent land occupations.

In June 1998, the government introduced a new land resettlement policy, where it aimed to acquire 5 million hectares and target 91,000 families, mainly smallholders. In order to secure funding for the programme the government hosted the land donor conference where multilateral institutions and international donors endorsed a two year inception period which was intercepted by farm invasions in early 2000 (UNDP 2002). The state or at least particular technical ministries still had some input at this time, but they were already being marginalised. Alongside land reform initiatives there were other initiatives like the lobby for a new democratic and people driven constitution by the National Constitutional Assembly (NCA).

2000-2010: Pro?

In February 2000, the Zimbabwe government sponsored a referendum which entailed a clause that made it mandatory for Britain to pay for land confiscated for land reform. The referendum was rejected by the people in a plebiscite. However, land reform was not the only issue as there were other issues around political governance. After the rejection of the draft constitution, Zimbabwe witnessed disorderly and violent land occupations of white owned commercial farms on a scale unprecedented in post-colonial Africa. Land invasions intensified in March, April, and May and by the time the late Vice-President, Joseph Msika, gazetted 804 farms to mark the start of Fast Track on 15 July 2000 (GoZ 2000) about 1000 white farms had already been occupied in invasions led by 'war vetcrans' (Marongwe 2008) and Chipinge was no exception.

Fast Track land reform entrenched the possibility to write 'specifically about land [in Zimbabwe], without reference to the high modernism or customary projects of previous decades' (Alexander 2006,180). In other words this heralds a new era where one has to integrate politics in the analysis of the state and the peasantry in post-2000 Zimbabwe. The state's authority was to be 'grounded in political loyalty and patronage, not expert knowledge and bureaucracy, a transformation that set in train a series of struggles that deeply divided the state *and society*' (Alexander 2006,187 *my emphasis added*). Given the narratives of ruling party politicians during Fast Track, which were based on the history of liberation struggle and the need to politically reclaim land, it is fair to say that for a moment the land question became disarticulated from the narrow concerns of agricultural production. The land question was now conceived as a political question justified by history and the need to resolve the inequities of colonialism and revive a waning ZANU-PF party and not conceptualised in the narrow sense of agriculture production. The political and social processes through which land was occupied or redistributed, provide a framework to understand the new farmers (see Marongwe 2008).

Recently, in a major study, Scoones *et al.* (2010, 8) challenge the myth that '[t]he beneficiaries of Zimbabwean land reform have been largely political cronies'. This is in response to a kind of popular image in the Western media where elites are deemed to have gotten all the farms. By 'crony' they mean quite specifically an elite group tied into the ruling party. I agree with them that it is not an accurate description of all Fast Track beneficiaries. Even if we accept that elite cronies were a minority, and concentrated in A2 farms and even more specifically the high value farms near urban areas it is still the case that some A1 beneficiaries, are still clients of patronage networks linked very directly to a partisan state and ruling party, willing to use coercion to exclude other 'ordinary' people who are understandably not a focus of Scoones et al.'s (2010) seminal study.

Claims to land are not only political (see Fontein 2006, 2009) and as Alexander (2006) argues claims to new resettlement land carry different meanings for different people ranging from historical, chiefly, political, need-based and ancestral claims. These factors in some instances influence how and why people got land. I am inclined to share Scoones et al.'s (2010) view that Fast Track processes possibly differed from place to place and from farm to farm hence the ideas sketched here are abstract but necessary to provide a theoretical entry point for further concrete investigations through field research as elaborated below.

Research sites and methodology

Scoones et al. (2010.8) rightly argue that theoretical debates on Fast Track have not been informed by the situation on the ground but rather by ideological persuasion and political affiliation. They stress that '[o]fficials ... have rarely been to the rural areas and if so only on flying visits and almost certainly never to the new resettlement areas' (ibid). Consistent with the above, Moyo et al. (2009, 4–5) and Matondi (2011) argue that the assertion that political patronage predominated in the land allocation process lacks empirical evidence.

As a result, my article is based on field research on three A1 resettlement schemes in Chipinge district. Data was collected at four levels, national, district, farm scheme and farm household. The three farms provide an opportunity to capture the diversity in terms of how people got land as they depict three different methods of acquisition. A questionnaire survey targeting 70 households was administered in the three schemes as tabulated in Table 5.

The questionnaire was organised into ten broad categories.[5] I flesh out the components that informed this article. These are (1) The origin of beneficiaries; (2) prior or current employment of household members; (3) method of land access, land size; (4) political and social networks; (5) motivations to seek land (6) social and cultural positions and other. I administered the questionnaire to the head of the household because the household was the unit of allocation in the redistribution or confirmation of land allocations under Fast Track.

I then carried out individual household interviews in the form of open-ended interviews and life histories. This enabled me to capture the origin of farmers, how

Table 5. Study sample (N=70).

Name of Farm	Farm Size (ha)	Land acquisition method	Number of resettled households	Number of household surveys
Glen View Lot 1	165	DLC[2]	17	11
Wolfscrag	810	Land invasion	44	28
Wedgehill	826	DLC	54	31
Total	1,801		115	70

---

[5]The categories were designed for my broad Doctorate thesis

they got land as a way to augment the annual data collected using the questionnaire. Life histories shed light on the claims and motivations of new farmers in getting land under the controversial Fast Track which is difficult to capture quantitatively. I targeted A1 farmers who owned different sizes of land, women and men, in order to capture different views across strata and gender in different periods between November 2010 and March 2011. Off farm I interviewed communal farmers and ordinary villagers in the low veld who failed to acquire land during Fast Track.

I also interviewed Agritex officers, Lands officials, District council employees, local NGO workers and other key informants who have been residing in Chipinge before and after Fast Track. This revealed subjective and objective interpretations which were important in explaining the nature of land and agrarian change in Chipinge.

Apart from field data, I used media articles, farmers' union reports, material from key informants, and government documents like local, provincial and national land allocation and progress reports.

For data analysis I used the Statistical Package for Social Sciences (SPSS) and the narrative approach for qualitative data so as to provide concrete details of the findings, what some scholars call the 'realist tale approach' (Neuman 2003). Next I delve further into my empirical findings.

How did they get the land?

As Scoones *et al.* (2010) argue there is no single story in the way land was taken and my study sites provide no exception. Empirical evidence shows three trajectories of land occupation namely land invasions, replacement of original occupiers and allocation through a formal process. Any generalisation can be misleading therefore I focus on the empirical specificities. Next, I give an overview of the Fast Track institutional framework put in place then the three trajectories on my study sites.

The institutional framework put in place to oversee Fast Track was the District Land Committee (DLC) and, in most cases, it had the final say in confirmation of land occupations or allocation of land to applicants. In Chipinge it was composed of Zimbabwe National Liberation War Veterans Association's (ZNLWVA) district chairperson, ZANU-PF district chairperson, Head of the Department of Agritex, Head of the Department of Lands, Chiefs' representative, District Administrator (DA), and a Security Sector representative. In a separate study Marongwe (2008) observed a similar pattern. In Chipinge, the DLC would sit every Tuesday from the year 2000 to allocate land, confirm applications and deal with boundary disputes and this committee was still functioning in early 2011. Zimbabwe has 24, 000 village heads, 271 Chiefs and 400 headmen (see Matyszak 2011,14, Chakaipa, 2010,48) and in Chipinge just like in many parts of the country chiefs and headmen are few but there are close to 500 village heads in the district. Some scholars usually recognise traditional authorities as composed of chiefs, headmen and village heads (see Chakaipa 2010, Matyszak 2011) but the Traditional Leaders Act [Chapter 29:17] recognises messengers appointed by chiefs and headmen as part of the traditional hierarchy. On my resettlement schemes the messengers' duties overlapped with those of village heads such as collecting fines and other charges and coordinating the village assembly known as *bandla* in the local lexicon. I also recognise their families and special advisors to the *bandla* who are not recognised in policy but are incorporated in the traditional hierarchy usually from the ruling lineage to help deal with issues relating to residential, grazing and agricultural land disputes among other social problems which give them a special social status and access to privileges that ordinary people cannot easily get.

By war veterans I refer to those who have been collaborators,[6] detainees and liberation war fighters at the battlefront during Zimbabwe's liberation war for independence. This is a broader definition that ZANU-PF uses but it should be noted that the ZNLWVA is only open to people who were trained and carried guns, and they are resolute about enforcing that definition. However, during land invasions war veterans established a marriage of convenience with detainees and collaborators on the ground in order to strengthen their numbers and effectively mobilise. How was land taken on the ground?

Wolfscrag

In 2000, a group of war veterans led the occupation of Wolfscrag farm. The former owner specialised in timber production mainly growing gum trees but also produced coffee and macadamia. The leading group of war veterans attempted to mobilise fellow veterans, particularly in the low veld but they were reluctant citing their allegiance to ZANU Ndonga, rather than the ruling party, ZANU-PF. However, the war veterans made an alliance with ZANU-PF local leaders and invited other war veterans loyal to the ZANU-PF ruling elite from the adjacent Buhera district[7] to join in the land invasion. Apart from the ZANU-PF structures, the war veterans also invited the traditional leaders displaced historically from the same farm in 1974. So a nexus of local war veterans, mobilised war veterans from Buhera and traditional leaders mobilised some people to join the occupation of the farm. A few farm workers also joined the invasion. Whereas some ordinary people joined the invasion the composition of beneficiaries is dominated by war veterans and members of the traditional ruling lineage. One would expect beneficiaries on this farm to be more diverse because it was acquired through a land invasion but this was not the case. The highly politicised process and the autochthonous nature of the invasion in a way limited possibilities of having a more diverse composition of ordinary beneficiaries.

The occupiers were divided into sub-groups with each led by a war veteran who then pegged plots for them. Mr X, the former white farm owner, reported to the police but no action was taken. War veterans and the identified traditional leaders allocated themselves land on one side of the farm with gum trees and macadamia that had survived fire outbreaks whereas others were given smaller pieces of land across the hill. The war veterans established a 'committee of seven' to govern the farm. In early 2000, the committee of seven were the main organisational structures on occupied farms. The committee of seven differed in composition on farms across the country. In Masvingo it involved 'a base commander, treasurer, women's and youth representatives alongside ordinary members' (Scoones *et al.* 2010,192). On this particular farm it was chaired by a war veteran and composed of other veterans and traditional leaders.

---

[6]Those who provided logistical support to trained guerillas at the battlefront during the liberation war

[7]Buhera district is 100 percent communal, it had no white owned large scale commercial farms thus no Fast Track could take place within this district. Aspiring beneficiaries had to acquire land from other districts.

In 2001, Agritex officers came to confirm the plots; about 32 households were allocated between 3 and 15 hectares each under the A1 model. The Agritex move was largely a confirmation exercise of the pre-existing allocations rather than a radical change of the land structure. Some of the war veterans and other ZANU-PF leading members had left the farm with some getting A2 farms elsewhere and others moving to invade more lucrative farms. A few farm workers confronted the DA at an entrance to the farm and forced him to allocate plots to them. The DA put cards in his Agritex sun hat and the farm workers had to randomly pick, with the number picked being the number of the plot. A new group of occupiers was chosen by the DLC and joined the original invaders. At this stage there are a few civil servants who manipulated the process but the number of civil servants is not the only benchmark of a patronage based acquisition. The war veterans and traditional authorities' actions were sanctioned by the ruling elite for reasons explored latter. In 2004, farmers started receiving official offer letters from the DA's office. To date there are 44 households resettled officially. Given the complexities of kinship and social networks among these clients some with kinship ties to traditional leaders, war veterans and the civil servants got an opportunity to get land and slightly bigger farms. Around June 2010, the white farmer who had kept the adjacent part of the farm was displaced by a top ZANU-PF official based in Harare but with origins in Chipinge in a new wave of elite capture.

Glen View Lot 1

Noting the escalation of farm invasions in 2000, the former white owner of Glen View Lot 1 offered a portion of his farm to the Ministry of Lands for resettlement. Glen View Lot 1 used to specialise in cattle breeding with high milking technology investment on the farm. Much of the land offered was meant for grazing. The Agritex officials subdivided the farm into A1 farms ranging from 3–10 hectares and demarcated common grazing land. Even though Glen View Lot 1 was acquired through state administrative procedures open to patronage abuse there were fewer civil servants who benefited because there was little investment on the farm as part of the land offered was previously grazing land. The DLC selected 17 households to settle on the farm. From 2004, farmers started receiving offer letters from the DA. They allocated bigger farms to a few civil servants and others who were politically connected whereas those with weaker political ties within ZANU-PF got the smaller farms with fields between 3 and 5 hectares. The latter were not socially excluded like those in the communal areas but were adversely incorporated. Political authority was not just used to gain access to land but to get larger farms than other clients of patronage. However, conflicts ensued between those excluded in the land allocation process and the new farmers. A group of about 200 people invaded the land meant for grazing and started erecting shacks and cultivating the land. Those officially allocated land have complained relentlessly to the traditional leaders and DLC but to no avail. One of the formal beneficiaries at Glen View Lot 1 complained:

> I am in the committee of seven here, and have been sent to the DA's office on many occasions on the issue of squatters occupying our grazing land. The DA has promised to address our issue but to no avail. At least the Agritex officers understand us on this one because they know our farming plans which are being interrupted by these squatters (Resettled farmer, Glen View Lot 1 Farm, 2011).

One of the squatters, a former communal farmer from Manesa, complained: 'How can they peg land for goats and donkeys when we the people do not have land? That is an expression of callousness. It is similar to the white farmers' mentality of prioritising the life of a goat to that of a human being.' (Squatter, Glen View Lot 1, 2011). The local police said they had an instruction from above not to evict the 'squatters'. I observed some ZANU-PF local campaign managers registering the names of the squatters in some A4 exercise books for purposes of forwarding the names to the DA who was supposed to ensure they were registered to vote in a 2011 anticipated national election.

Wedgehill

The previous owner used to specialise in production of macadamia for marketing to the Southern African Macadamia Association (SAMA) and had started venturing into coffee production at the time of the invasion and had gum tree plantations that acted as firebreaks. The farm was highly mechanised with macadamia shelling machines, irrigation equipment and other high-technology machinery. On the other hand, there were good houses for farm workers. However, at the height of jambanja, a term used to describe the violent, chaotic invasion of farms from 2000, the farm owner fled to Australia taking with him what he could and allegedly entered into a lease agreement with Southdowns estate. Former Wedgehill farm workers and those from adjacent Southdowns estate invaded and started clearing the land for cultivation of maize, uprooting the young coffee plants and cutting some gum trees for firewood and for sale. However, the farm worker honeymoon was short lived as the Ministry of Lands designated the farm as vacant and sub-divided it into A1 farms ranging between 3 and 15 hectares. The farm was then allocated by the DLC mainly to war veterans, former prison officers and army officers resulting in the eviction of farm workers. The composition of beneficiaries on this farm shows that what was at stake was not just the mode of acquisition but the nature of investment on the farms such as macadamia trees, plantations and other infrastructure which resulted in civil servants and war veterans manipulating the administrative structure in their favour.

Who got the land?

Let me emphasise that I did not ask directly about people's political party affiliation but some of my respondents from communal areas freely expressed that they were excluded from Fast Track because they were opposition supporters and castigated the partisan nature of the programme as detailed latter. The local history of opposition to the ruling party since 1980 makes some people more open about their political association. On the other hand, some war veterans overtly declared their allegiance to ZANU Ndonga, ZANU-PF and other parties. I have a biographical connection to the district, family links and social networks that made some respondents comfortable to talk about sensitive topics. It is still difficult to deduce everyone's political party affiliation except those in high ranking positions and willing to do so. Moreover, political party affiliation can shift over time. Many scholars factually argue that people were compelled by circumstances and would pretend to belong to ZANU-PF to get land hence it is difficult to pin down beneficiaries' political party affiliation (see Matondi 2011). My emphasis is that this cannot be celebrated as it does not reflect a vibrant political society where individuals can freely exercise their constitutional right to freedom of assembly, association and expression without jeopardising their access to land. Whereas one may never be able to deduce the political affiliation of every beneficiary but voting patterns alluded to earlier help to show that ZANU-PF has substantial electoral support in the resettlement schemes. Nevertheless, the pinnacle of my argument is not that new beneficiaries were ZANU-PF but that Fast Track was a highly partisan process where political loyalty and patronage to the ZANU-PF ruling elite largely explain how new beneficiaries acquired land across my study sites.

For both government officials and scholars it is difficult to disentangle categories of settlers to come up with a summary that best represents who got what. Any summary of categories of settlers that give precedence to occupation or social status is bound to overlap as people have multiple identities. Abstractions are unavoidable in this case but the argument is which categories are more useful for a particular context. For example, chiefs as a singular category may be important in Mashonaland central province where chiefs in some areas claimed a 5 percent quota on every farm (Matondi 2011,61) based on their traditional status rather than their other multiple identities. A business person can double up as a civil servant whereas the latter can be employed in the security services and vice-versa. Scholars are unlikely to agree on a single template that best represents who got what as it has implications on interpreting both the Fast Track beneficiary selection process and its outcome which remains intellectually contested. If such a template is not carefully informed by empirical data in terms of how people got land in a particular context it can mislead interpretations.

In the seminal Masvingo study, Scoones et al.'s (2010, 52–53) summation of categories of settlers that got land is based on the occupation of the household head prior to getting land which is one possible way. Based on this criterion they argue that '68.2 percent of new settlers were... a diverse ordinary group, with about half of all new settlers coming from nearby communal areas' (Scoones et al. 2010, 238). It could be useful to reproduce the categories used by Scoones et al. (2010) for direct comparison but emphasis should be determined by local dynamics at least until such a time scholars have gathered empirical data district by district to give a clearer national picture. Table 6 shows the singular settler categories that take precedence in summarising the Masvingo data.

Scoones et al. (2010, 53) define ordinary as 'not members of the other categories [they identified above] and as 'largely asset and income poor'. Due to the problems of multiple identities as discussed above Scoones et al. (2010, 54) spread war veterans

Table 6. Settler profile across resettlement schemes in Masvingo.

Category	Percentage
Ordinary	68.2
Civil servant	16.5
Security services	3.7
Business person	4.8
Former farm worker	6.7
Total	*99.9*

Adapted from Scoones et al. (2010:53)

across all categories and in the process argue that all war veterans on A1 villagised farms in their study sites were ordinary beneficiaries. This is because they were unemployed and 'many had long dropped their war veteran identity' and were just small-scale farmers in the communal areas (Scoones *et al.* (ibid). Classifying war veterans under those without formal employment, a definition Scoones *et al.* (2010) use to define the ordinary, would suggest that war veterans got land just like ordinary people during Fast Track. Yet in Chipinge some war veterans reinvented their identity as it became political currency in gaining access to land through mobilisation, land invasions and the [in] formal allocation process during Fast Track. It is therefore difficult to place unemployed war veterans in the ordinary category. Their status as war veterans was the significant factor during Fast Track and to a large extent determined 'who got what when and how' rather than their other multiple identities hence it takes precedence in my summary of categories. Even the government recognised war veterans as a special interest group and reserved a special 20 percent quota for them on A1 resettlement schemes (GoZ 2001), a priviledge not accorded to other ordinary people. Consequently, official land audits identify war veterans as a special interest group in summarising who got what during Fast Track (see Buka 2002, Utete 2003). The ordinary should not include members of distinct interest groups such as war veterans, traditional authorities, civil servants, security services and business people.

While there can be debates around what ordinary really means I posit that classifying land reform beneficiaries according to their previous professions and describing them as ordinary based on not having 'formal employment' (Matondi *et al.* 2008, Marongwe 2008) or of 'low income and assets' (Scoones *et al.* 2010) or simply from adjacent communal areas (Moyo *et al.* 2009) without considering the distinct social status one holds is inadequate if one is to explore the nuances of political patronage in gaining access to land. The ordinary can still be rich in political capital and be clients of patronage networks and beneficiaries of patronage relations linked to the ruling elite.

Traditional authorities in Chipinge played a crucial role and were dominant in what appears to be clear cases of restitution hence I single them out. On the other hand, business persons were insignificant in the process and outcome of beneficiary selection in my study sample. On treating farm workers and security services as singular categories I agree with Scoones *et al.* (2010). Former farm workers can be categorised as ordinary but due to contested views on whether they benefited during Fast Track they are worth singling out (see Chambati and Moyo 2004 and Sachikonye 2003). Security services[8] are civil servants but given the militarisation of state institutions (see Raftopolous 2003) in Zimbabwe they are worth mentioning as a separate category. Consequently, I adopt the categories in Table 7 in summarising my data on who got what and my results show three dominant types of beneficiaries: war veterans, traditional authorities, and civil servants.

Overall, those that dominate, among beneficiaries, are civil servants mainly from the Ministry of Lands, Agriculture and Local Government. How did they come to dominate and why? Those working for the government had easier access to information regarding the Fast Track process. Civil servants in Chipinge district

---

[8]Security services are composed of police, army, Central Intelligence Organisation and prison officers.

Table 7. Profile of beneficiaries as of 2010.

Category	Percentage of beneficiaries
Civil servants	35.0
Traditional authorities	22.2
War veterans	16.7
Security Services	15.0
Other 'Ordinary'	6.9
Former Farm workers	4.2
Total	**100.0**

Source: compiled by author from survey data (N=70).

would submit their applications to their respective head of department who would not only forward the application but recommend beneficiaries to the DLC. As indicated above, the Agritex and Lands representatives sat in the district committee which was an added advantage to employees in their department. Civil servants are usually the agents through which state programmes are implemented and their loyalty is important in maintaining state power. Following the disorderly onset of Fast Track it also made sense to incorporate civil servants from the technical planning offices of Agritex and Rural Council to give some semblance of technocratic legitimacy and order to Fast Track. In this process the civil servants had an opportunity to self-serve.

In addition, civil servants benefited because they were no longer considered an overt political threat. The allocation of posts to head government departments at district level had become highly politicised and partisan after 2000 as the thin line between ZANU-PF and the state was eroded. One had to be vetted by the security services before being appointed to such a post to ensure s/he was not a risk to state 'security', which usually means not a risk to ZANU-PF. One of the employees in the Agritex department emphasised:

> Our District Agricultural Extension officer has a diploma from an unknown institution yet most of his subordinates have degrees from universities [mainly Zimbabwe Open University]. There was a human resources audit that recommended that he be demoted but ZANU-PF district leadership came to back him up because he is a strong party cadre. So he is hanging in there on a party ticket not on the basis of competence (Agritex officer, Chipinge, 2011).

Within this context, personal, state and party interests became conflated. For example, the late Mr Samuel Zuze, another magistrate deemed to be politically correct by most war veterans I spoke to, presided over the case of a white farmer, Mr Mike Jahme, and convicted him in January 2010 under the Gazetted Land (Consequential Provisions) Act because he was refusing to vacate his former Newcastle farm. The Chipinge magistrate is the one who had been allocated Jahme's farm as shown in the offer letter below but he did not excuse himself.

Even though the offer letter below is an A2 offer letter my point here is that government workers had become so compromised to an extent that personal, state and party interests became intertwined. Also note that the Minister in the offer letter above reserved the right to withdraw the offer anytime which effectively means the beneficiary is obliged to reassert his legitimacy on the land. The A1 offer letter is

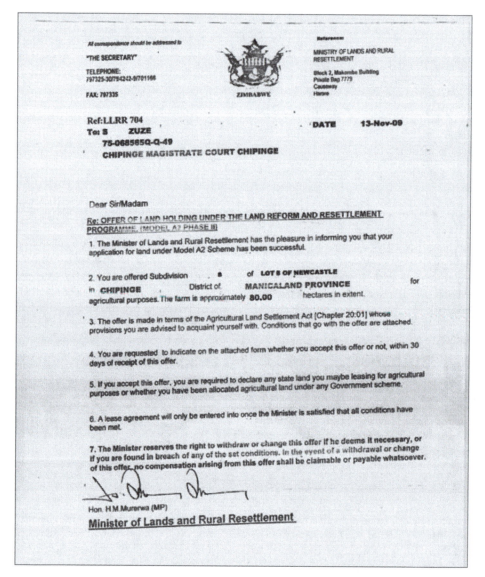

Figure 1.   Magistrate Zuze's offer letter.
Source: obtained by author.

similar in indicating that the government can withdraw the offer anytime without compensation for any improvements made on the resettlement farms. On the ground, war veterans continuously demonstrated that civil servants should exhibit 'correct' political behaviour which meant acting in the interest of the ZANU-PF ruling elite.

Once deemed to be politically correct the beneficiaries were expected to continuously show political loyalty to the ruling elite. Civil servants perceived to be colluding with the opposition were in many cases subjected to violence by the local war veterans in a tactical manner that served to terrify other civil servants. The civil servants could be land reform beneficiaries but were going to be kept in check through the use of brutal force. For example, Walter Chikwanha, a Chipinge magistrate, was assaulted in August 2002 by war veterans for passing a judgement in favour of MDC (Movement for Democratic Change) supporters as one of my key informants, a teacher, who has spent the past four decades in Chipinge recalled:

> They took him [Walter Chikwanha] from the court [Chipinge magistrate court] to our government offices [Government district departments are located in one complex in Chipinge town] and started beating him with open and closed fists, some with a log and it was so barbaric, he started screaming in pain until he was just bleeding. They accused him of supporting the MDC because he had given bail to some MDC people. It was a lesson to all of us government workers of what it would mean to act against ZANU-PF and we were all scared. They [war veterans] were so inhumane … they started walking with him around town toward Chipinge hotel singing ZANU-PF songs and shouting pashi nema sell-outs! (Down with sell-outs!). From there people were terrified at work and knew these guys meant business (Key informant, Chipinge, 2010).

More recently, war veterans invaded the magistrate court in April 2010, singing songs in praise of ZANU-PF and threatening to assault magistrate Thomas Masendeke who had ordered an eviction of war veterans and other settlers from a farm they had invaded. Thomas was later transferred to another city, Mutare, just like Walter Chikwanha for personal safety. The war veterans-led crowd also went to sing at the DA's office giving him a warning against keeping such MDC oriented people in the district (also see Chimhete 2010).

This typifies McGregor's (2002, 9) observation that the state had already politicised the bureaucracy thereby largely undermining 'the scope of professionalism within the public service through physical and psychological harassment' led by war veterans in her study of Matebeleland and Midlands. Rather, in post 2000, some civil servants were now an ally in promoting personal and ruling elite interests using state institutions.

The civil servants who got land in Chipinge attend ZANU-PF political meetings and rallies for they have to show continuous loyalty as a way to safeguard their land. It is increasingly difficult for the civil servants who got land to act outside the interests of the ZANU-PF ruling elite. Some civil servants were a potential active constituency for the opposition but land distribution transmuted them into a potential constituency for the ZANU-PF ruling elite thereby laying foundation for an alliance based on patron-client relations.

Apart from civil servants, traditional authorities and war veterans were major beneficiaries of land reform. Why? After independence government pursued decentralisation in local government such that The Chiefs and Headmen Act [Chapter 29:01] of 1982 removed most powers traditional authorities had prior to 1980 (Chakaipa 2010,48) and placed them in elected democratic structures. However, in reality villagers would still report to traditional leaders and I am inclined to share Fontein's (2009,8) view that, '[t]his value system is like a religion or a way of life, the chief as custodian of peoples' heritage'. Consequently, in early 2000, the ZANU-PF ruling elite shifted and developed a strategy to co-opt traditional leaders in their ideological and mobilisation campaign for political hegemony. The traditional leaders had their powers restored through the Traditional Leaders Act [Chapter 29: 17]. This Act increased functions of a chief from three to 22; a headman from three to 11 and village heads were allocated 16 functions but it does not specify the number of functions for messengers. Alongside the co-option of traditional authorities, Sadomba (2011,157) rightly argues that there were signs of cooptation of some war veterans by the ZANU-PF elite in early 2000. Sadomba (2011,194) reveals that President Robert Mugabe invited Chenjerai Hunzvi then chairman of ZNLWVA at a central committee meeting on 18 February 2000 and asked him to lead the election campaign and offered him a special position on the central committee of ZANU-PF.

In my case study, war veterans articulated with the traditional leaders to invade Wolfscrag farm because they knew that they would buy into the project given their history of ancestral claims to the land. It is important to note traditional leaders were not just passive beneficiaries. As Fontein (2009,1) argues in Masvingo, some local clans 'often made very specific claims to land which appealed to autochthonous knowledge of the land, invoking memories of past occupations and the burial of ancestors in the land' in order to gain legitimacy. This is the same case with Chiefs on Wolfscrag who had been displaced in 1974 and made autochthonous claims as I show in the motivation cases later. It was more a case of restitution though it is not an official land reform policy in Zimbabwe. Let me point out that the different autochthonous imaginations on Wolfscrag are not as stark as in Fontein's (2009,5–6) study in Lake Mutirikwi because his study area as he confirms is made 'of the variety of historically different landscapes' and also contains 'Zimbabwe's most important national heritage site, Great Zimbabwe, and several even older rock art sites' (Ibid).

In addition, traditional authorities were represented in the DLC and, together with ZANU-PF representatives and war veterans, played a major role in deciding who got what. Such co-option of war veterans and traditional leaders would lend 'legitimacy to the government's ongoing anti-colonial rhetoric' and 'appeal to a rural constituency' (Fontein 2006,238) as 'part of a continuing struggle of liberation against the forces of imperialism' (McGregor 2002,10) in order to legitimise the ruling elite's anti-settler and anti-neo colonial rhetoric (Sadomba 2011,145) 'at a time of political discontent' (Chaumba et al. 2003,599). Yet the redistribution agenda and neo-colonial discourse was meant to hide authoritarian, exclusive and repressive politics (see Raftopolous 2003). In Chipinge a partisan discourse of 'Pamberi na Mugabe' (Forward with President Mugabe) also formed part of claims to land by war veterans and traditional leaders because historical and traditional claims were not enough to ensure one's tenure but political loyalty reasserts legitimacy in ongoing contestations over resettlement land in Chipinge.

Land was used as a source of patronage and beneficiaries (clients) are supposed to show loyalty to the patron (ZANU-PF ruling elite) and failure to do so may result in one being threatened or actually losing the land. Mr Z, the village head at Glen View Lot 1, was visited by his cousin from Harare[9] in the run up to the 27 June 2008 Presidential run-off election and was questioned by the war veterans for hosting people from Harare who were labelled MDC supporters as he elaborates:

> They questioned me about my cousin who came to see me … everyone from the town was said to be MDC. They threatened to evict me but I explained to them. In any case it is difficult to evict the owners of the land but if I was another person I could have been chased. However, for safety of my cousin I told him to return to Harare. My father started ZANU-PF with the likes of [Ndabaningi] Sithole in Mt Selinda so there is no way I can be evicted (Village head, Glen View Lot 1, 2010).

---

[9]Harare is the capital city of Zimbabwe where the MDC party enjoys significant support

The village head raises both traditional and political authority to safeguard his land which shows that a beneficiary with weaker traditional or political ties could have possibly been evicted.

The role of chiefs and war veterans was not bifurcated into traditional and political as Fontein (2006,241; 2009,13) seems to suggest in a separate study in Masvingo. Rather the village heads and headmen played a political role alongside war veterans on the farms, they belonged to the ZANU-PF political party committees and helped to mobilise new farmers to go and cast their vote in the 2000, 2002 and 2005 elections urging and sometimes intimidating them to vote for President Mugabe's party. On each farm, every head of household was allocated a political post in ZANU-PF's ward committee that would meet every Tuesday where the ideology of the ruling party was drilled. Political support was more overt in the 27 June 2008 run-off election when the traditional leaders gave a directive to the resettled farmers to assemble at the village head's homestead then proceed from there to go and vote as a group. This was another way to legitimate their stay on resettled land through continuously showing political loyalty. It was not just a farming landscape but a heavily politicised and Zanunised landscape with campaign posters of President Mugabe, for the 27 June 2008 run-off election, still visible on some rocks and homesteads. A military base was set up at Wolfscrag farm in early 2011 to carry out political intimidation campaigns in the form of pungwes (all night political education campaigns) where neighbouring communities were forced to attend and threatened with death if they were to vote for the opposition in an election local ZANU-PF structures anticipated in 2011. I was advised by a security agent to wind up my fieldwork because of the anticipated election. The ZANU-PF governing elite was aware of the political capital of war veterans and traditional leaders in terms of both legitimising the land reform project and in terms of ongoing political mobilisation beyond the 2000 and 2002 elections.

I have discussed why war veterans, the third dominant group, alongside civil servants and traditional leaders managed to get land but how did they manage to get land? They mobilised villagers in the communal areas to go and invade white-owned farms and as a result were in the forefront and subsequently became the first line of beneficiaries. War veterans also played a significant role in the initial unofficial pegging and allocation of land to occupiers as illustrated in my Wolfscrag example. However, not all war veterans got land through occupation in Chiping. Some formally submitted their applications to the ZNLWVA at district level. The district leadership would vet and submit the names to the Chipinge district ZNLWA chairperson who sat in the DLC. The government also had a quota system for the war veterans. This process was difficult for the ordinary people to penetrate.

Not all war veterans endorsed the ruling party's political project of mobilising rural support but some played into the hands of the ruling party by becoming agents of an authoritarian state. Some war veterans who paid allegiance to Ndonga had grievances about land but they did not share the same partisan approach in realising their claims to land; in fact some refused to join the invasions because they say they did not want to give legitimacy to the ZANU-PF project. A ZANU Ndonga member queried, 'If ZANU-PF cared about land reform they should not have destroyed Churu farm' (ZANU Ndonga supporter, Mutema, 2011). In the early 1990s, Reverend Sithole had offered his Churu farm in the outskirts of Harare for poor households and by 1993, close to 20 000 people had been resettled on the land. The ZANU-PF governing elite was not happy to have land beneficiaries under the benevolence of an opposition party leader. They used the Minister of Lands to acquire the farm in April 1993 on grounds that it was being underutilised for agricultural production. The beneficiaries were dispersed violently by anti-riot police. Even at the national level some of the war veterans refused to be part of the ZANU-PF governing elite agenda and they formed the Zimbabwe Liberators' Platform (ZLP) in 2000 whose objective was to 'refocus on the original aims and objectives of the liberation' and to assist war veterans so that they would not become 'vulnerable to manipulation by unscrupulous politicians' among other things (Kubatana 17 December 2007).

Some war veterans who supported opposition parties, notably the MDC and ZANU Ndonga in Chipinge, alleged victimisation in the land allocation process as an ardent MDC supporter illustrated:

> The ZNLWVA district chair dogmatically supports ZANU-PF and when he sits in the land committee his primary duty is to ensure that those real veterans who do not overtly carry out ZANU-PF's sting operations [meaning occupations, political violence] are excluded from the process. So even if you were saying war veterans benefited they are ZANU-PF aligned war veterans (MDC supporter, Mutema, 2011).

A member of the ZNLWA defended the association's partisan position: 'Those who were supporting the MDC must go and get their land in Britain. The land is for sons of the soil not sell outs' (War veteran, Mutema, 2011). At a national level ZANU-PF politicians openly castigated opposition supporters and denounced them as unfit to get land under Fast Track (see Marongwe 2008).

Not all beneficiaries who can arguably be classified as 'ordinary' people are exempt from the politics of patronage with ZANU-PF as the vehicle. Why? For some ordinary families in the communal areas of Chipinge to get land, they had to apply to the chairperson of ZANU-PF in their respective ward who would submit the names to the ZANU-PF district chair who would then forward the names to the DLC where he had a seat. Materially, as Scoones *et al.* (2010, 53) observed, they can be defined as 'largely asset and income poor' but it does not mean that the poor cannot be clients within the ruling party's patronage networks. Some of these people benefited using local political networks and others were close companions of the local ZANU-PF leadership, so it is farfetched to dismiss the political patronage thesis in this sense. Why is it that other ordinary people who wanted land were not prioritised? An aspiring beneficiary explained:

> I also wanted a piece of land because this area [Mutema] is dry but I could not submit my name to the ZANU-PF [ward] chairperson because I am a known opposition supporter since the days of the great Ndaba [former president of ZANU Ndonga Ndabaningi Sithole]. The chairperson was saying the land is for ZANU-PF people only. Others encouraged me to submit my application directly to the Ministry of Lands but without any connection I realised it was a futile exercise. So I have decided to wait because our day will come (Opposition supporter, Mutema, 2011).

An MDC supporter who was excluded from land reform also stated:

> I am a known MDC supporter; I was busy putting posters for the MDC campaign in this whole area. When Fast Track started I was caught by surprise. Everyone wants land but the same war veterans who were intimidating us, are the ones who were dishing land so it was impossible to break through (MDC supporter, Manesa, 2011).

The above account resonates with a separate study by ZHR NGO Forum (2010) which concluded that it was ZANU-PF oriented war veterans and youth militias who dictated the allocation of land yet they were also involved in violence and intimidation against non-ZANU-PF supporters.

A communal farmer still subsisting in the poor sandy soils in the low veld area of Tanganda also expressed disgruntlement with the overall process:

> I am a communal farmer who was on the waiting list for land reform, but when this thing started, the waiting list was subverted and it was these guys who were in control, who got the land. If they had followed the list and not prioritised the new people registered in year 2000 by the party [ZANU-PF] then it could have been better (Communal farmer, Tanganda, 2011).

The exclusionary process possibly shaped people's reactions to the partisan land redistribution programme as indicated by a communal farmer in Manesa area:

> Because we were left out in this process we are still suffering from food shortages. It is very difficult to produce on this sandy soil and it rarely rains in this low veld. We just hear stories that those who were resettled are enjoying having three meals a day and lots of money. But we know how we will punish the politicians as we did in the previous election (Communal farmer, Nyunga, 2011).

As noted before, it is important to note that ZANU-PF lost elections in the lowveld but won elections in the highveld with concentrated resettlement schemes. An MDC Chipinge central election agent told me that in the 29th of March, 2008 election 'the resettlement farms cost us [the MDC to win Chipinge central]'. The posting of results outside the polling stations helped people to show where each party had support. Most scholars have confused the ZANU-PF March 29, 2008 loss of support in communal areas with a reversal of ZANU-PF's electoral support in resettlement schemes (see Scarnecchia *et al.* 2009). Ranger (2009:1) argues that the peasantry did not support Mugabe 'despite land re-distribution' in the March 29, 2008 election. Readings of results by polling stations can help clarify this misinterpretation otherwise ZANU-PF's substantive rural support was concentrated in the resettlement schemes.

The following cases based on qualitative interrogations reveal how some of the ordinary people became an extension of the patronage system.

Case A

Mr T comes from Ngaone in Chipinge district and occupies 10 hectares of fertile land on Glen View Lot 1. He used to work as a storekeeper and tailor at the Wattle Company in Chipinge town. Mr X resigned from his job and settled on the farm in 2002. Is this not an ordinary person who benefited from Fast Track? Further interrogation revealed that he stays on the farm under an informal arrangement with his son to safeguard the property. Mr X's son is a soldier currently stationed at KG 6, a military barrack in Harare and he applied for land through the army. As I have already indicated, the military was represented in the committee that made allocation decisions. Though Mr X is tilling the land, the offer letter is in the name of his son who serves at the headquarters of the army.

Case B

Mr Y occupies 15 hectares of land at Wedgehill farm and he has no formal employment and he has low income and poor assets. Yet Mr Y is a village head in the communal areas who got land because he is part of the ruling lineage. When the farm was allocated the DLC prioritised those in the ruling lineage, civil servants and war veterans. Some traditional authorities were prioritised over other ordinary families.

Case C

A former farm worker got five hectares of land at Wolfscrag farm. She is 76 years old and all her life she has worked at a farm near to her new home. How did she get the land? First, she joined land invaders but was excluded from the official confirmation of land. She got the land not because she was a priority of the DLC or as a way to show the impartiality of the committee but because she and a few others confronted the District Administrator and forced him to formalise their occupation. It may appear as an exceptional case but it shows the determination of the war veterans and DLC to push out ordinary people.

Why then do my results differ to a certain degree from those of Scoones *et al.* (2010)? There are five possible explanations. First, my findings on who got what during Fast Track differ because of different contexts. Whereas much of Masvingo province lies in agro-ecological zones 3 and 4 designated for extensive and semi-intensive farming, land reform dynamics in Chipinge were key in natural region 1 and 2 meant for specialised, diversified and intensive agricultural production. There were high value crops in Chipinge like macadamia, coffee, gums and wattle with good investments in infrastructure on farms. So much was at stake as one official from the Agritex department highlighted:

> We could not let others come and feast ... there was so much at stake, for example a macadamia plantation can sustain a family for 25 years by simply harvesting and selling. That is a fortune for a generation. So there is no way we could have let communal farmers and others just enjoy (Agritex official, Chipinge, 2011).

Second, there is a dominant culture of political and partisan control of processes in Chipinge given the local history of ZANU Ndonga.

Third, local dynamics like superimposition of processes from above such as the displacement of farm workers at Wedgehill farm affected the composition of beneficiaries. The state administrative procedures that were put in place to redistribute land on Wedgehill farm were open to patronage affecting the composition of beneficiaries on the farm, a situation which could have been different if the farm workers had been allowed to stay.

Fourth, in their summary of settler categories Scoones *et al.* (2010) do not single out the category of traditional authorities and spread war veterans across other categories.

Fifth, a slightly different methodology that focuses on those excluded from Fast Track brings out different dynamics. Based on my three study sites it will be creating a new myth to say that there was no substantial patronage in the administrative allocation of land and in the highly politicised farm invasion given the evident role of war veterans, civil servants and traditional authorities in land acquisition and the composition of beneficiaries thereof. Myths and realities may differ from place to place, which makes generalisation a poor way of understanding what is happening in resettlement areas in post-2000 Zimbabwe.

What was the motivation of seeking land?

Claims to land are multiple and overlapping and thus difficult to disentangle. For some, land provides an opportunity to expand production and accumulation; for others it is a process of reclaiming ancestral land in clear cases of restitution; for some it has political and historical significance; for others it is a source of livelihood in an era of crisis of livelihoods and for some it is a source of food security for their households or all of the above. Though motivations are diverse, in some cases, the vehicle to fulfil such motivations was politicised and took place through patronage connections. I illustrate these different claims using voices from fieldwork in the cases below.

Case A: Food security

I used to live in Musani communal area in the low veld where we subsisted on poor sandy soils. My family could hardly produce 50 bags of maize per hectare and we could only afford one decent meal a day. So when we heard the comrades encouraging people to join the land invasions as loyal party members we quickly joined so that we could get land that is highly productive. Now we can produce about 2 tonnes of maize a hectare and enjoy three decent meals a day with our children even having more meals. So the main driving force was to eradicate hunger in our household and that is what we have achieved (land reform beneficiary, Wedgehill farm, 2010).

Case B: Restitution

Our ancestors were the original traditional leaders in this area. So in 1974 our family was evicted by the white farm owner because we were laying claims to some parts of the farm for rain making ceremonies and other rituals yet the owner wanted to utilise all the land. So when our grandfather was chased he left to the mountainous areas of Samutsa where you can hardly produce anything. So when we heard of independence in 1980 we thought we would return to our land but we were let down. But when the second independence came in 2000, we seized the opportunity to return to our land. We had a traditional ceremony where we confirmed our headmen. So for us it is rather re-occupation than occupation. We have always wanted to come back home and we thank Gushungo [President Mugabe] for allowing this (Land reform beneficiary, Wolfscrag farm, 2010).

Case C: Straddling of livelihoods

I am formally employed in the Chipinge Rural District Council as an internal auditor but my salary is of no value. The money I was getting was not enough to fend for my family. So I decided to seek land as a way to supplement my meagre earnings through producing our own food. I was in good books with the Chief (Head of Department). Now I do not have to buy everything from the market and my family is much better. I combine my formal wage and what I get from the farm in order to eke a better livelihood. If my salary was enough maybe I would not have thought of seeking land but the life nowadays you need to be dexterous in order to survive. It is a tough economy (Land reform beneficiary, Wedgehill farm, 2011).

Case D: Political and historical

To me this is the completion of the liberation war. I joined the war to fight for land which had been stolen by the British imperialists. We were educated in the camps about the need to use the gun in order to reclaim our soil. When independence came, us guerrillas never really felt the independence because the land was still in the hands of the enemy and the elite politicians were dining and wining with them. Do not think that was exciting for us because we felt we had wasted our time fighting for nothing. Every time I would pass through this area my heart would ache seeing a white man still in charge. In 2000, we felt politically triumphant and whatever we do with the land it is now ours. We have completed the war through the Third Chimurenga (Land reform beneficiary, Glen View Lot 1, 2011).

Case E: Social safety net

I was nearing retirement and I saw Fast Track as an opportunity to get a retirement home. In fact, this is my second pension. The first pension was not of much value. Now my children and grandchildren have a place to call home where they can fend for themselves. As a pensioner without a home I really needed this land that is why I applied to my head of department [Agritex] and thank God I got the land (Land reform beneficiary, Glen View Lot 1, 2011).

The above voices point to the fact that not all claims to land in Chipinge district were political. However, what comes out clear is that some beneficiaries used patronage networks embedded in ZANU-PF to realise their different aspirations. For example; in case A, the beneficiary utilised her ZANU-PF party membership; in case B the beneficiary with autochthonous claims pays his allegiance to President Mugabe; in case C the motive to diversify livelihoods was realised through connections with the head of department who sat in the land committee a similar phenomenon with case E. As for case D I have explored how ZANU-PF manipulated anti-settler concerns by war veterans to legitimise Fast Track as an ongoing liberation movement.

I agree that there were some claims not based on party politics but this still does not underplay the importance of how the ZANU-PF ruling elite manipulated such to further a political agenda to stay in power. By realising their claims through Fast Track the above beneficiaries became embroiled in a web of patronage networks. The manipulation of local grievances that war veterans and chiefs had, has led some scholars to say the process was wrong but the issues were right (see Mamdani 2008) which relegates and underplays the impact of the state's authoritarian, repressive and exclusionary politics.

2000-2011: A case of patronage

Whereas the land redistribution programme was started as pro-poor in the 1980s and drifted toward pro-elite in the 1990s, neither is adequate to describe Fast Track. The fact that beneficiaries are dominated by civil servants, war veterans and traditional authorities on my three study sites show that there is more to the process than meets the eye. I therefore argue that Fast Track in this context was a pro-ruling party process with the majority of beneficiaries having access to political networks and partisan state networks. I illustrate my argument with the four points below.

First, the DLC in charge of allocating land to new beneficiaries or formalising occupations was composed of ZANU-PF functionaries such as the district chairperson of the party. This is consistent with Marongwe (2010, 8) who argues in reference to A2 farms that, 'political considerations determined the outcome of beneficiary selection' because of the partisan and militarised composition of Land Identification Committees. According to Alexander (2006) land committees were put in place in order to ensconce politicised and partisan interests. The process was evidently exclusionary of other political parties. Neither ZANU Ndonga nor the MDC were part of the decision-making process. As already indicated, the war veterans and ZANU-PF local agents were the domineering characters. Moreover, it can be argued that the civil servants in the DLC, being head of departments, were serving with the approval of the ZANU-PF ruling elite. I have demonstrated how magistrate Samuel Zuze abused his position and how the Head of Agritex kept his job despite his low qualifications because of support from the ZANU-PF ruling elite and McGregor (2002) and Hammar (2003) also explain how certain civil servants' appointments became highly politicised in post-2000 Zimbabwe.

Second, the replacement of farm workers by new occupiers on Wedgehill farm by the DLC indicated that Fast Track was far from targeting the 'ordinary' people without political connections. The farm workers were replaced mostly by civil servants who were well connected.

Third, most 'ordinary' families scattered across the 30 wards of Chipinge district had to put their application to the ZANU-PF chairperson or its representative of that ward where it had no clear structures. I have already alluded to how difficult it was for some overt opposition supporters to hand in their applications or be considered for land reform. The process was advantageous to most people connected to ZANU-PF local elites, who could easily hand in their applications, and get a favourable recommendation from the party's district chair, who played a central role in the Tuesday DLC meetings.

Fourth, even during the land occupations my case of Wolfscrag shows that the involvement of the ZANU-PF local leadership in the mobilisation of the masses to invade the farm gave an impression that land occupation was a ZANU-PF driven project. This was buttressed by the central role played by ZNLWA members who are perceived allies of the ZANU-PF ruling elite. This perception prompted some war veterans aligned to ZANU Ndonga to shun the process as I have indicated. Moreover, some traditional authorities became appendages of the ZANU-PF ruling elite given the symbiotic relationship revived from the 1990s, and very notably in a political alliance since 2000, signified by their role in advancing party interests. All these factors had an effect on who eventually got land.

My observations above are not to entirely dismiss that for some Fast Track beneficiaries the idea was not to shore up ZANU-PF's waning political support. Fontein (2006,244) has argued that:

> The ruling party's far narrower nationalist imagination of recent years has not necessarily obliterated or limited the nationalist imagination of others. For ... chiefs, war veterans who have felt themselves excluded since independence, the current politics of exclusion may seem more like a renewed opportunity for their inclusion.

My emphasis is that their inclusion was necessitated by an authoritarian ZANU-PF governing elite using land as a source of patronage to remain in power. Within this matrix, the different actors could be manipulating each other in order to shape each other's interests but the ruling elite maintains coercive power and the legal right to expel Fast Track beneficiaries who fail to show political loyalty.

Conclusion

I therefore conclude that, on my three A1 study sites, due to corrupt administrative practices in land allocation and the politicised and autochthonous nature of the land invasion, civil servants, war veterans and traditional authorities were the major beneficiaries of Fast Track. Though these beneficiaries had diverse claims to land that were not entirely political the vehicle to realise such was linked to political loyalty, connection and patronage under the ZANU-PF umbrella. Political patronage extended beyond gaining access to land as land reform beneficiaries have to continuously show political loyalty through 'correct' political allegiance in highly politicised resettlement landscapes. The authoritarian and partisan nature of the state excluded many farm workers, white commercial farmers, ZANU Ndonga and MDC supporters and some ordinary people in the communal areas in Chipinge who never had a chance. A new agrarian structure has emerged but one shaped mainly by socio-political dynamics rooted in the ZANU-PF ruling elite.

References

Alexander, J. 1994. State, peasantry and resettlement in Zimbabwe. *Review of African Political Economy*, 21 (61), 325–345.

Alexander, J. 2003. 'Squatters', veterans and the state in Zimbabwe. *In*: Hammar, A. *et al.*, eds. *Zimbabwe's unfinished business: rethinking land, state and nation in the context of crisis.* Harare: Weaver Press, pp. 83–117.

Alexander, J. 2006. *The unsettled land: state-making and the politics of land in Zimbabwe 1893-2003.* Oxford, Harare and Athens: James Curry, Weaver Press, Ohio University Press.

Buka, F. 2002. A preliminary audit report of land reform programme. Buka report. Harare. Government Printers.

Central Statistics Office (CSO). 2002. Zimbabwe's preliminary report. Harare: Central Statistical Office.

Chakaipa, S. 2010. Local government institutions and elections. *In*: Visser, J *et al.*, eds. *Local Government reform in Zimbabwe: a policy dialogue.* Community Law Centre. University of the Western Cape,pp. 31–68.

Chambati, W. & Moyo, S. 2004. *Impact of land reform on farm workers and farm labour processes.* Harare: African Institute for Agrarian Studies.

Chaumba, J., Scoones, I. & Wolmer, W. 2003. New Politics, New Livelihoods: Agrarian change in Zimbabwe. *Review of African Political Economy*. 30 (98), 585–608.

Chimhete, C. 2010. Illegal settlers threaten Chipinge magistrate. *The Standard*, 26 April.

Commercial Farmers Union (CFU). 2000. Farm invasions update. Available from http://www.zimbabwesituation.com/cfu_15May.html [Accessed on 29 January 2010].

Dekker, M. and B. Kinsey. 2011. Contextualizing Zimbabwe's Land Reform: Long-term Observations from the first generation. *Journal of Peasant Studies*, 38(5) this collection.

Fontein, J. 2006. Languages of land, water and 'tradition' around Land Mutirikwi in southern Zimbabwe. *Journal of Modern African Studies*, 44 (2), 223–249.

Fontein, J. 2009. We want to belong to our roots and we want to be modern people. New farmers old claims around Lake Mutirikwi, Southern Zimbabwe. *African Studies Quarterly*. 10 (4).

Government of Zimbabwe (GoZ). 1986. Land use in Zimbabwe. Ministry of Lands, Agriculture, and Rural Resettlement. Harare, Zimbabwe.

Government of Zimbabwe (GoZ). 1998. Land reform and resettlement programme phase 11. Harare, Zimbabwe.

Government of Zimbabwe (GoZ). 2000. Extraordinary government gazette notice 233A of 2000. Harare: Zimbabwe.

Government of Zimbabwe (GoZ). 2001. Land reform and resettlement programme: revised phase 11. Ministry of Lands, Agriculture and Rural Resettlement: Harare.

Hammar, A. 2003. The making and un (mas) king of local government in Zimbabwe. *In*: Hammar, A *et al.*, eds. *Zimbabwe's unfinished business: rethinking land, state and nation in the context of crisis*. Harare: Weaver Press, pp. 83–117.

Kubatana. 2007. Zimbabwe Liberators Platform (ZLP). Available from http://www.kubatana.net/html/sectors/zim026.asp [Accessed on 13 August 2010].

Kinsey, B. 1999. Determinants of rural household incomes and their impact on poverty and food security in Zimbabwe. A report prepared under the Zimbabwe Rural Household Dynamics Project. University of Zimbabwe/ Free University, Amsterdam.

Mamdani, M. 2008. Lessons of Zimbabwe. *London Review of Books*, 30 (23), 17–21.

Marongwe, N. 2008. Interrogating Zimbabwe's Fast Track land reform and resettlement programme: a focus on beneficiary selection. PhD thesis. University of *the* Western Cape.

Marongwe, N. 2010. Institutional and process issues of settler selection in Zimbabwe's land occupations driven Fast Track land reform. Submitted to Development and Change, 8 April 2010.

Matondi, P., B., Khombe, T.C., Moyo, N.,R., Matondi, M and Chiweshe, M. 2008. The land reform and resettlement programme in Mangwe district, Matebeleland South Province. University of Zimbabwe: Centre for Rural Development.

Matondi, P. B. 2011. Zimbabwe's Fast Track land reform programme (forthcoming). London: Zed Books.

Matyszak, D. 2011. *Formal structures of power in rural Zimbabwe*. Harare. Research and Advocacy Unit.

McGregor, J. 2002. The politics of disruption: war veterans and the local state in Zimbabwe. *African Affairs*, 101 (402), 9–37.

Moyo, S. 2007. The land question in southern Africa: a comparative review. *In*: Ntsebeza, L *et al.*, eds. *The land question in South Africa*. Cape Town: Human Sciences Research Council Press,pp. 220–245.

Moyo, S. 2009. *Agrarian reform and prospects for recovery*. Harare: African Institute of Agrarian Studies.

Moyo, S., Chambati, W. and Murisa, T. 2009. *Fast Track land reform: trends and tendencies report of the baseline studies*. Harare: African Institute of Agrarian Studies.

Moyo, S. 2011. Land concentration and accumulation after redistributive reform in post settler Zimbawe. *Review of African Political Economy*, 38 (128), 257–276.

Neuman, W.L. 2003. *Social research methods qualitative and quantitative approaches*. Boston: Allyn and Bacon.

Palmer, R. 1990. Land reform in Zimbabwe, 1980–1990. *African Affairs*, 98, 163–181.

Raftopoulos, B. 2003. The state in crisis: authoritarian nationalism, selective citizenship and distortions of democracy in Zimbabwe. *In*: Hammar, A *et al.*, eds. *Zimbabwe's unfinished business: rethinking land, state and nation in the context of crisis*. Harare: Weaver Press.

Ranger, T. 2009. Lessons of Zimbabwe. *Concerned African Scholars*, 82. Available from http://concernedafricascholars.org/bulletin/82/ranger/ [Accessed on 28 May 2011].

Sachikonye, L.M. 2003. From growth with equity to Fast Track reform: Zimbabwe's land question. *Review of African Political Economy*, 30 (96), 227–240.

Sadomba, Z.W. 2011. *War veterans in Zimbabwe's revolution: challenging neo-colonialism, settler and international capital*. Oxford, Harare: James Curry, Weaver Press.

Saxon, T. 2011. Apostles invade Chipinge farm. *The Zimbabwean*, 08 Jun.

Scarnecchia, T and 33 others. 2009. Lessons of Zimbabwe. *London Review of Books*, 31 (1).

Scoones, I., Marongwe, N., Sukume, C., Mavedzenge, B., Murimbarimba, F. & Mahenehene, J. 2010. *Zimbabwe's Land Reform: myths and realities*. Harare: Weaver Press.

Sender, J. & Johnston, D. 2004. Searching for a weapon of mass production in rural Africa: unconvincing arguments for land reform. *Journal of Agrarian Change*, 4(1/2), 107–142.

Sithole, M. 1993. Is Zimbabwe poised on a liberal path? The state and prospects of the parties. *A Journal of Opinion*, 21 (1/2), 35–43.

United Nations Development Programme (UNDP). 2002. *Land Reform and resettlement: assessment and suggested framework for the future: Interim mission report*. New York: UNDP.

Utete, C. 2003. Main report to the President of the Republic of Zimbabwe. Vols 1 and 2. Harare: Zimbabwe.

Zamchiya, P. 2007. 'How many farms is enough?'. *In*: S. Moyo *et al.*, eds. *The day after Mugabe: prospects for change in Zimbabwe*. London: Africa Research Institute, pp. 78–82.

ZHR NGO Forum. 2010. Land reform and property rights in Zimbabwe. Harare, Zimbabwe. Available from http://www.hrforumzim.com/special_hrru/land-reform.pdf [Accessed 2 February 2011].

Phillan Zamchiya is a Doctorate candidate in International Development at Oxford University in the United Kingdom. He holds an MPhil degree (cum laude) from the Institute of Poverty, Land and Agrarian Studies (PLAAS) based at the University of *the* Western Cape. He has published a number of book chapters on policy options on land and agrarian reform in Southern Africa. His academic interests are in politics, land and agrarian change. Phillan Zamchiya is also a former President of the Zimbabwe National Students Union (ZINASU).

Land, graves and belonging: land reform and the politics of belonging in newly resettled farms in Gutu, 2000–2009

Joseph Mujere

The return to ancestral lands has been at the centre of the land reform rhetoric in Zimbabwe. This argument is premised on the fact that many communities were displaced from their ancestral lands during the colonial period hence they saw the Fast Track Land Reform Programme (FTLRP) in 2000 as an opportunity for them to 'return' to their old homes. This article explores and analyses the issues surrounding land disputes in one village model A1 after the land reform programme and the role played by the regime of traditional authorities in determining how belonging is negotiated. It also analyses the conflicts between autochthons and migrants over the control of the new resettlement areas and over the authority of village heads and chiefs. Claims to land based on ancestral graves and autochthony are also analysed in view of the power of political authorities in land allocation. This paper offers an analysis of the intricacies of land reform in the newly resettled areas and examines the interface between politics and traditional authority on how belonging is negotiated in these contexts. The article is largely based on qualitative interviews with resettled farmers in Gutu, informal interactions as well as personal observations.

Introduction

Zimbabwe's accelerated land reform and resettlement programme which is also known as the Fast Track Land Reform Programme (FTLRP) was launched in 2000 when war veterans and peasants invaded white owned commercial farms. The early phase of this land reform programme was chaotic and marked by a lot of violent displacement as war veterans and peasants occupied farms spontaneously. Studies of this land reform have thus emphasised the chaotic manner in which the process was carried out, the new forms of livelihoods it provided the landless peasants and also its impact on agricultural productivity and on the displaced former farm workers (Sachikonye 2004, Chaumba *et al.* 2003, Moyo 2000). Little attention has however been paid to institutional authority in the resettled farms, especially the regime of traditional authority. The redistributive nature of the land reform has brought new

This work was in part supported by the Livelihoods after Land Reform small grants fund (http://www.lalr.org.za/zimbabwe/zimbabwe-working-papers-1). I wish to thank Ian Scoones, Joost Fontein, Francesca Locatelli and the anonymous reviewers for their helpful comments on the earlier versions of the article.

trajectories of belonging which need to be problematised. The newly resettled farms have seen the conflation of people from different areas and different traditional authorities which transformed the resettlement areas into some kind of a melting pot. Moreover, the influx of people from different areas into the newly resettled farms, especially in the more communal A1 Model has engendered new politics of belonging which have widened the division between those who see themselves as autochthons or the owners of the land and those they view as outsiders. Furthermore, chiefs and headmen have also sought to reconfigure their traditional boundaries and expand their authorities into these newly established resettlement areas which they consider to be their lost ancestral lands. Consequently, *matongo* (old homes), *mapa* (ancestral burials), rivers, mountains and other traditional markers of boundaries have assumed a new significance in the land reform aftermath. The land reform has thus brought with it not only new forms of livelihoods but also new and often contested forms of belonging as well as the intensification of boundary disputes.

Using the case of Gutu District in the north of Masvingo Province this article argues that the return to ancestral lands has been one of the major issues at the centre of the land reform in Zimbabwe. A number of traditional authorities have seen the land reform as an opportunity to 'return' to their ancestral lands, which were alienated during the colonial period and to redefine their boundaries. In most cases traditional authorities have used old homes and graves to lay claim to the areas opened up by the land reform. The article also explores the role of traditional authorities in the resettled areas and how their powers have intersected with or have been undermined by the government sponsored village committees and political elites.

Since not much has been written on the regime of traditional authority in the newly resettled areas oral history was the mainstay of this research. The article tries to recover the voices of traditional authorities in the land reform discourse and examine how they have negotiated their boundaries and recast their authority in the new resettlements. The article is largely based on oral interviews with traditional authorities, newly resettled farmers and government officials. Structured and semi-structured interviews were held with the aim of capturing the experiences of people who benefited from the land reform programme together with those who consider themselves to be the autochthons of the area.

The first section conceptualises the interplay between land and politics of belonging and gives an overview of the FTLRP, land restitution and the politics of belonging. It discusses land reform in the context of land reclamation, autochthony and the re-establishment of traditional authorities in the newly resettled farms. The second section uses ethnographic data gathered in Gutu to explore the intricacies of land reclamation, boundary disputes and the authority of traditional authorities in the aftermath of the FTLRP. It also analyses the role of village committees in the management of resettlement areas and how this has led to the marginalisation of traditional authorities. It observes that although the FTLRP has arguably left out traditional authorities in the management of the new resettlement areas, the traditional authorities have largely been quite resilient in their efforts to reclaim their lost lands, renegotiate traditional boundaries and recast their authority in these areas. The land reform altered local traditional boundary dynamics and also fuelled traditional authorities' desires to reclaim their lost lands.

Land reform and the politics of belonging: an overview

Belonging is a relational concept which entails among other things attachment to a group, place or other category. Local notions of belonging especially among peasants in Africa tend to revolve around religion, autochthony and ownership of land and among other local attachments. It is therefore about being locally embedded. Belonging also raises the question of access to resources such as land. The land and belonging nexus has thus seen the rise of debates over autochthony-based claims to land and the conflicts that arise from counter claims. It is therefore important to explore this nexus in the light of the land reform movements in Africa. According to Geschiere and Nyamnjoh (2000, 425) ethnicity and autochthony, 'are equally capable of arousing strong emotions regarding the defence of home and ancestral lands, but since their substance is not named they are both more elusive and more easily subject to political manipulation'. Leonhardt (2006, 70) contends that, 'autochthony is not a coherent body of principles on which rights are based. It is a mystification of ancestry, a method used for the purposes of magically extracting wealth from the state.' It is also based on the contestable claim of being the first comers and being the sons and daughters of the soil as opposed to strangers, aliens or late comers. Political elites thus use autochthony and narrow citizenship policies for political expedience.

It is all the more striking that in spite of it being seemingly embedded or primordial, autochthony is a very fluid form of identity. This makes the process of defining who is an autochthon and who is a stranger a very difficult task. Since identities are fluid, claims to autochthony are often met with counter claims. In the end autochthony is by no means cast in stone as 'strangers' can also claim autochthony thereby turning the former autochthons into strangers. It is in this light that Geschiere and Jackson (2006, 1) argue that, 'belonging turns out to be always relative: there is always the danger of being unmasked as "not really" belonging, or even of being a "fake" autochthon'.

The autochthon-stranger dialectic is largely played out on conflicts over control of natural resources such as land. Lentz (2007, 37) argues that land and land rights play an important role in the politics of belonging in Africa due to the fact that rights to land 'are intimately tied to membership in specific communities'. Hence the scarcity of land has the effect of increasing the need to identify those who 'really belong' and those who are late comers and therefore have limited rights to the land. Control over land therefore becomes a sign of the extent to which one belongs (Lentz 2006, 1). According to Lewis (2004), 'the vagaries of belonging(ness) are crafted in the context of formal rights and entitlements'. The scramble for resources and political manipulations have therefore led to the crystallisation of the divisions between autochthons and allochthons in Africa thereby crippling the processes through which immigrants could be integrated into a community and enjoy the same rights as the autochthons (see Kopytoff 1989).

Using the case of the Anglophone region of Cameroon, Konings (2001, 117) argues that, land was not the only reason for the development of antagonistic relations between autochthons and strangers. The latecomers' success in agriculture, trade and other entrepreneurial activities also contributed to the strained relations as the autochthons became jealous of the success of the immigrants. Politicians exploited these localised conflicts to further their own agendas of entrenching themselves. According to Lentz (2007, 47), 'in many cases, it is young men who invoke powerful discourses on autochthony, much more so than their fathers, who continue to insist that well-intentioned strangers should not be refused land if they need it for subsistence'. This is often a result of petty jealousies over the success of the immigrants and also competition over resources.

Belonging also entails rootedness or being attached to a place. This involves an attachment to place, being an indigene or having roots in a certain place as opposed to being a stranger. Malkki (1992), however, suggests that the idea of being rooted needs to be problematised as it fails to appreciate people's ability to construct new notions of belonging when they get 'uprooted' or migrate. As she argues, 'there has emerged a new awareness of the global social fact that, now more than perhaps ever before, people are chronically mobile and routinely displaced, and invent homes and homelands in the absence of territorial, national bases – not in situ, but through memories of, and claims on, places that they can or will no longer corporeally inhabit' (Malkki 1992, 24). This captures the need for scholars to escape the pitfalls of always attaching people to place and nations to territory.

Shipton (2009) used the case of the Luo in Kenya to show that African ideologies about land and attachment have often clashed with governments' land reform policies aimed at titling land and making it possible to use it as collateral in applying for agricultural loans. It is this Western model of land tenure and credit which Shipton sees as threatening the Luo people's attachment to their ancestral lands and the future of their belonging. Particularly important is the debate on the applicability of the concept of freehold tenure in Africa and whether land can be bought or pledged as collateral for a loan and consequently forfeited if the debtor defaults. Shipton (2009) asserts that the freehold-mortgage system of credit does not suit the Kenyan people and arguably many other rural African communities whose tenure is communal. As he puts it, the mortgage 'threatens to separate people in rural areas from home, from kith and kin, and from ancestral graves, with all that these mean' (Shipton 2009, ix). The nature of a number of African communities' attachment to land thus makes any land reform aimed at making formerly communal land freehold difficult to implement since this ultimately makes it possible to mortgage the land and by extension 'mortgag[e] the ancestors'.

In a recent study on land reform and land restitution in South Africa, Deborah James (2007) shows the salience of ancestral graves on people's claims to land and demands for restitution. The study illustrates the salience of graves and old homes in claims to land and negotiation of belonging in Africa. In her study, attachment to graves and old homes allows the Doornskop community to articulate their belonging and to reclaim their ancestral lands in the post apartheid land reform in South Africa (James 2007,16). Land restitution is thus here based on 'claiming graves' and old homes. As she puts it, 'where stories of lives lived on the soil have been told or features of the landscape invoked to endorse claims to particular farms or clusters of territory, these tend to highlight moments in the experience of the claimants' or tellers' own childhoods or to emphasise the burial sites of their parents rather than stressing long-term occupancy of the land' (James 2007, 16). It is quite evident that James's study seeks to show that restitution is easier for those people with titled land. 'The most obvious beneficiaries of land reform', she argues, 'were to be former title-holding landowners who had been relocated to the homelands during apartheid's "Black Spots" removals. Their property rights have been clearer cut and rather easier to assert and retrieve through restitution than those of other claimants' (James 2007, 8). Those claims that are based on ancestral graves without backing of any legal documents are however complicated by the fact that there are always counter claims by other communities who are also able to point to their own ancestral graves in the same areas claimed by others.

Just as in many African societies belonging in Zimbabwe, from the pre-colonial period, has often been about attachments to ancestral lands, graves and old homes *(matongo)*. Spierenburg's (2004) work on migrations and land reforms in Zimbabwe examines the conflicts over land between autochthons and strangers *(vatorwa)* which emerged as a result of the government initiated and donor funded Mid Zambezi Rural Development Project in the Dande communal areas. The aim of the project was to redistribute land and ensure increased production (Spierenburg 2004). Spierenburg (2004) examines the Dande people's resistance to the government land reforms in the 1980s and 1990s and the role of the *Mhondoro* spirits (mediums of royal ancestral spirits) in the allocation of land and integration of the migrants in their spirit province. The *Mhondoro* spirits were in the forefront in expressing discontent with the land reforms. The mediums claimed that traditionally land belonged to the royal *Mhondoro* spirits and before any strangers could be allowed to settle they had to be presented to the *Mhondoro* spirit mediums. The *Mhondoro* would then do the rituals of accepting the immigrants and integrating them into the community. The owners of the land were thus the *Mhondoro* spirits whose mediums had the power of allocating it, a role which was now being usurped by the government technocrats imposing land reform. The land shortages generated by this land reform project led to conflicts between the autochthons and the migrants as the former saw the migrants as benefiting from the project at their expense and 'stealing' their land. These conflicts often spilled into witchcraft accusations or threats of use of witchcraft as the autochthons were often disgruntled with the strangers being allocated more land.

Traditional authorities and the FTLRP in Zimbabwe

Zimbabwe's Fast Track Land Reform Programme has attracted a lot of scholarly attention since its beginning in 2000. Much focus has however been on the *jambanja* (chaos) in which it was carried out: the violence, the massive displacements, and also the new livelihoods that it availed to the peasants (see Rutherford 2001, Chaumba *et al.* 2003, Sachikonye 2004).[1] There has however been little debate on people's claims to what they consider to be their lost ancestral lands. In a number of cases starting with that of Chief Svosve in Marondera District, traditional authorities have made attempts to re-establish their control over those lands from which they were alienated during the colonial period, in some instances even claiming restitution. As noted by Scoones *et al.* (2010,165) chiefs took advantage of the new areas of land being opened up by the land reform to 'exert control over these new territories, often claiming that past ancestors had occupied the areas and that the land reform had been a process of restitution, and that they, through evidence of burial sites and past settlement, along with religious sites where spirits resided, had new authority'. This has often led to land and boundary disputes between traditional authorities as claims and counter claims on particular pieces of land emerged.

---

[1]The FTLRP in Zimbabwe has received greater scholarly attention elsewhere and it is not my intention to recount it here save to provide a nuance that has so far not been given much attention.

There has been a general assumption that the FTLRP has led to the marginalisation of traditional authorities as the management of the new resettlement areas has been the responsibility of village and district land committees as well as war veterans and 'Base commanders'.[2] This has meant that traditional authorities have been left out on the margins in as far as the management of the newly resettled farms is concerned. Be that as it may, some traditional authorities have been making efforts to recast their authority in the areas, renegotiate boundaries and reclaim alienated ancestral lands. The FTLRP has thus provided traditional authorities with the opportunity to pursue an agenda akin to land restitution as they have been making a number of claims both substantiated and unsubstantiated over the new resettlements which they view as their *matongo* (old homes). According to Marongwe (2003,174), 'despite the fact displacement of communities took place several decades ago, there was still an expectation that the government would facilitate their return to their original home areas under the current land reform programmes'. Thus the non-recognition of such historical land claims by the policymakers coupled with the communities' desire to return to their traditional land alienated during the colonial period was one of the major factors that fuelled land occupations (Marongwe 2003,171). The return to ancestral lands is arguably at the centre of most land reform programmes allowing restitution in Third World countries (see James 2007). Communities' claims to ancestral lands have thus shown the centrality of *matongo* and *mapa* (ancestral burial sites) in the Shona people's negotiation of belonging in the newly resettled farms. Mazarire (2008, 2009) has argued that the struggles over land in Chishanga in Masvingo Province in Zimbabwe have revolved over the people's return to their ancestral *gadzingo* (traditional headquarters). A *gadzingo* was at once a clan's traditional headquarters, a burial ground for important figures, a place of ancestral veneration as well as a stronghold and a place of refuge during attacks (Mazarire 2009,7). This thus made ancestral lands such as the *gadzingo* an important feature in Shona people's construction of their *nyika* (territory).

The colonial displacements which led to the carving out of large farms by a few white farmers alienated many locals from their homes. The completion of this displacement and emplacement circle with the FTLRP has nevertheless created a crisis of belonging leading to boundary disputes. This is further complicated by the fact that very few people actually 'returned' to their original homes during the FTLRP making the newly resettled farms a melting pot full of people who could at best be described as outsiders living among very few autochthonous people.

The FTLRP can also be argued to have led to the extension of the traditional authorities' power into the resettlement areas. Formerly, the government through the Ministry of Local Government Rural and Urban Development argued that the provisions of the Chiefs and Headmen Act (1982) stipulated that chiefs could only exercise authority in communal areas meaning that farms and resettlement areas were out of bounds for the traditional leaders (Mubvumba 2005, 46). However, the FTLRP brought a new trajectory to the role of traditional authorities in the resettlement areas providing chiefs and headmen with an opportunity to seek restitution, reclaim their ancestral lands or to recast their authority in these formerly alienated lands. It is also important to note that some traditional authorities fighting for restitution do so not because they wish to be productive on the land but because

---

[2]Base commanders were usually war veterans who led farm invasions and imposed a regime of control on the newly acquired farms.

of the fact that 'some people only wanted to get back land that belonged to their fore-fathers and not to necessarily commercially farm' (Matondi 2010, 7). Chakanyuka (2007) argues that the FTLRP provided the traditional authorities who are the custodians of cultural heritage sites with the opportunity to reclaim their lost ancestral lands and sacred places and in some instances demand restitution. As she puts it,

> for the government, FTLRP was meant to return land to the dispossessed indigenous people yet for most rural communities it was not only access to fertile land but also restorations of cultural links with the land of their ancestors. Hence from the beginning, the FTLRP was accompanied by claims based on ancestral links to the land. (Chakanyuka 2007, 4)

In some instances the land reform brought conflicts between those who view themselves as the autochthons and those they see as outsiders or strangers. Some settlers do not respect the traditional authorities in the area because they hail from different areas and therefore do not have a strong sense of belonging in the resettlement area. Such a scenario has brought the farmers who do not have a direct link with the former occupants of the land and therefore do not have any ancestral links with the land into a collision course with those who claim direct links with original owners of the area and thus feel they are custodians of the land, sacred places, and ancestral graves.

Village Committees and the marginalisation of traditional authorities in Gutu

Gutu is one of the most overcrowded districts in Zimbabwe. It lies in Masvingo Province which tops the list of the most overcrowded communal areas in Zimbabwe. It falls in Natural Regions (agroclimatic regions) 3, 4 and 5, with the largest part of the district being in Region 5 which is characterised by poor soils and low rainfall. Not surprisingly, most of the lands that were alienated during the colonial period fall in Natural Regions 3 and 4, which have comparatively better soils and better rainfall. The new resettlement areas in Gutu are in wards 1, 7 and 32. At present there are 27 traditional authorities in Gutu, nine of which are chiefs and 18 of which are headmen. The majority of the chiefs and headmen are of the *Gumbo Madyirapazhe* clan. Chief Makore and Chief Chiwara are of the *Moyo Duma* clan while Serima is of the *Gushungo* clan. Some of the traditional authorities were moved from their ancestral lands when their lands were alienated during the colonial period or when their chieftaincies were abolished by the colonial government during the time their areas were alienated.

When large tracts of land such as the Eastdale Estate, formerly owned by Willoughbys Consolidated Company, were alienated, this led to massive displacements of people. The affected people were either moved to new areas together with their traditional authorities or were placed under other traditional authorities. Chief Serima and Headman Denhere of the Gutu Dynasty are the main claimants of the resettlement areas in Eastdale Estate. Both traditional authorities claim authority over much of Eastdale based on the status quo before the colonial land alienation. They have sought to use graves, old homes and landscape features such as mountains and rivers to substantiate their claims. The area covering resettlements in the former Chatsworth Estate, which is in ward 32, is being claimed by Headman Gadzingo as well as Chief Serima.

Apart from the Chiefs and Headmen who are making efforts to extend their spheres of influence into the resettlement areas by claiming that they are their ancestral lands, there is also another group of traditional authorities whose chieftainships were abolished during the colonial period. These traditional authorities are seeking to reclaim their alienated ancestral lands and have their authority resuscitated. Among these groups making efforts to take advantage of the FTLRP are Musarurwa, who claims to have been removed from the Chindito area and had his headmanship abolished in the colonial period, and Makumire, who claims to have been removed from Eastdale and now wants to return and re-establish his headmanship. However, the government has put a moratorium on the resuscitation of abolished chieftainships and refuses to entertain any attempts to have these abolished chieftaincies reclaim their positions and land for fear of opening up the way for many such claims (Interview with N. Muzenda, 25 August 2009). This has left people like Musarurwa and Makumire with little room to manoeuvre. As Chakanyuka (2007, 88) puts it, 'the FTLRP policies were not sensitive to the demands by communities of returning back to the ancestral lands but were more concerned with landlessness'.

Before the FTLRP, the white commercial farms were out of bounds for traditional authorities as chiefs and headmen's authority was confined to communal and resettlement areas. Since chiefs and spirit mediums were left out of the administrative hierarchies of the commercial farms; they continued to be alienated from both their ancestral lands and their sacred sites like ancestral burials. The FTLRP therefore provided people with an opportunity to have access to their sacred places from which they were alienated during the colonial period and also to re-establish their authority on the land and with the people. In spite of such efforts, however, the government seems to have continued with its policy of sidelining traditional authorities, choosing instead to work with elected village committees and councillors. Hence, traditional authorities have had to compete for space with village committees as far as the administration of the resettlement areas is concerned.

Village committees, which are also known as 'seven member committees' because they are composed of seven office bearers, have been the mainstay of institutionalised management of the resettlement areas. These committees report to the District Lands Committee chaired by the District Administrator. The District Lands Committee has the responsibility of identifying 'vacant land' and allocating it to other beneficiaries. It also has the power to withdraw the offer letters given to plot holders in the event of a violation of the offer conditions or absenteeism from the plot. According to Scoones et al. (2010,165),

> as people settled into their areas, the *jambanja* period institutions of base commander and committee of seven still had important roles to play. In addition to negotiating with technocratic authorities of the state around land-use planning, they increasingly had to articulate with other forms of 'traditional' authority – notably chiefs and their headmen, as well as spirit mediums, rain messengers, church leaders and others.

The legitimacy of all these new forms of authority was contested at various levels. While the village committees saw themselves as the institutions in charge of the resettlement areas because they represented the District Lands Committee, the traditional authorities challenged this based on the fact that they owned the land before the colonial land alienation.

The new village committees strongly resemble the Village Development Committees (VIDCOs) created in the early 1980s. According to Goebel (2005,106), 'VIDCOs were to be secular and democratic, a departure from the traditional local institutions, which were based on ancestral religion and lineage membership.' She further adds that it was quite easy to impose this new structure in the resettlements because the areas were composed of people from many different home areas and also because traditional authorities usually remained in the communal areas (Goebel 2005, 106). By the same token the government has used the same ideology in the management of the village models in the newly resettled areas. According to Goebel (2005, 107), 'the village chairman is elected by democratic vote and can be deposed if performance is unsatisfactory. The job of the chairman is to channel grievances or issues of dispute from the settlers to the resettlement officer and take information given by the resettlement officer back to the settlers.' These village committees however work with almost complete disregard for traditional authorities. A member of the Uitcom 1 Village Committee in Gutu explained the village committees' relationship with traditional authorities thus:

> Village Committees are an elected body mandated by the government through the District Lands Committee to manage the resettlement areas. Chiefs and headmen's powers are confined to communal areas where authority is hereditary and lineage based. We are more concerned with development and productivity than the reclamation of ancestral lands, sacred places and the powers of traditional authorities (Interview with Mr Gumbo (pseudonym) 10 August 2009).

This illustrates the kind of importance placed on village committees and how this has affected the functions of chiefs and headmen in the new resettlement areas. Because village committees see themselves as functioning outside traditional authorities' spheres of influence they have continued to work with complete disregard for the authority of traditional leaders.

Silverdale, Uitcom, Allenbale and Lionsdale which are all located to the north of the district are each divided into three villages. Each of these farms is managed by a seven-member village committee. In most cases the village committees are composed of people who are well known in the village and often are first comers in the farms. In some instances, these people will also be influential war veterans or politicians. In the end they combine their political clout and the authority they wield as elected members of the village committees to command respect in the villages and determine policy. In the case of Silverdale, Mr Manenji Murodhini, being one of the first settlers in Silverdale farm has used his influence as a first comer to assume the unofficial position of the farm chairman (Interview 12 September 2009). He thus heads the committee which oversees the work of all the three village committees in Silverdale. This is however outside the norm as other farms such as Uitcom and Allenbale do not have a farm committee or a chairperson as they only work with a village committee. Yet because of his influence as a first comer in Silverdale and also as a person with some political clout, Mr Manenji Murodhini continues to maintain his position as de facto head of the farm. Village 1 is under R. Chizema, village 2 is under Gotore while village 3 is headed by E. Chara. All these village chairpersons happen to belong to a special group of first comers who claim to have participated in the *jambanja* and therefore claim to have a bigger sense of belonging than the late comers.

Similarly, Uitcom Farm is divided into three villages which are all managed by seven-member village committees which report to the District Lands Committee.

The majority of the people who settled in the three villages in Uitcom came from ward 18 which covers the Vhunjere and Zinhata areas as well as the Dewure Purchase Areas (Interview with E. Pugede 13 July 2009). War veterans, prominent ZANU-PF (Zimbabwe African National Union- Patriotic Front) activists and members of the ZANU-PF Youth Wing largely dominate the village committees and determine policy in the villages. For example, Mr Chamwaura, the Base Commander of Eastdale Estate, still holds a considerable amount of power and is consulted by village committees in cases of boundary and other land disputes (Interview with E. Pugede 13 July 2009). Mr Chamwaura has managed to combine his position as a liberation war veteran and the fact that he was the leader of the first group to occupy Eastdale to maintain a position of authority. Though the position of Base Commander is not in the formal government structures of management of resettlements he has managed to use his considerable political influence to accumulate wealth and to retain a measure of influence. Though the government is against multiple farm/plot ownership Chamwaura is thought to have more than one plot as well as a large herd of cattle 'inherited' from the former white farmer (Interview with E. Pugede 13 July 2009).

The village committees are mandated with spearheading developmental projects such as building of schools, clinics and cattle dip tanks among other projects. Recently the Uitcom village committees managed to make a lot of progress in erecting more permanent structures and teachers' houses at Chivake School in Uitcom 1, which was just a pole and mud thatched structure. It is quite apparent that all these projects have been mooted and managed by village committees and prominent war veterans with little or no involvement of traditional authorities who are still jostling to substantiate their claims to the resettlement areas.

Although village committees can identify 'vacant' land within their village and recommend to the District Land Committee to have that piece of land re-allocated, they do not have the authority to allocate land. In spite of this, it is not uncommon to find cases where village committees try to arm twist the District Lands Committees to allocate land to 'our son who has no piece of land' or 'a very loyal member of the party [ZANU-PF]' (Interview with N. Muzenda 25 August 2009). Consequently the Lands Committee ends up being asked to approve land allocations made and fiercely defended by village land committees or just chairpersons of these committees. Village committees' functions have also extended to include helping in settling disputes among plot holders. The disputes largely revolve around conflicts over the sharing of communal resources such as pastures and water and also boundary disputes.

From as early as the 1980s traditional leaders were lobbying the government to be given more authority over land. This culminated in the enactments of the Traditional Leaders Act (25/1998). The Act 'granted chiefs and headmen and village leaders of development institutions below the council, control over land allocation in communal areas, and a greater role in that one-time bastion of state-directed modernisation, the resettlement areas' (Alexander 2006,183). Section 5 of the Act, in particular, stipulates that the chief shall be responsible for taking traditional and related administrative matters in resettlement areas, including nominating persons for appointment as headmen by the minister (Traditional Leaders Act 25/1998). According to the Act, the minister is the one with the power to authorise a chief to exercise authority over a certain resettlement area provided that the declared area is a single resettlement ward and does not fall under the authority of more than one chief (Traditional Leaders Act 25/1998). In spite of all these pronouncements traditional authorities have however continued to play second fiddle to village committees who are responsible for the day to day running of the resettlement areas and are answerable to the District Land Committees (Mubvumba 2005, 37). Hence, in spite of the government's moves towards strengthening the position of chiefs and headmen, 'authority over land remained the subject of heated contestation among a wide range of local and state institutions' (Alexander 2006,183). Thus chiefs and headmen often find it difficult to assert their authority in resettlement areas and continue to agitate for the re-establishment of the traditional boundaries that existed prior to colonial land alienation.

The push towards gender equity in the country has also seen an increased participation of women in the village committees in Silverdale, Uitcom and Eastdale. According to Mr Pugede (Interview 13 July 2009), though men, especially war veterans, ZANUPF activists and pioneer settlers, tend to dominate village committees, a significant number of women are being elected into Village Committees. For example, Mai Mada (pseudonym), in one of the three villages in Uitcom, chairs the village committee and claims that she is respected by her peers in the committee as well as other villagers. Mai Mada is a former *chimbwido* (female liberation war collaborator) and was among the first group of people to take up land in Uitcom (Interview with A. Matangi 10 August 2009). Though this is a unique case in the area it nevertheless illustrates the extent to which the village committee as a management unit has helped women transcend barriers placed by traditional institutions such as chieftainship (Goebel 2005). In the communal areas where traditional authorities are the major institutional authority women still have little chance to hold positions of authority, which democratic institutions in the resettlement areas such as Village Committees are providing. As Marongwe (2003,182) observed elsewhere, 'some farms, particularly those close to communal and resettlement areas, had fairly even numbers of male and female occupiers. In other cases, wives and husbands participated together in the farm occupations.' The FTLR thus arguably provided women with not only new livelihood portfolios but also opportunities to assume positions of authority in the Village Committees and other structures thereby ensuring a gendered access to land.

By leaving out traditional authorities such as village heads, headmen and chiefs in the management of the resettlements the Land Ministry disempowered the traditional authorities, which are often portrayed as the 'conservative guard' of the old and unproductive system in the communal areas. This was also coupled with the enactment of legislation which took away chiefs' powers to redistribute land (Alexander 1994,333). Such enactments have made chiefs feel like they have been left out in the management of resettlement areas and replaced by technocrats and democratic institutions such as village committees. It is quite clear that the government is not keen to involve traditional authorities in the agrarian reform. This explains the government's reluctance to allow the land reform to take a land restitution angle which would have allowed those people who could substantiate their claims to ancestral lands to be resettled in such lands (Mubvumba 2005, 18). By contrast the South African model allows for restitution when such claims can be substantiated (James 2007).

In spite of the seeming friction between different groups in the management of resettlement areas the differences between these groups are not always cut and dried. For example, a war veteran can also be an autochthon as well as a member of a village committee, thus being in a position to articulate the views of one or the other group depending on the occasion. Hence it is quite difficult to separate these groups and make them clearly identifiable as members tend to straddle between groups making use of them to make claims to land, authority and resources. Arguably some people are able to articulate a particular kind of claim more strongly based on their positions as chiefs, members of the village committee, agricultural extension officers or councillors (J. Fontein, Personal Communication 8 December 2009). For instance, a chief or headman can effectively make claims to autochthony based on links to graves and old homes while a village chairperson is able to derive his/her authority from the fact that he/she is a democratically elected officer who represents the District Land Committee. However, this leaves out the common people who fall outside these high profile positions. The new settlers often have to fit into categories such as autochthon/stranger which at times makes little sense to them as some just identify themselves as simple farmers without the added baggage of being identified as either an autochthon or a stranger or being dragged into this belonging milieu.

Traditional authorities, disputed territories and belonging in Gutu

The discussion has thus far focused on the role of village committees as the core management institution for resettlement areas and how this has further alienated traditional institutions from the day to day running of resettlements. This section focuses on traditional authorities' efforts at asserting their authority in the newly resettled areas and how this has impacted on belonging and boundary politics. In spite of the marginalisation that they are suffering at the hands of government imposed institutions such as village committees, traditional authorities have been quite resilient in their attempts to assert their power in resettlements in Gutu.

The return to ancestral homes, redefinition of traditional boundaries and negotiation of belonging are issues that have dominated traditional authorities' efforts at re-asserting their powers in resettlement areas. The land reform in Gutu has also stimulated debates about autochthony and disputed territories as traditional authorities are being engaged in disputes over the extent of their territories and also making efforts to reclaim what they consider to be their lost lands. According to Moyo and Yeros (2007, 110), 'during the fast-track process, the role of traditional leaders in beneficiary selection was often overridden by war veterans, but still chiefs did influence policy to the effect of extending their territorial control into contiguous resettlement areas'. This was normally based on traditional authorities' quest to reclaim what they consider to be their alienated ancestral lands.

The much publicised case of Chief Svosve and his people's reoccupation of their ancestral lands in Marondera in 1998 shows the vital, yet neglected, trajectory of restitution and return to ancestral lands in the land reform discourse in Zimbabwe (see Alexander 2006, 184; Marongwe 2003). As Marongwe notes (2003, 184) 'restitution claims based dispossessions, were a strong rallying point for participants in the [farm] occupations'. As a result, claims to ancestral lands such as the Tangwena people's claims over Kairezi in Manicaland (Moore 2005), Sekuru Mushore's claims over the Nharira Hills in Marondera, Mashonaland West; and the Ndau people's claims over Chirinda forests, among other historical claims, have been a major highlight in the FTLRP (Marongwe 2003, 186). These people who have historical claims to such lands often view other non-local farmers as 'foreigners' or 'strangers' with little if any claim to the land. This resonates with Lentz's (2007) argument that scarcity of land tends to increase the need to identify those who are insiders and the strangers who have limited rights to land. Control over land therefore becomes a sign of the extent to which one belongs. It is important to note that though the land reform in Zimbabwe has not incorporated restitution, traditional authorities have continued to make efforts to regain control over their formerly alienated lands.

In 2001 Mr Jara Mudziwaniswa, the village head of the Mudziwaniswa Village in Headman Mawungwa's area, left his home together with other people from his village that included his two young brothers. Their intention was to return to their ancestral lands in Harawe area in Chief Chikwanda's area. They allocated each other pieces of land and began to engage in agriculture. Mr Mudziwaniswa and his brothers were already in the process of negotiating with Chief Chikwanda over the re-establishment of their village when they were told by the police to vacate the farm as it had not yet been gazetted as one of the farms to be reposed. Narrating their eviction from their ancestral lands Mr Jara Mudziwaniswa (Interview 17 September 2009) stated that:

> Our houses were burnt and crops destroyed by the police. In fact we were forcibly removed from our ancestral lands and now we are back in this barren land with little hope of ever reclaiming our lost lands. Of course later on the farm was gazetted and people resettled but since it is in Masvingo District we stood little chance of returning to the farm as preference was being given to people who resided in the district. However, we still have an emotional attachment to the area because it is the land of our forefathers where they lay buried.

Though Mr Jara Mudziwaniswa and his people's return to their ancestral lands was short lived they still cherish the few months they were reunited with their lost ancestral lands. It also reveals the saliency of the materialities of graves, old homes and traditional boundary markers such as mountains and rivers in land reclamations during the FTLRP (Fontein 2009a).

Belonging is indeed contested and negotiated between and among different communities. It also involves negotiation over territorial boundaries and control of the land within the said boundaries. Lentz (2006, 2) sums up this argument by asserting that, 'if claims to land are linked to group membership, the reverse is also true: control over land has been and is still used as a way of defining belonging and ...' this places land at the centre of how belonging is contested and negotiated especially at the boundaries between communities. By the same token in the south-eastern parts of Gutu District in Zimbabwe the contestation over land and belonging has been a central feature in the boundary dispute between Chief Makore and Chief Chikwanda, which dates back to the colonial period. Colonial land alienations and the resultant re-configurations of reserves saw Chikwanda losing all his land and having to be accommodated by other traditional authorities. According to Mtetwa (1976, 313),

> as a result of this continued attack on Chikwanda Reserve, its size by 1913 had shrunk to accommodate only the people of Makore, Rupiri and Mutema and not Chikwanda's own people. Thus, the NC [Native Commissioner] said that Chikwanda Reserve was a misnomer since the people of Chikwanda were outside the reserve and suggested that the reserve should be called Makore.

Thus the colonial land alienation which left Chikwanda with virtually no land for his people threw the Chikwanda people into perennial land disputes with other traditional authorities, especially Makore who is to the north of Chikwanda in Gutu District.

Chief Makore's land claims present a very interesting dimension in traditional authorities' land reclamations and attempts to re-constitute their territories and reconfigure their traditional boundaries. Makore is a chief in Gutu District but belongs to the *Moyo Duma* clan and occupies the area to the south-west of Gutu District bordering the Masvingo and Ndanga Districts. He claims that his chiefdom used to stretch as far south as the Mbebvume River in Masvingo District prior to the colonial land alienations. However, he lost a large portion of his lands to white farms such as N. Richards, J. Bolland, Springfield, Wepener, Portigieter, Standmore, Dromore, Elands Kop, Bonair, Lynington, and Chidza, among others. These areas used to be occupied by people under Chief Makore before they were displaced to pave way for the creation of the white commercial farms. Makore also argues that mountains and sacred sites such as Musanawengwe, which is in Springfield Farm, together with Nyoni and Zishumbe Mountains are well known Makore ancestral *mapa* (royal burial sites). The Zishumbe *mapa* is actually where the Makore founding father Risipambi was interred after being dried according to the *Duma* customs. The Makore people are thus keen to regain custodianship of these important ancestral *mapa* (Interview with J. Tangemhare Councillor Ward 27, Makore, 21 August 2009).

The FTLRP has, however, seen Chief Chikwanda, who is also of the *Duma* clan, expanding his boundary into areas which formerly belonged to Chief Makore. This has been occasioned by issues which include the fact that Makore did not quickly make an attempt to repossess their ancestral lands in significant numbers when the FTLRP was in full swing in 2000 and 2001. On the contrary, Chief Chikwanda, who was facing serious land shortages, encouraged his people to occupy farms including those in the areas which formerly belonged to Makore. He also settled his representatives in the area and used that to claim the area and impose his authority. Makore's claims to these lost ancestral lands have also been further complicated by the fact that Makore's chiefdom is in Gutu District yet the contested areas are demarcated as belonging to Masvingo District in terms of the district administrative boundaries. Scoones *et al.* (2010,166) argue that Chief Chikwanda's people claim that the area belongs to Chikwanda because Chief Makore rules in Gutu administrative district while Chief Chikwanda rules in Masvingo District where the contested areas lie. Consequently, it has proved a very difficult task for Chief Makore, in spite of the availability of material evidence in forms of graves and ruined old homes supporting his case, to reclaim his ancestral lands which now fall in another district and are now being claimed by Chief Chikwanda. As a result of this, Chief Chikwanda managed to install his village heads and headmen in the areas contested by Chief Makore and expelled the headman who had been appointed by Chief Makore, thus effectively taking control of the disputed areas (Scoones *et al.* 2010, 166).

The settlement of people with little regard for the Makore ancestral lands around Zishumbe and other *mapa* has angered Chief Makore, who has blamed the constant droughts in his area on the desecration of these sensitive ancestral *mapa*. According to Fontein (2009a, 22), in 2006,

> the senior Chief Makore (the acting chief's father), blamed poor rains and failing harvests on 'those fast track people' who 'do not know the land'; and was particularly angry about a local councillor who has occupied Zisubwe [Zishumbe] hill, Makore's sacred *mapa* where their founding ancestor Risipambi lies (dried) buried, and near to his birth place.

Chief Makore thus sees the alienation of Makore people from their ancestral lands and the desecration of the same by 'those fast track people' as contributing to calamities such as droughts.

The simmering boundary dispute between Chief Makore and Chief Chikwanda has been raging since 1982 with Makore claiming that Chief Chikwanda is refusing to give him back his ancestral lands alienated during the colonial period. This boundary dispute reached new levels in July 2008 with some youths allegedly working for Chief Chikwanda assaulting acting Chief Makore (Phineas Makore) who was admitted at Gutu Mission Hospital after sustaining serious injuries. The case was reported to the police leading to the arrest of 12 villagers suspected to have participated in the assault (*Herald* 9 July 2008). The incident had been triggered by the fact that Chief Makore had installed village heads in areas which were also being claimed by Chief Chikwanda. Chikwanda argued that the area fell under his dominion while Makore was claiming it as his lost ancestral lands (*Herald* 9 July 2008). The situation between these two *Duma* Chiefs has been so tense that Chief Makore ended up asking for police protection, fearing similar attacks from Chief Chikwanda (*Herald* 9 July 2008). The long standing boundary dispute between the two chiefs illustrates the complex nature of the politics of belonging in the post FTLRP era and the challenges faced by traditional authorities in re-establishing traditional boundaries and reasserting their authority in their former white commercial farms. It also shows the limits of attempts to reclaim ancestral lands based on pointing to ancestral graves and old homes.

There have also been similar boundary disputes in the resettlements in Silverdale, Uitcom and Eastdale Estates. Chief Serima who is of the *Gushungo* clan claims to be the sole traditional authority in all the new resettlement areas to the north and north-west of the district. This area covers resettlements such as Eastdale, Silverdale, Lionsdale and Uitcom, among others. Mazarire's argument about the centrality of ancestral lands is especially useful in understanding various land claims that are being made by traditional authorities in Gutu (Mazarire 2009). Serima uses known Serima ancestral burial sites and *nhare* (strongholds) such as Chikwidzire, Gonwe and Songorera, all in Eastdale Farm, to substantiate his claims over the resettlement areas. The Songorera *nhare* and *mapa* features prominently in Serima oral traditions for it was the centre of Serima defence against the Mfecane invaders (Interview with Mr Chikanya, 15 August 2009). It is on the basis of the presence of such sacred sites in the Eastdale Estates that Chief Serima claims to be the sole traditional authority in the resettlement areas in the former Eastdale Estates. Chief Serima's claims are however contested by other traditional authorities such as Hwenga, Denhere, Makumire and Musarurwa.

Mr Nicros Makumbe of the Mudavarwe clan also claims the same area arguing that the area originally belonged to the Mudavarwe clan. He argues that Dhombo Mountain was a *mapa* for the Mudavarwe as it was used for the burial of important people in the clan (Interview with Mr N. Makumbe, 12 September 2009). He claims that when other people were driven out of the area to Chiguhune his family together with other members of the Mudavarwe clan decided to stay put and work for the white farmers who took over their land. Thus when the land occupations began in 2000, Mr. Makumbe chose to remain in his ancestral lands and thus took up land in Uitcom Village 1 in the vicinity of where his parents used to live. Though he was allocated a piece of land close to Chivake River in Uitcom Village 1 Mr Makumbe instead asked his nephew to run this plot on his behalf as he chose to stay in the area designated for Chivake School arguing that 'this is where my forefathers used to live' (Interview with N. Makumbe, 12 September 2009). From his arguments, it is clear that Mr Makumbe belongs to a group of farmers whom Fontein (2009b) refers to as 'new yet autochthonous farmers who easily relate to the "history-scapes" and real landscape features of the area such as graves, trees, springs, rivers, hills as well as *matongo* (old homes)'. Such farmers use their links with these areas which are their *matongo* to legitimise their occupation or re-occupation on the area. Such a position enable these 'returning' autochthons to articulate certain claims to the land and also to assert their authority as people who have direct, if not imagined, links with the original owners of the land. They can easily point to old homes as well as graves of their ancestors in legitimising their claims to autochthony. Yet it should be made clear that it is not only because of the simple 'presence' of graves, old homes, hills and springs, among other landscape features, that traditional authorities have been able to substantiate their claims to the land but it is more importantly about the materiality of such features (see Miller 2005, Fontein 2008). It is about the history and culture that these features are able to articulate and also about how such articulations impact on the people. These features thus impact on the imaginings of the local people in as much as the people make use of them to construct belonging. In other words it is as much about how local people use landscape features in laying their claims as it is about the affordances of these features.

Traditional authorities are pushing for the establishment of village heads (*maSabhuku*) in the resettlement areas that would be answerable to the chiefs and headmen and would ensure that the authority of the traditional leaders is not neutralised by that of the village committees. At present, though no village heads have been installed in the resettlement areas in Eastdale, Uitcom and Silverdale, Chief Serima has appointed representatives in the villages who work towards ensuring that traditional customs such as the *chisi* holiday and sacred sites are respected. Chief Serima sees this as a precursor to the appointment of substantive village heads under his authority. The appointment of village heads is already causing a lot of conflicts among rival traditional authorities in Gutu especially given the contestations over boundaries. In Ward 7, Gooile Hoop, Willands and Mazongororo farms are hotly contested by Chief Serima, Headman Gadzingo (Dzimba Madondo) and Headman Makumbe. All the three traditional authorities make historical claims to the area and also use old homes, mountains and graves to legitimise their claims. Interestingly all have also made attempts to install village heads in the area or at least their representatives. However, though Chief Serima has been more vigorous in his claims and in his attempts to install his *maSabhuku* (village heads) he has made little progress to this end. It is believed that the reason why Serima has failed to quickly install his *maSabhuku* and bring the area into his sphere of influence is that he has been asking the aspiring village heads to give him one head of cattle each so that he would recognise them as village heads under his authority. This has not gone down well with the aspirants who see it as an excessive demand bordering on extortion. As a result of this, Headman Makumbe and Headman Gadzingo have decided not to make such demands in order to appease the disgruntled aspirants and gain their allegiance in the battle for the control of the area (Interview with Mr. D. Mberikwazvo, Gooile Hoop Farm, 12 December 2009).

Chief Chiriga of the Gumbo Madyirapazhe clan which is dominant in Gutu has also asserted his authority in Clare and Lonely farms arguing that the farms belonged to his clan. According to Scoones *et al.* (2010, 165), 'his territorial control is enforced through kraal heads (*sabhukus*) and chief's aides. The chief also works together with coordinators (former war veterans) and committees of seven.' It is quite evident that the installation of village heads is one of the greatly contested issues in the land reform discourse as it has the effect of reconfiguring traditional boundaries as well determining who belongs where. This explains why the appointment of village heads has raised so much debate and so many disputes. The land reform in Zimbabwe has thus led to the emergence of new forms of authority in the resettlement areas which have often resulted in intense contestations over legitimacy and control (Scoones *et al.* 2010,160).

In some cases, funerals become theatres of struggle as boundary politics are often played in the resettlement areas. For instance, when somebody dies, the bereaved family has to inform the immediate traditional authority who is often a headman or chief before they can bury their relative in his *ivhu* (soil). In the case of Eastdale, Silverdale and Uitcom, people inform either Chief Serima or Headman Denhere depending on which traditional authority they fall under. The choice to inform one traditional leader over the other thus strongly suggests where one feels they belong and has serious consequences in boundary politics. In the event that representatives of both traditional authorities attend the same funeral in a disputed area, tension often arises over who has the authority to speak at the funeral as the traditional authority of the area (Interview with J. Mumanyi 15 August 2009). Funerals thus assume a very important position in the belonging matrix. In fact they becomes a sort of ritual for asserting belonging. In his recent publication, Chabal (2009) asserts that burials reinforce a collective sense of belonging and strengthen an individual's attachment to the community. As he argues, 'the link to the ancestors, wherever they are buried, is an integral part of the meaning of origin, and of the texture of identity, which cannot be disregarded' (Chabal 2009, 29). Belonging is here linked with attachment to a physical place which draws its meaning from people's attachments with the ancestral graves. A critical point noted by Chabal (2009) is that, 'a properly executed burial reinforces the collective sense of belonging, without which the person is not fully human and the community is not fully complete' (Chabal 2009,49).

Similarly, Geschiere and Nyamnjoh (2000, 435) argue that in Cameroon many Cameroonians consider burial locations as a very important criterion for belonging. In essence the basic test for one's belonging will be to ask them to show where their ancestors are buried. A failure to do so would be interpreted as meaning that the person belongs elsewhere, in this case where the bones of his/her ancestors are interred (see Geschiere 2009). This is the reason why among the Igbo in Nigeria, rural-urban migrants 'face powerful expectations to be buried at home in their ancestral villages and perform elaborate and expensive funeral ceremonies for their dead relatives' (Smith 2004, 569). The value of this is basically the maintenance of migrants' ties with their rural homes, each funeral they attend being a reminder of where they really belong. This also means those urban migrants in the end have dual or multiple notions of belonging. Hence funerals and graves have so much significance in the negotiation of belonging in Africa. Belonging is here linked with attachment to a physical place which draws its meaning from people's attachments to the ancestral graves. As Shipton (2009, 20) argues, 'graves are the symbolic focal points of human attachments to place: the living and dead, the social and the material, all connect here'. Graves also form the material evidence of who has been living on the land longer and therefore has historical links with the land.

Rainmaking is another ritual which sees the power dynamics and politics of belonging being played out in newly resettled areas. In Gutu rainmaking ceremonies are normally done at the beginning of every rainy season at the clan's main *mapa*, which is usually located in the ancestral *gadzingo*, and the smaller *mapa* in the wards. Normally all people in the ward are asked to contribute millet for the brewing of the traditional beer used for the rainmaking rituals. In the case of Makore, people visit Masvitsi Hill, which is also Makore's *mapa*, to ask for rain from the spirits of the chiefs buried there. As Chabal (2009, 29) puts it 'the relation of land and sense of origin are both rooted in the location where ancestors are buried and propitiated'. This places *mapa* and other sacred places at the centre of people's construction of belonging. The other chiefs in Gutu like Serima and Chiwara also have their different *mapa* where they conduct their rainmaking and other rituals. Each family is supposed to contribute millet or *rapoko* to be used in brewing the beer used in the rituals. However, often such ceremonies also lead to conflicts over boundaries in resettlement areas as contributing millet for a ceremony organised by a rival claimant to the area often denotes one's allegiance. Consequently, some people prefer not to contribute towards such ceremonies for fear of being labelled as having allegiance to a certain traditional authority. Thus, together with funerals, rainmaking ceremonies provide stages where belonging and boundaries are negotiated. These rituals and their performance provide traditional authorities with a platform to assert their authority as well as creating tensions between rival claimants to a territory.

Apart from contestations over boundaries and reclamation of lost ancestral lands the land reform has also brought about dual belonging as some of the new settlers straddle between communal and resettlement areas. Cohen and Atieno Odhiambo's (1992) ethnographic study of the conflicts surrounding the burial of SM, a prominent Luo Kenyan lawyer, illustrate the problems that emanate from dual belonging and the extent to which people can go to prove where one really belongs. In the case of Gutu this dual belonging or straddling between old homes and new ones is largely caused by the fact that some people see their positions in the resettlement areas as very insecure because of the unclear tenure system. For example Mr Matangi, who owns a plot in Eastdale, still maintains his old home in Guzha Village under Chief Nyamande. As he puts it, '*handina kusiya ndapisa musha wangu kwaGuzha*' (I did not burn my homestead when I left Guzha). I just left my eldest son to take care of the homestead as I am trying to improve my livelihood here in the farms' (Interview with A. Mutangi, 10 August 2009). Such farmers who continue to hop between communal areas and resettlement areas seem to be trying to use communal areas as a safety net in case they lose their plots in the resettlement areas. Some farmers however argue that they are content with the conditions in the resettlement areas and have cut ties with their old homes in communal areas. On the contrary, Manenji Murodhini who is the Farm Chairman of Silverdale, claims that he has cut ties with his old home in Headman Mazuru's area and now feels that he belongs to Silverdale, though he stills occasionally returns to his old home for funerals and other family gatherings (Interview, 12 September 2009). Manenji Murodhini's stance however seems to be influenced by the fact that he was among the first comers in Silverdale and is an important figure in the community, rendering dual residence or dual belonging difficult to maintain for him. The practice of straddling therefore seems to be less common among those farmers who consider themselves to be pioneers in their villages and are also important figures in the community than among late comers who do not have a strong sense of belonging.

It is quite evident that though not widely acknowledged, ancestral claims to land have played a significant role in the politics of belonging in the newly resettled lands. As James (2007) has observed for South Africa and as the Gutu case discussed here has shown, though land restitution is an issue widely debated in land reform discourses it is very difficult to implement especially when it is not based on formal title. Claims for land restitution based on ancestral claims not aided by title deeds or any other formal ownership are often met with equally convincing counter claims. It is therefore vital to juxtapose Zimbabwe's redistributive land reform with other models such as the South African one which recognises the rights of those displaced during the Apartheid era and therefore allows for restitution.

It can also be argued that the FTLRP has also ensured gendered access to land as it has seen an increased number of women owning land and participating in the management of the village committees in the resettlement areas. Communal areas, with their emphasis on traditional institutions, limited women's access to land as they could only claim ownership of land through their husbands. This explains why in spite of some women leading village committees in the newly resettled lands their claims to land are seldom based on restitution as women's access to land does not fit comfortably with claims based on ancestral claims unless such a woman is possessed by an ancestral spirit.

It is a common practice to see the so-called new farmers straddling between both the newly resettled areas and their old homes as they are not yet sure of the security of tenure in their new pieces of land. The case of the Matangi family discussed above has shown these inherent fears within the new farmers regarding severing their ties with their former homes. They are afraid of being left homeless if for some reason they get evicted from the resettlement areas. The kinship ties that still exist between these new farmers and their relatives in the communal areas also demand that they continue to have contacts with them. Such people arguably have a dual sense of belonging as they have one foot in the new resettlement and another in their old homes.

Conclusion

What emerges here is that while for the government the land redistribution programme has been about taking land from the minority white farmers and giving it to the landless black majority, for the traditional authorities it has also meant a return to their ancestral lands. Traditional authorities in many parts of the country have seen this as an opportunity to reclaim lost ancestral lands, graves, mountains, and sacred places and also to re-establish their *nyika* boundaries which had been greatly altered during the colonial period. It is quite clear that although the Traditional Leaders Act gives chiefs and headmen authority over land in both communal and resettlement areas, there have continued to be some contestations between traditional authorities and democratic institutions such as village committees over control of land in resettlement areas. Without clarity on this matter there will continue to be contestations over which group has authority over land allocations and determining boundaries. It is therefore necessary for the government to come up with a clear policy position on institutional authority in resettlement areas to avoid these contestations.

The FTLRP has also brought new challenges for traditional authorities in Gutu who have continued to be locked in disputes over the re-alignment of their traditional boundaries. Any analysis of the politics of belonging in the resettlement areas must therefore appreciate and address issues surrounding reclamation of ancestral lands and boundary politics. For the Eastdale Estates, Silverdale and Uitcom, the boundary conflicts pitting Chief Serima against Headman Denhere are illustrative of the intricate and fluid nature of traditional boundaries and the centrality of graves, mountains, and rivers, among other features, in the negotiation of such boundaries. Makore's and Chikwanda's claims over resettlements to the south west of Gutu bordering Masvingo District are complicated by the fact that the area now falls in Masvingo District therefore making it quite difficult for Chief Makore to reclaim what he considers to be his ancestral lands. It is quite evident that the FTLRP has brought about new trajectories of belonging and also broached questions about disputed territories and contested boundaries. These claims and counter claims reveal the intricacies of land reclamation in the aftermath of the FTLRP and the importance of graves, old homes, mountains and rivers in traditional authorities' efforts to reclaim their ancestral lands and re-establish their authority. Hence, restoration of links with ancestral lands, re-establishment of boundaries and the re-assertion of traditional authority has remained at the heart of a number of local communities' conceptualisations of land reform in Zimbabwe.

References

Alexander, J. 1994. State, peasantry and resettlement in Zimbabwe. *Review of African Political Economy*, 21(61), 325–345.

Alexander, J. 2006. *The unsettled land: state-making and the politics of land in Zimbabwe 1893–2003*. Oxford: James Currey.

Chabal, P. 2009. *Africa: the politics of suffering and smiling*. London: Zed Books.

Chakanyuka, C. 2007. Heritage, land and ancestors: the challenges of cultural heritage management in the face of the fast track land redistribution programme in Zimbabwe 1999–2007. Thesis (MA). Heritage Studies, History Department, University of Zimbabwe.

Chaumba, J., I. Scoones and E. Wolmer. 2003. From *jambanja* to planning: the reassertion of technocracy in land reform in south-eastern Zimbabwe? *Journal of Modern African Studies*, 41(4), 533–554.

Cohen, D.W. and E.S. Atieno Odhiambo. 1992. *Burying SM: the politics of knowledge and the sociology of power in Africa*. London: Heinemann.

Fontein, J. 2008. The power of water: landscape, water and the state in Southern and Eastern Africa: an introduction. *Journal of Southern African studies*, 34(4), 737–756.

Fontein, J. 2009a. Graves, ruins and belonging: towards an anthropology of proximity. Curl Lecture, Institute of Archaeology, UCL, 17 September.

Fontein, J. 2009b. We want to belong to our roots and we want to be modern people: new farmers, old claims around Lake Mutirikwi, Southern Zimbabwe. *African Studies Quarterly*, 10(4), 1–35.

Geschiere, P. and F. Nyamnjoh. 2000. Capitalism and autochthony: the seesaw of mobility and belonging. *Public Culture*, 12(2), 423–452.

Geschiere, P. and S. Jackson. 2006. Autochthony and the crisis of citizenship: democratization, decentralization, and the politics of belonging. *African Studies Review*, 49(2), 1–14.

Geschiere, P. 2009. *The perils of belonging: autochthony, citizenship, and exclusion in Africa and Europe*. The University of Chicago Press.

Goebel, A. 2005. *Gender and land reform: the Zimbabwe experience*. Montreal and Kingston: McGill-Queen's University Press.

Hadzoi, L. 2003. Continuity and change in the powers of chiefs, c. 1951–2000: a case study of Gutu District. BA Honours Dissertation. History Department, University of Zimbabwe.

James, D. 2007. *Gaining ground? Rights and property in South African land reform.* New York: Routledge-Cavendish.

Konings, P. 2001. Mobility and exclusion: conflicts between autochthons and allocthons during political liberalization in Cameroon. *In*: M.R. de Bruijn, R. van Dijk and D. Foeken, eds. *Mobile Africa: changing patterns of movement in Africa and beyond.* Leiden: Brill. pp. 169–194.

Kopytoff, I. 1989. *The African frontier: the reproduction of traditional African societies.* Bloomington: Indiana University Press.

Lentz, C. 2006. Land rights and the politics of belonging in Africa: an introduction. *In*: R. Kuba and C. Lentz, eds. *Land and the politics of belonging in West Africa.* Leiden: Brill, pp. 1–34.

Lentz, C. 2007. Land and the politics of belonging in Africa. *In*: P. ChabalU. Engel and L. de Haan, eds. *African alternatives.* Leiden: Brill, pp. 37–58.

Leonhardt, A. 2006. Baka and the magic of the state: between autochthony and citizenship. *African Studies Review*, 49(2), 69–94.

Lewis, G. 2004. 'Do not go gently. . .': terrains of citizenship and landscapes of the personal. *In*: G. Lewis, ed. *Citizenship: personal lives and social policy.* Bristol: The Policy Press, pp. 1–39.

Malkki, L. 1992. National Geographic: the rooting of people and the territorialization of national identity among scholars and refugees. *Cultural Anthropology*, 7(1) (Space, Identity, and the politics of difference), 24–44.

Manganga, K. 2007. An agrarian history of the Mwenezi District, Zimbabwe, 1980–2004.Thesis (MPhil). PLAAS, School of Governance, University of Western Cape.

Marongwe, N. 2003. Farm occupations and occupiers in the new politics of land in Zimbabwe. *In*: A. Hammar, B. Raftopolous, S. Jensen, eds. *Zimbabwe's unfinished business: rethinking land, state and nation in the context of crisis.* Harare, Weaver Press, pp. 155–190.

Matondi, G.H.2010. Traditional authority and Fast Track Land Reform: empirical evidence from Mazowe District, Zimbabwe livclihoods after land reform in Zimbabwe. Working Paper No. 11.

Mazarire, G.C.2008. The Chishanga waters have their owners: water politics and development in southern Zimbabwe. *Journal of Southern African Studies*, 34(4), 757–784.

Mazarire, G.C.2009. The invisible boundaries of the Karanga: considering pre-colonial Shona territoriality and its meanings in contemporary Zimbabwe. Paper Presented to the ABORNE Conference on 'How is Africa transforming border studies?' School of Social Sciences, University of the Witwatersrand, Johannesburg, South Africa, 10–14 September.

Miller, D. 2005. *Materiality.* London: Duke University Press.

Moore, D. 2005. *Suffering for territory: race, place, and power in Zimbabwe.* Durham: Duke University Press.

Moyo, S. 2000. The political economy of land acquisition and redistribution in Zimbabwe, 1990–1999. *Journal of Southern African studies*, 26(1), 5–28.

Moyo, S. and P. Yeros. 2007. The radicalized state: Zimbabwe's interrupted revolution. *Review of African Political Economy*, 34(111), 103–123.

Mtetwa, R.M.G.1976. The political and economic history of the Duma people of South-Eastern Rhodesia from the early 18[th]c to 1945. Thesis (DPhil). Department of History, University of Rhodesia

Mubvumba, S. 2005. The dilemma of governance: government policy on traditional authority in resettlement areas 1980–2004: the case of Guruve. B.A. Special Honours in History, History Department, University of Zimbabwe.

Rutherford, B. 2001. Commercial farm workers and the politics of displacement in Zimbabwe, colonialism, liberation and democracy. *Journal of Agrarian Change*, 1(4), 626–651.

Sachikonye, L.M.2004. Land reform and farm workers. In: D. Harold-Barry, ed. *Zimbabwe: the past is the future.* Harare: Weaver.

Scoones, I, *et al.*, 2010. *Zimbabwe's land reform: myths and realities.* Oxford: James Currey.

Shipton, P. 2009. *Mortgaging the ancestors: ideologies of attachment in Africa.* New Haven and London: Yale University Press.

Smith, D.J. 2004. Burials and belonging in Nigeria: rural-urban relations and social inequality in a contemporary African ritual. *American Anthropologist*, 106(3), 569–579.

Spierenburg, M.J. 2004. *Strangers, spirits, and land reforms: conflicts about land in Dande, Northern Zimbabwe*. Leiden and Boston: Brill.
Traditional Leaders Act, 25/1998
Traditional Leaders Act, Chapter 29: 17 section 30 (Government of Zimbabwe, 2006)

Joseph Mujere is a PhD Candidate in the School of History, Classics and Archaeology at the University of Edinburgh. He received his MA in African History from the University of Zimbabwe. His PhD research focuses on the interface between migration, land and the politics of belonging in colonial and post-colonial Zimbabwe. Email: j.mujere@sms.ed.ac.uk.

Local farmer groups and collective action within fast track land reform in Zimbabwe

Tendai Murisa

Fast track land reform led to the restructuring of agrarian relations in Zimbabwe. This paper explores the emerging forms of local agency on selected A1 farms in Goromonzi and Zvimba. It analyses how A1 beneficiaries have formed local farmer groups (LFGs) and the extent to which these have contributed towards relieving farm production challenges that include the unavailability of productive assets, limited household labour and unavailability of inputs. Through case studies of local farmer groups this paper manages to examine the internal dynamics of local agency, the nature of participation and the extent to which these formations actually provide a relief to members and provide the first line of defence of the newly found land rights. The findings provide important clues regarding local agency in a context where lineage forms of organisation do not exist and in most cases 'strangers' from different places have been resettled next to each other.

Introduction

Since 2000 Zimbabwe's agrarian landscape has been fundamentally reconfigured through a major convulsion centred on farm occupations and the eviction of the former, largely white, owners of these farms. This paper focuses on a key dimension of post-2000 land redistribution that has not received much attention to date – that of emerging forms of collective action and organisation on the former commercial farms, and associated social and political dynamics, but focuses in particular on local farmer groups. A key focus of most of these farmer groups has been securing access to inputs such as hybrid seeds and fertilisers, which have been in short supply as a result of the wider economic crisis in Zimbabwe since 2000. Some groups have organised the collective production of specialised crops such as wheat, while others have arranged for the pooling of labour for certain key farming tasks, the shared use of farm assets such as tobacco curing barns, the hire of tractors and other equipment, and cooperative marketing. Other motivations for the formation of these local farmer groups have included defending newly acquired land rights in a context where the land tenure regime remains poorly defined. These groups have also engaged with various tiers of government around the provision of a variety of support services, and some have attempted to nurture the participation of their members in collective decision-making.

This article explores why and how Fast Track Land Reform Programme (FTLRP) beneficiaries on four A1 resettlement schemes in Goromonzi and Zvimba districts established local farmer groups (LFGs) and engaged in various forms of collective action via these groups. The farmer groups were formed in a context where not only the lineage framework is largely absent, but also where non-state actors such as non-governmental organisations (NGOs) and the national small farmers' union have been hesitant to support these new farmers. Left to organise themselves, beneficiaries formed farmer groups and engaged in various collective action strategies aimed at gaining access to external resources and support, enhancing their productive capacity, and defending their newly-acquired land rights.

The article begins with a brief review of the wider literature on farmer groups and other associational forms in rural Zimbabwe, with a particular emphasis on resettlement schemes. It then presents case study material from four FTLRP farms, and concludes with a discussion of the wider significance of these cases.

Local farmer groups and collective action in Zimbabwe

Various studies reveal that Zimbabwe's countryside is made up of a mosaic of associational forms, including loose, unstructured mutual-aid networks such as faith-based groups, credit associations, women's groups, labour-sharing groups, as well as the more structured peasant or farmer organisations, which might be either localised or national in scope (Bratton 1986, 358). It was estimated in 1982 that 44 percent of households in communal areas belonged to such associational forms, and in 2002 it was estimated that there were 3000 local smallholder farmers' organisations in Zimbabwe (Bratton 1986, 371, Moyo 2002, Sibanda 2002). Bratton's (1986) study of communities belonging to local farmer groups in Hwedza District in Mashonaland East demonstrated that these institutions often go beyond their stated goals to include sanctioning of social behaviour, and organising consumption, especially in times of distress. A local level study of rural responses to economic reforms in the late 1990s, in Shamva district, found out that over 50 percent of the smallholders in the district belonged to a local farmer group (Arnaiz 1998).

Moyo (2002, 1) argues that these local associational forms serve a variety of purposes and assume a multitude of roles, but are mostly formed in response to the negative effects of state policies and market penetration. In his view collective action by farmers is most common where both the state and market have a strong presence, and least common where both are weakly represented. In contrast, Bratton argues that in Mashonaland East, '"group development areas" have risen since 1972 in response to a government programme to deliver extension advice on a group basis' (Bratton 1986, 371). The state owned Agricultural Finance Corporation's expansion into communal areas coincided with the mushrooming of 'credit and cash groups' (Bratton 1986,372). These are indications that group formation can arise in response to positive opportunities too.

There is evidence in the Zimbabwean literature that the formation of groups to engage in collective action has occurred through the initiatives of external agents (especially NGOs). These have promoted interventions such as diversification into non-farm income sources and introduced methods that intensify the use of available land through the adoption of high yield seed varieties and synthetic fertilisers (Helliker 2008). In the process these interventions served to pacify demand for structural transformation through land redistribution (Helliker 2008). In fact, both Moyo and Yeros (2005) and Helliker (2006) view the majority of local associations formed since independence as subordinate partners of intermediary NGOs, which have steered them into localised 'development projects' that focus on the immediate objectives of securing food security, but within a framework that does not challenge the structural causes of this insecurity.

Some farmer associations function at the national level. In the period just after independence many rural households were recruited into local structures of the national smallholder farmers union – the National Farmers Association of Zimbabwe (NFAZ) – and by the end of 1988 the union had established 4,000 local clubs countrywide (Bratton 1994,15). In 1994 a new smallholders union was formed from a merger between the NFAZ and the Zimbabwe National Farmers Union (ZNFU). The ZNFU represented the interests of approximately 9,500 black farmers who owned private farms (ranging from 20 to 200 ha and averaging 80 ha) in the then African purchase areas (Bratton 1994,14). The NFAZ on the other hand represented the majority of the smallholders located in communal areas. At the time of independence the ZNFU and NFAZ were probably the only self-managed national smallholder unions on the continent, with 9,500 and 85,000 members respectively.

The resultant Zimbabwe Farmers Union (ZFU) has been slow to establish its structures in resettlement areas. Studies by Barr (2004) in areas resettled during the first phase of land reform between 1980 and 2000 did not find any ZFU activity, and by the end of 2009 it had not established its presence in fast track areas either. There are several reasons that help to explain the ZFU's antipathy to fast track land reform, including their dependence on the donor community (which was mostly opposed to the farm occupations) and the fact that it continues to be led by elite sections of black farmers.This antipathy to resettlement was apparent, however, even before the farm occupations and the FTLRP. Instead of targeting their neediest members, those concentrated in the communal areas, the ZFU leadership gave priority to programmes aimed at securing tractors and pick-up trucks for members in the small-scale commercial farming sector. They failed to address major differences in infrastructural development between the privately owned farms of their wealthier members and those in communal and resettlement areas, the land tenure challenges of resettled farmers, and the overcrowding that communal area based farmers suffered (Bratton 1994,28). Furthermore, although the union claimed to represent all black farmers, 50 percent of its members were poor, land-short, cattle-less and experiencing food deficits – yet the leadership of the union advocated that land be redistributed to 'capable farmers' rather than the 'needy' (Moyo 1999,17). By the end of 2009 the ZFU had not established any structures within newly resettled areas.

Studies by Dekker (2004) and Barr (2004) in pre-2000 Model A resettlement schemes found that despite official insistence on maximising production, the immediate concern of many resettled households was to forge new social relations. The social composition of the 'communities' in which they found themselves was markedly different from communal areas, in that kinship relations were largely absent, and settlers also lacked access to organisations with a broader mobilising platform such as the ZFU. There were rare instances of family relationships, either by birth or through marriage between near neighbours (Dekker 2004), and some people would move to the new areas with friends or acquaintances from their places of origin. In the majority of cases, however, people were strangers to one another (Barr 2004,1753). In the absence of pre-existing ties and forms of traditional leadership, settlers invested in new social relations by creating new local organisations (Barr 2004,1754). The social synergies thus created were utilised to improve production capacity, share risks and attend to other social obligations such as mobilising support during funerals and weddings. Barr (2004) concluded that in comparison with their customary area counterparts, resettled households have a tendency towards 'increased socio-civil activity' (2004, 1756).

The land occupation era, leading to the FTLRP, saw the emergence of a variety of organisational forms. The most common was the committee of seven, which was in most instances led by a war veteran who occupied the post of Base Commander (Chaumba et al. 2003, 8). The committee of seven was associated with replication of an 'army barracks-like' form of organisation in which curfews were established, females were separated from the males and all visitors had to report to the Base Commander. The committees of seven were also responsible for ensuring that farm production commenced as soon as land beneficiaries had been allocated their individual plots (Sadomba 2008, 115). Traditional authority functionaries such as chiefs and headmen rarely formed part of such structures, but they were consulted on some matters, such as traditional cleansing ceremonies and beneficiary selection. Although spontaneous in essence the land occupations received national support from the war veterans association and ZANU-PF (Zimbabwe African National Union Patriotic Front) and this partially explains the uniform nature and structure of the committee of seven (Chaumba et al. 2003).

In 2003 the government issued a directive on local government which stated that 'in terms of the Traditional Leaders Act (Chapter 20, 17) all resettlement areas shall be placed under the relevant traditional chiefs or headmen' (GoZ 2003,4). The Traditional Leaders Act (TLA) empowers the chief to nominate village headmen, and since 2003 the chiefs have been arbitrarily appointing headmen from among the land beneficiaries in newly resettled areas despite the absence of a genuine historical claim to that position by the individual being promoted. The village headman chairs the new Village Development Committee (VIDCO). Chaumba et al. (2003,10) describe these new village authority structures as 'a sudden emergence from seemingly nowhere'. However, in reality they are a slightly different version of the VIDCOs that had been established in the mid-1980s.[1] Members of the VIDCO are directly elected into office by villagers. A new post of war veteran representative was created in the aftermath of the conversion of the occupation era committee of seven into a village development council. Other posts in the village council include officers responsible for village development, security, women's affairs, health and the youth.[2]

Local associational life is a feature of FTLRP farms. A household survey of FTLRP beneficiaries in 2005/06 found that approximately 26 percent of resettled A1 households belonged to local farmer groups (Moyo et al. 2009). Scoones et al. (2010, 207) also made similar observations: '... social networks replicating those found in communal areas have emerged in various forms ... and these include work parties,

---

[1]The Prime Minister's Directive of 1984–1985 provided for the creation of a hierarchy of representative bodies at village, ward and district levels. The local development committees – the Village and Ward Development Committees (VIDCOs and WADCOs), also composed of elected members – were charged with the responsibility of defining local development needs (Mutizwa-Mangiza 1985). These development committees were described as 'democratic institutions of popular participation to promote the advancement of development objectives set by government, the community and the people' (Alexander 2006, 108).

[2]Based on interviews with headmen at Dunstan and Buena Vista farms, September 2008.

funeral assistance and religious based interactions'. The authors note that 'religion and church affiliation have emerged as a vital component in the construction of social relations and networks on the new resettlements (Scoones *et al.* 2010, 71). Findings from my field work in Goromonzi and Zimbabwe concur with Moyo *et al.* (2009) and Scoones *et al.* (2010). Approximately 40 percent of the A1 land beneficiaries in Goromonzi and Zvimba belong to local farmer groups.

The local farmer groups examined in this article have emerged in a context of very thin agricultural markets, a constrained state and isolation from the wider networks of civil society such as the larger farmer unions. These factors contributed to the emergence of 'Janus' faced local associations, focused not only on enhancing their productive capacity, but also, of necessity, engaged in complicated processes of negotiating for resources and support from state institutions. This is one of the major differences with the groups that emerged in resettlement areas of the 1980s and 1990s.

What were the key objectives of the local farmer groups that beneficiaries formed in FTLRP contexts, and what forms of collective action did they engage in? To what extent did they succeed in achieving their objectives? What problems and challenges did they confront, and what strategies did they adopt to address these? Case studies of farmer groups on FTLTP farms in two districts in the high potential high veld of Zimbabwe provide provisional answers to these questions, and show that although they have significantly contributed towards easing farm production constraints through sourcing of inputs, pooling together labour and savings, they still face major challenges in establishing sustainable systems of production.

The study area: Goromonzi and Zvimba districts

Goromonzi is 32 km south east of the country's capital, Harare and lies within NR IIa and NR IIb.[3] The total population of Goromonzi district is estimated to be 200,000 (GoZ Census, 2000). The district is further divided into 25 smaller administrative units called 'wards' or 'intensive conservation areas' (ICAs). Bromley, where the four case studies are situated, is one of these ICAs. The area normally enjoys reliable rainfall conditions and rarely experiences severe dry spells in summer (Jiri 2007). The region is suitable for intensive farming of crops such as maize, soya beans, sugar beans, ground nuts, potatoes, tobacco and wheat. Livestock production was very limited. Moyo (2000, 68) found that the majority of the large-scale farmers in Bromley made the switch to horticultural production in the 1990s and they were producing vegetables and flowers for export. The switch to horticulture led to an actual decrease in the amount of land devoted to crop production. There were 38 large scale commercial farms prior to fast track. Currently there are 9,382 A1 units and 2,319 A2 units in the area (Goromonzi AREX Office 2008).

Banket is an ICA within the Zvimba district which is part of Mashonaland West Province. The population of Zvimba is estimated to be 220,000 (GoZ Census, 2000) and the district lies in natural region IIb. Land use patterns before the land redistribution included staple cereals (maize and wheat) and livestock production. Dairy production was more prevalent among the large-scale commercial farms.

---

[3]Zimbabwe is divided into five natural regions or agro-ecological zones. The first three (I–III) are suitable for intensive agriculture, while the other two (IV–V) are only suitable for limited extensive agriculture (heavily reliant on irrigation), livestock and game ranching.

Other such farm land use patterns included flue cured tobacco and soya beans (Muir 1994). There were 41 large-scale farms in Banket prior to resettlement and 16 of these have been subdivided into A1 farm units and 25 into A2 farms.

The study sites were, prior to land reform, part of Zimbabwe's green belt in terms of food production. The GoZ had until the late 1990s discouraged disruption to the large-scale farms operating in the area, for instance Bromley had never experienced any land redistribution prior to 'fast track'. Furthermore these areas are serviced by a reliable road network and are on the electricity grid and closely located to the capital city of Harare.

A total of six farmer groups from Goromonzi and Zvimba districts were studied using both quantitative and qualitative methods. A household questionnaire targeted at members of farmer groups was administered in the same areas where the initial household survey was carried out by researchers from the African Institute for Agrarian Studies (see Moyo *et al.*, 2009 and Moyo this collection) had been carried out. A total of 137 households resettled at Dunstan, Lot 3A of Buena Vista, Dalkeith and Whynhill were asked to respond to the questionnaire (see Table 1).

The selection of the four former large-scale farms and the groups that have emerged in these areas was influenced by the need to show internal variation among the groups, and the preliminary criteria for selection included the differences in the ways that the farm was occupied and the manner in which local authority emerged, the process of the formation of the groups, land use patterns among the members and the nature of activities chosen by the groups. These variables provide initial indications of the differences within these local groups and the potential for further variation among the groups. The selection criteria also considered how the analysis of each case study would contribute towards a more illuminating process of understanding the emergence of local farmer organisations in A1 settlements.

The case studies describe in more detail the processes of formation and the nature of the activities carried out by the local farmer groups. They also help to explain the manner in which these groups are contributing to the formation of new communities and the ways in which they are ensuring that the newly resettled areas remain visible to the GoZ.

Membership levels within these groups vary from as low as 10 to as many as 75 (see Table 1 above). The groups serve a variety of purposes ranging from lobbying

Table 1. Local farmer groups.

Name of Group and Farm	District	Preferred Land Use	Year of Formation	No. Interviewed	Total in Group
Budiriro-Dunstan	Goromonzi	Maize	2003	10	10
Muswiti-Dunstan	Goromonzi	Maize	2004	9	15
Salt-Lakes-Dunstan	Goromonzi	Tobacco	2005	11	16
Tagarika-Lot 3A of Buena Vista	Goromonzi	Maize	2003	30	40
Chidziva-Dalkeith	Zvimba	Wheat &Maize	2004	59	75
Zhizha-Whynhill	Zvimba	Wheat & Maize	2004	18	31
Total				137 (73.26%)	187 (100%)

Source: Murisa (2008), Goromonzi and Zvimba, LFG Households Survey.

for improved distribution of inputs, mobilisation of savings, joint use of infrastructure and defending the newly acquired land rights. The groups under study have come up with their own constitutions which spell out the objectives of the group, activities, and conditions of membership and roles of office bearers. Table 2 provides an overview of how the local groups under study were established and the areas of collective action.

Cases

Small farmer groups at Dunstan Farm

Dunstan farm is located in Goromonzi district. It is 3200 hectares in extent and it was converted into an A1 settlement of 115 households in 2001 after occupation. In 2003 a local farmer group called Budiriro (Shona for development/progress) was established with 64 members.[4] The leadership of Budiriro faced challenges in convening meetings and difficulties in agreeing on the objectives of the association.[5] Based on advice from the local extension officer, the group was split into smaller groups with an average of 13 members on the basis of common land use activities.[6] In 2004 the group was subdivided and four groups for those interested in growing maize were eventually formed: Budiriro, Muswiti (named after the river that passes by the farm), Dunstan and Shingai (Shona for 'resilience'). Table 3 provides a summary of the groups that existed at Dunstan in 2008. Of the 115 official beneficiaries 81 (70.4 percent) belong to farmer groups.

The formation of the farmer groups in Goromonzi took place mostly in response to the suggestions of the local extension officer, except in the case of Salt-Lakes group, whose origins derive from the contract farming agreement some beneficiaries have entered into with a multinational tobacco trading company of the same name. Seven households had consistently grown tobacco since 2002, and these pioneered the establishment of Salt-Lakes[7] local farmer group in 2005 following an approach from a Salt-Lakes Ltd local representative offering to finance their farming inputs and to market their crop. Under the agreement all the tobacco produced by group members is sold to the company on the basis of prevailing tobacco auction floor prices.[8] The group had 16 members in 2008, and on average each member devoted two hectares of their six-hectare plots to growing tobacco.

Fourteen households who subsequently became interested in tobacco growing could not join the Salt-Lakes group because of the conditions of the agreement with the company. These households formed the Gutsaruzhinji group in 2006, which also focuses on tobacco growing but does not engage in contract farming.

Other crops grown by group members include tobacco, sunflower (on a very small scale), and sweet potatoes and there is very limited livestock rearing. The majority of farmer groups at Dunstan Farm have been established in order to

---

[4]Based on interviews with Extension Officer, September 2008 and focus group discussions held at Dunstan Farm, September and October 2008.

[5]Dunstan focus group discussion notes 01/08, September 2008

[6]Interviews with Bromley area Extension Officer and Budiriro Maize Group Chairperson, June 2009.

[7]Salt Lakes is a multinational tobacco company that entered into an agreement with a consortium of mainly A2 farmers growing tobacco. Within the consortium one of the leading A2 farmers also acts as an agent for A1 farmers interested in growing tobacco.

[8]Interview with Chairperson of Salt-Lakes LFG, September 2008.

Table 2. Synopsis of origins and activities of Local Farmer Groups.

Farm	Size of farm (ha)	Beneficiaries on farm	Associational form	Number of Members	Key actors involved in formation of group	Focus of collective action
Dunstan	3,200	115	6 x farmer groups	81	Five farmer groups: local extension officers One farmer's group: tobacco trading company that contracts with the group	Access to government subsidised inputs and credit via Agri-Bank loans Joint savings accounts and rotating savings One group: co-operation in tobacco production and marketing in terms of contact with trading company
Buena Vista, Lot 3	533 350 ha is irrigable, of which 18 ha has been cultivated to date	40	Tagarika Irrigation Scheme Co-operative	40	War veteran who led farm occupation	Irrigated wheat production by co-operative Application to government for irrigation equipment Pooling of resources for purchase of irrigation equipment Hire of tillage services Access to agricultural credit via Agri-Bank loans Defence of land rights against incursions by neighbouring A2 farmer/politician
Dalkeith	600	79	Chidziva Farmer's Association	79	Son of spirit medium and local extension officer	Irrigated wheat or seed maize production by farmer's association Access to government subsidised input supply programme Defence of land rights against incursions by neighbouring A2 farmer

(continued)

Table 2. (Continued).

Farm	Size of farm (ha)	Beneficiaries on farm	Associational form	Number of Members	Key actors involved in formation of group	Focus of collective action
Whynhill	1038	45	Zhizha Farmer's Co-operative	33	Local extension officer	Irrigated wheat production by co-operative
			Whynhill Farmer's Association	12		Access to government subsidised input supply programme
						Defence of land rights against incursions by neighbouring A2 farmer

Table 3.   Small farmer groups on Dunstan Farm.

Name	Preferred Land Use	Year of Formation	Total in Group
Budiriro	Maize	2003	10
Muswiti	Maize	2004	15
Salt-Lakes	Tobacco	2005	16
Dunstan	Maize	2004	10
Shingai	Maize	2004	16
Gutsaruzhinji	Tobacco	2006	14
Total		*81*	

Source: Murisa (2008), Goromonzi and Zvimba, LFG Households Survey.

improve access to farming inputs. Since their formation all the groups have registered formally with the local extension office. The registration of farmer groups with the extension office is mostly unique to Goromonzi and is used as a condition for accessing the government's input subsidy scheme. The groups submitted their constitutions and lists of members as part of the registration process,[9] and managed to secure limited quantities of inputs for their members.

In addition to the focus on farm inputs, all the groups based at Dunstan have introduced savings schemes though the opening of joint savings accounts with the local branch of Agri-Bank. Since 2005, this bank has been issuing short term loans for the purchase of inputs under the GoZ's Strategic Grain Reserve Facility[10] that encourages loans to LFGs that maintain savings accounts.[11] The leadership of the groups are tasked with ensuring that all members repay the loan. Members of Muswiti and Salt-Lakes were among the first recipients of the loans. In December 2006 Muswiti farmers group was issued with a loan of ZW$ 5,040,000.00 (approximately US$ 1,000.00) for the purchase of inputs which was to be repaid by September 2007.[12] Each member received two 50kg bags of seed maize and two 25kg bags of fertiliser. The group managed to repay the loan at the end of June 2007.

The relationship between Agri-Bank and farmer groups in Goromonzi thrived from 2005 until the end of 2007. During this period groups such as Salt-Lakes managed to acquire loans to purchase farm inputs for two consecutive seasons, but during the 2007/08 farming season the bank could not issue any new loans because the revolving fund had been wiped out by the hyperinflationary pressure on the local currency.[13]

Hyperinflation also harmed the savings accounts maintained by the groups. On two occasions members of Budiriro and Muswiti groups lost the real value of their

---

[9]Interview with Goromonzi District Extension Officer, August 2008.
[10]There were a number of subsidy initiatives developed mostly by the GoZ's Ministry of Agriculture and also the Reserve Bank. The Strategic Grain Reserve facility was developed and targeted at smallholders (and to lesser extent A2 farmers) who were willing to borrow loans at concessionary rates to purchase inputs. The Agri-bank instead of giving loans would proceed to purchase the inputs on behalf of the farmers and the Marondera branch insisted on lending only to smallholders within groups.
[11]Interview with Agri-Bank Branch Manager, November 2008.
[12]Agri-Bank 2006, Loan Agreement Form with Muswiti LFG.
[13]Interviews with Mr Hlophe, Agri-Bank Branch Manager, September 2008.

savings because they had left their money in the bank for too long.[14] The groups still maintain a savings account, but as of 2008 Budiriro and Muswiti moved towards a rotating savings scheme (colloquially referred to as *kukandirana*), whereby the agreed amount saved in each month is given to one group member, on a rotating basis. Funds from these initiatives have been deployed towards a number of social functions and farm investments by members. One member used the money to establish a small chicken rearing project.[15] In 2007 two members of Muswiti purchased a small water pump, which they use jointly for the irrigation of their vegetable gardens.[16]

The key differences between Salt-Lakes and the other farmer groups at Dunstan derive from its agreement with the tobacco trading company which supplies the group with inputs, whereas the others mostly rely on state agencies (including the state owned Agri-Bank). Tobacco growing requires specialised skills, and some of these were lost during the process of acquiring the farm as a result of the displacement of farm workers. To reduce the skills constraint, the Salt-Lakes farmer's group organises the hiring of specialised labour for activities such as the treatment and transfer of seedlings and the curing of tobacco. The group also coordinates the joint use of productive assets such as tobacco curing barns and grading sheds. Since its formation in 2005 the members of the group have received tobacco farming inputs from the contracting company, Salt-Lakes Limited, sufficient to cultivate two hectares per member. In the first year the company also provided extension support to assist with the technical aspects of growing tobacco. Sales records since 2005 show that each member has been able to deliver an average of three tonnes of tobacco. However, some members struggled to deliver more than one tonne, and the more successful delivered much more (e.g. the female chairperson of the group delivered four tonnes in 2005 and six tonnes in 2007).

Tagarika irrigation scheme at Lot 3 of Buena Vista

In 2004 all of the 40 beneficiaries resettled on A1 plots within what used to be Lot 3A of Buena Vista Farm established an irrigation cooperative, to take advantage of the irrigation potential of the farm. The establishment of the cooperative was led by the former leader of the occupations on the farm, a veteran of the liberation struggle trained in the former Soviet Union who had worked in an agricultural collective[17]. The irrigation system that existed prior to land reform had the capacity to irrigate 350 hectares of the 533 hectares within the farm.

The Tagarika Irrigation Scheme Cooperative was formally established in 2005, when it comprised all beneficiaries on the former large-scale farm. The process of establishment entailed the adoption of a constitution and the election of an executive committee chaired by the former leader of the occupations. According to the constitution the purpose of the cooperative is 'to improve the lives of the members and to increase production at the farm through the resuscitation of the irrigation system and to venture into crop horticulture under irrigation' (Tagarika

---

[14]Interviews with Budiriro and Muswiti chairpersons, September 2008.
[15]Budiriro, Muswiti, focus group discussions, September 2008.
[16]Muswiti, 01/08 focus group discussion notes, September 2008.
[17]Interview with Lot 3A Extension Officer, September 2008.

Constitution, 2005, 2). One of the initial motivations for the formation of the group was the need to ensure equitable access to irrigation equipment.

The group has faced challenges in terms of translating the objectives of the constitution into reality. The constitution lists fishing, resuscitation of irrigation equipment and horticulture as some of the activities that the group should engage in. The group has created four sub-committees responsible for maintenance of irrigation equipment; marketing and production; property and security; and welfare and health. The sub-committee for property and security is responsible for the maintenance of all the assets owned by the group. It maintains a register of assets and regulates the use of the limited irrigation infrastructure that is currently functional.

The group has struggled to revive the irrigation system to its full capacity despite having appealed successfully to the GoZ in 2005 for the replacement of some missing water pumps. When the group was formed, it established a smaller sub-committee tasked with engaging the Ministry of Agriculture's department of mechanisation. It managed to secure two 125 horse power water pumps. The pumps could not be operated, however, because some valves and sprinklers were missing. In 2006 the members of the group each contributed 50kg of maize for sale, and the proceeds were used to purchase the outstanding parts.[18] Less than half of the required valves and sprinklers were purchased from the proceeds of the sale.[19]

In 2005 the group opened a savings account with the Agricultural Bank (Agri-Bank) and was able to secure a loan for the 2005/06 agricultural season for the purchase of farm inputs for wheat production. Although one of the group's key objectives is to organise farm production and the marketing of crops on a collective basis, members of the group have mostly been producing crops such as maize, groundnuts, sunflowers, sweet potatoes and very few produce tobacco on an individual basis. The only instances of joint production of wheat were in the 2005 and 2006 winter seasons. The most visible intervention made by the group so far has been through securing inputs (maize seed and fertiliser) and repairs to irrigation equipment.

The sub-committee for property determines access to jointly owned equipment for individual use, and is also responsible for the hiring of tillage services. The joint hiring of tillage services has not been implemented in a uniform manner since the group was formed. In the 2005/06 and 2006/07 agricultural seasons the group secured the tractor services of the GoZ owned District Development Fund (DDF) to prepare three hectares for each member. In 2007/08 a private contractor was invited to prepare an average of five hectares per member. The service was provided to only 14 members, including the entire executive committee.[20] Explanations for such a discrepancy differ; according to a member of the sub-committee, those who were excluded had not paid for the services, while participants in the focus group discussion argued that they had been made to understand that the 50kg of maize they had provided in the previous season would cover the cost of hiring the tillage services.[21]

---

[18]Interview with Tagarika Chairperson, September 2008.
[19]Interview with Tagarika Chairperson, September 2008.
[20]Tagarika Focus Group Discussion Notes, September 2008.
[21]Tagarika Focus Group Discussion Notes, September 2008.

Despite the fact that there is a large dam on the farm that could irrigate up to 350 hectares, the beneficiaries have not yet exploited the full irrigation potential. They have thus far managed to irrigate only approximately 18 hectares of land in the area close to the dam, for joint[22] wheat production. They have mobilised subscriptions from members to purchase the outstanding parts required for the underground irrigation system to cover the whole farm but they are yet to acquire adequate sprinkler heads and mainline underground hydrant taps and caps.[23] The group has pursued several options for purchasing the required equipment. In early 2008 they applied to the GoZ for a loan to finish installing the irrigation equipment. An assessment team led by the acting Engineer for Goromonzi district came up with an inconclusive appraisal which stated that the:

> ...nature of the [irrigation] project is so massive and the options for financing include: (i) resettled farmers mobilise funds to buy the outstanding equipment and the local council will provide electricians to fit the equipment or (ii) the Ministry of Mechanisation to assist in buying the remaining equipment and farmers pay for it through agricultural commodities (Goromonzi RDC 2008, 1).

By the end of 2008 the GoZ was yet to commit itself to supporting the venture beyond the water pumps that they had already provided, and the chairman of the group conceded that it was proving difficult to continue asking members to contribute towards the purchase of these given the manner in which the prices kept changing.

Meanwhile the GoZ had in 2007 introduced the Agricultural Mechanisation Programme through the Reserve Bank and the Ministry of Mechanisation. Some land beneficiaries (mostly A2 farmers) were issued with brand new tractors, ploughs, ridgers and harrows. The Tagarika leadership applied for a tractor to be issued to the cooperative but they were advised that the tractors were for land beneficiaries with more than 50 hectares of arable land, in essence excluding all A1 farmers[24].

The group has also intervened to defend the land rights of its members. In 2006, 15 of the land beneficiaries at Lot 3 of Buena Vista were facing possible eviction. A local politician from ZANU-PF, resettled on a neighbouring A2 plot without irrigation capacity, approached the group's leadership to ask for shared use of the dam. The members agreed, but later on the A2 farmer came back with an offer letter from the GoZ for a portion of Lot 3 adjoining to his farm. He argued that he was in a better position to utilise the dam.[25] The leadership of the group approached the provincial offices for clarification on the offer letter and demanded to know why the farm was being subdivided and where the affected 15 households were to be resettled. They also threatened the A2 farmer that they would occupy his portion of the farm if he pursued his claim.[26] It was only after the intervention of the provincial administrator that the claim to Lot 3 by the A2 farmer was withdrawn.

In 2004 the chief in Goromonzi appointed one of the beneficiaries as headman of the farm, without consulting the former Base Commander and now head of the Tagarika Cooperative. This was seen as an attempt to marginalise those who had

---

[22]They have used the 18 hectares close to the dam as a collective farm where they produce and market the wheat as a group.
[23]Interview with Tagarika Chairman, September 2008.
[24]Interview with Tagarika Chairman, September 2008.
[25]Interview with Bromley Ward 21–22 Extension Officer, June 2009.
[26]Tagarika, Focus Group Discussion Notes, September 2008.

been at the forefront of the occupation of the farm and their efforts in establishing local authority. The members of the original committee of seven mobilised the beneficiaries to resist the appointment.

The ambitious objectives of the Tagarika Cooperative have clearly not yet been achieved, and secure and regular access to farming inputs and tillage services remains a problem. By the end of 2008 the irrigation system was still not functional and there was very limited collective action in terms of organising farm production. However, the leadership has managed to establish a sense of belonging among the cooperative's members and the successful defence of the land rights of the 15 members whose plots were under threat from the neighbouring A2 farmer has strengthened members' loyalty to the group.

Chidziva farmer's association on Dalkeith Farm

The Manjinjiwa lineage group under the Matibiri-Magaramombe chieftainship was forcibly moved from the area converted into Dalkeith and Noordt Gate farms in the early 1950s. When land occupations began across the whole country in February 2000, the Manjinjiwa spirit medium is said to have gone into a trance and directed lineage members to occupy Dalkeith farm.[27] The clan's leaders negotiated with the farm owner for a portion of the farm which they considered sacred, and he acquiesced to their demands and also offered them accommodation within the farm workers' compound. In August 2000 the farm owner was served with a Section 8 notice that informed him that the farm had been designated for compulsory acquisition. He was required to vacate the property within two months and was not permitted to remove any property from the farm. The owner contested compulsory acquisition through the courts, but was unsuccessful. The farm is 600 hectares in extent, and was used to produce tea tree oil (on 36 hectares), rearing of crocodiles and also the growing of wheat, maize and tobacco by its previous owner.

Starting in 2001, the 79 land beneficiaries at Dalkeith farm decided to combine their individual plots to grow irrigated wheat in winter as a collective. However the process of growing the winter crop was marred by tensions and the erratic mobilisation of labour. Members of the group felt that the allocation of tasks had unfairly burdened some of the households. It was thereafter felt that there was a need for a committee to facilitate the collective production of wheat. Initially the village head wanted a smaller sub-committee to be established and to operate within the confines of the recently established Village Development Committee (VIDCO) that he chaired but some of the younger members advocated for a fully fledged farmers' association that would focus on farm production.[28] They also argued that the association should focus primarily on securing farm inputs, which were difficult to obtain at the time. In late 2002 one of the sons of the spirit medium was tasked by the lineage elders with pioneering the formation of a farmers' group based on his previous agricultural experience and modest education. Local extension officers were invited to provide assistance in the establishment of the institutional structures of the group.

Three key motivating factors behind the formation of the group include the shared social identity as a lineage group, the existence of a dam on the farm that

---

[27]Interview with Village Head, August 2009.
[28]Interviews with Dalkeith Farm VIDCO Treasurer, August 2009.

could provide water for irrigated cropping and the potential of increased incomes through the marketing of wheat. The Chidziva Farmers Association was established to 'improve the farming capacities of the members and to contribute towards better lives through collectively seeking for farm inputs, markets and introducing other income generating projects' (Chidziva Farmers' Association, Constitution 2004,1).

The initial focus of the association was on irrigated winter wheat, and specifically on securing inputs and coordinating both household and hired labour. The leadership of the group successfully requested assistance for farm implements from the GoZ after the co-existence deal with the former farm owner had collapsed. The government issued them with two brand new 125 horse-power water pumps and some of the pipes required for irrigation and the GoZ insisted that the group should produce wheat for resale to the Grain Marketing Board. However by 2008 the group members were yet to use the new equipment due to the non-availability of electricity.[29]

Farming inputs have mostly been secured through the government's input subsidy programme. In the 2006/07 agricultural season the leadership of the group attempted to break away from dependence on the government scheme by entering into an agreement with a maize seed company called Pannar Seeds, to grow seed maize on individual plots with the intention of joint marketing as a collective. The company provided the association with the necessary inputs, namely seeds and fertiliser. The group's leadership also expected the company to provide them with tillage services but the company insisted that they could only supply farm inputs. The inputs provided were inadequate and could only support approximately 40 members.[30]

From early on the group tried to diversify into other farming activities. After resettlement the group's leadership had entered into an arrangement of co-existence with the former owner. He was allowed to continue with his tea tree oil and crocodile rearing business on 3 hectares of the farm, and in return the former owner provided tillage services to the new land owners. The arrangement did not last for long, however. In late 2003 the leadership of the association contrived to evict the former farm owner from the land on which he produced tea tree oil. They convinced the village head that they could continue with the tea tree oil enterprise to the benefit of all the members[31] but the expulsion of the former owner was carried out without the knowledge of other members of the group[32]. Although the association's leadership retained the same labour pool that was working in the tea tree business, they struggled to find a market for the oil, and in 2006 made the decision to cut down the trees.

In the aftermath of the collapse of the co-existence arrangement with the former owner, the leadership of the association approached one of the neighbouring A2 farmers to help them in looking after the crocodiles, and entered into a verbal agreement with him. Within a space of 12 months disagreements arose over the distribution of proceeds from the sale of the crocodile skins.[33] The leadership of the group alleges that the A2 farm owner was behind the attempts by the Zvimba District Land Identification Committee (ZDLIC) to remove them. In 2006 officials from the Ministry of Lands informed the leadership at Dalkeith that the farm had

---

[29]Interview with Chidziva Chairperson, September 2008.
[30]Interview with Chidziva Chairperson, September 2008.
[31]Interviews with Dalkeith Farm VIDCO Treasurer, August 2009.
[32]Chidziva Focus Group Discussion Notes, September 2008.
[33]Interview with Chidziva Farmers' Association Treasurer, August 2009.

been re-zoned into an A2 farm.[34] The Chidziva leadership lodged an appeal against their eviction with the Minister of Local Government (who was also the MP of the area), and threatened that if the matter was not resolved they would approach the President. This strategy proved successful in warding off the threat of eviction

The leadership at Chidziva has established cordial relations with the local extension office, and the association continues to be prioritised in the distribution of inputs via government input supply schemes. The leadership has also made attempts to join the Zimbabwe Farmers' Union (ZFU). In 2006 they submitted membership forms to the national ZFU office and assumed that their membership was confirmed.[35] However there is no documentation confirming their membership and they are yet to realise any tangible benefits.

Although not clearly mentioned in its constitution, the farmer's association seemed to have the necessary elements for joint farm production due to its common lineage identity (which suggests potential ease of mobilisation of labour, and trustful relationships) together with the availability of irrigation water that can be more efficiently utilised on a larger scale than on small individual plots. Joint farm production has, however, been constrained by the fact that water pumps for irrigation purposes require electricity, which has not been available to date. In addition, some of the members have refused to cede their fields for collective production. The Chidziva Farmers' Association has thus a mixed record of limited success and many failures in their collective action efforts.

Collective agriculture at Whynhill Farm: the case of Zhizha Cooperative

Whynhill farm measures 1038 hectares and is located in the Banket area within Zvimba district. It was officially converted into an A1 settlement in 2001. Initially there were 83 A1 beneficiaries but the number was reduced to 45 when 380 hectares of the farm were offered to a former ZANU-PF councillor as an A2 farm. The majority of A1 beneficiaries are involved in crop production, the most common crops being maize, cotton, tobacco and soya beans. Twenty-eight households are involved in market gardening, in hedged, 0.2 hectare gardens that are situated close to the dam. The majority of those with gardens are growing leafy green vegetables, mostly for their own consumption, but seven reported that they also sell crops to neighbouring villagers. Livestock production is very limited in comparison to the period prior to resettlement; only 11 households own cattle and the average size of each herd is six. Only one A1 farmer has established a pig project by taking advantage of the pig handling facilities left behind by the former owner that are located on his plot.

In 2003 the A1 beneficiaries at Whynhill came together to form the Zhizha ('fresh harvest') Farmers Cooperative to ensure equitable and optimum usage of inherited irrigation equipment through joint production and marketing of wheat.[36] The extension officer responsible for the greater Banket area was part of the team that demarcated A1 plots, and during further visits he made suggestions for all land beneficiaries to come together and form a production cooperative that would utilise the dam on the farm and available irrigation equipment and also benefit from the

---

[34]Interview with Chidziva Farmer Group, Chairperson, September 2008.
[35]Interview with Chidziva Chairperson, June 2009.
[36]Interview with Zhizha Secretary, Whynhill Farm, August 2008.

government farm inputs programme.[37] The Village Development Committee (VIDCO) established in the aftermath of resettlement was instrumental in mobilising all the households to commit to the idea of the group. Elections for the executive, comprising the chairperson, secretary, treasurer and two committee members, were held in March 2003. Besides overseeing the group's meetings, the chairperson is responsible for ensuring that members have access to farm inputs. After two years of operation the leadership established three new sub-committees: one for the coordination of the group's labour requirements during the growing of wheat, another for maintenance of irrigation equipment and the third responsible for marketing. In 2006 the VIDCO chairperson was added to the structure as a committee member and the VIDCO made a similar provision by accommodating the chairperson of the farmer group into the VIDCO structures. This was done in order to ensure better coordination among the structures and to minimise duplication of activities.[38] Initially the group comprised of all the 45 A1 households, but in 2007 12 of the members split off to form the Whynhill Farmers Association.

Activities of Zhizha Cooperative

The group functions as a collective cooperative during the winter season. Members combine their six-hectare plots of land, obtain wheat inputs as a group and use the available irrigation capacity to water the crop. At the end of the season the group is responsible for the marketing of wheat and they share the proceeds from the sale equally among the members.

One of the first activities of the group was to follow up on a request for irrigation equipment that had been made by the VIDCO in late 2002. In 2003 government hired a contractor to install three new water pumps and to repair available irrigation equipment. Since 2003 the group has been utilising the donated irrigation equipment jointly to grow wheat. In the first three years they grew wheat on 40 hectares, the inputs being obtained from the state through the local extension office in Banket. The first three years of winter production were quite lucrative for the group and marketed output averaged 60 tonnes per year.[39]

However, collective wheat production has been on the decline since 2007, when the group managed to plant only 16 hectares; in 2008 this was reduced to three hectares. At least five reasons are behind the decline. First, the A2 farmer settled within Whynhill farm has in the past few years encroached into the irrigable area, and some of the A1 plots that were once part of the pool of fields combined for winter production now belong to the A2 farmer. Second when the group split in 2007 it lost some land in the process. Third the group did not receive payment for the 2006 crop on time, and they were not provided with subsidised inputs in the following year. Fourth when the group split into two, those who left the cooperative took with them one of the water pumps and some of the pipes, which further reduced irrigation capacity. In addition, the A2 farmer resident on the farm managed to connive with the first chairperson of the group to appropriate a water pump for her own use.

During the summer agricultural season (November to March) members revert to individual farm production and the leadership of the groups focuses on sourcing

---

[37]Interviews with Extension Officer, September 2008.
[38]Interview with VIDCO Chairperson, August 2009.
[39]AREX, 2008, Whynhill Farm Wheat Deliveries Records.

inputs. Since its formation, the group has only managed to secure inputs for members twice, in the 2004/05 and 2007/08 farming seasons. Extension officers responsible for input allocation alleged that the group did not submit their application on time for the 2006/07 allocations,[40] while the leadership of the group claims that inputs intended for the group members were allocated to the A2 farmer.[41]

Zhizha is a fully fledged cooperative engaged in collective production and marketing of commodities. However it faces many internal problems which arise out of weak leadership, limited participation of the members in the daily affairs of the group and weak government support. In 2005 the Zhizha Chairperson agreed to share equipment with the neighbouring A2 farmer without consulting other executive committee members.[42] When the members learnt of the arrangement, approximately 36 percent resigned to form another group. The remaining members tried for more than two years to remove their chairperson from the post but were afraid to do so because of his perceived good relations with the local extension officers as the latter are instrumental in ensuring access to inputs. Even after he was eventually deposed, the group still failed to extricate itself from the equipment sharing agreement with the A2 farmer.[43] The group has, since its formation, struggled to secure adequate supplies of labour during the period of joint wheat production. Less than 40 percent of the members have consistently contributed labour to the group's activities.[44] The problem is so acute that the group's biggest costs involve the hiring of external labour to complement the labour contribution from the members.

Collective action by local farmer groups in land reform contexts – Lessons from the case studies

As the case studies illustrate, a key focus of collective action by local farmer groups on these farms has been to improve the productive capacity of the land reform beneficiaries. This has not, however, been the sole objective of these local farmer groups and cooperatives – they have also defended the land rights of their members, and attempted to remain visible to government departments, their main source of material support. The discussion that follows examines in more detail the strengths and weaknesses of the different strategies pursued by these groups in their efforts to improve the livelihoods of their members.

Farmer groups and sourcing of inputs

The farmer groups in Goromonzi attempted to become independent of the government's inputs subsidy scheme through mobilisation of their own savings to purchase inputs. This has contributed towards nurturing trust and loyalty among members, and is contributing towards a nascent philosophy of independence from the state.

[40]Interview with Extension Officer, August 2009.
[41]Interview with Zhizha Secretary, August 2008.
[42]Zhizha Focus Group Discussion Notes, September 2008.
[43]Interviews with Zhizha Secretary, September 2008.
[44]Interview with Zhizha Chairperson, August 2009.

However the hyperinflation that characterised Zimbabwe prior to the adoption of the multi-currency system in 2009 severely constrained these efforts. Most of the savings were very low; for example in 2007 contributions from members of the groups based at Dunstan farm were less than US$4.00 per month. The groups had an average membership of around 12, making difficult the mobilisation of significant amounts. The larger groups in the Zvimba cases have been unable to engage in such activities due to insufficient levels of trust. It seems that strategic interventions such as rotating savings and labour pooling only begin to add value after a certain threshold of membership has been reached, but experience within LFGs in Zvimba has shown that as the number of members increases the levels of trust and loyalty to the group's objectives also decrease. This was compounded by the perceived failure of the leadership of the Zvimba groups to transparently allocate received inputs. Allegations of corruption within the leadership were rife at Chidziva and Zhizha.

Furthermore the inputs subsidy programme has fallen victim to bureaucracy and patronage. The process of acquiring inputs is very elaborate and includes the registration of farmer groups and their requirements with the district extension office three months before the beginning of the planting season. During the period between the formal application and the actual receipt of inputs the groups engage in a variety of, at times costly, discrete lobbying activities to be prioritised in the allocation of the inputs. The fact that often times the rationale for establishing these groups has been to secure a government subsidy programme has had unfortunate effects on the levels of commitment to the group.

Joint production and marketing

Only two groups, Zhizha and Salt Lakes, sustained collective action in joint production and marketing. Collective farming and marketing of wheat was introduced among the Zhizha members in order to take advantage of existing irrigation equipment and government's irrigation rehabilitation programme. None of the members had any prior experience of joint production nor were they given an opportunity for training in this new form of social organisation. Most of the land beneficiaries are from the Chirau and Kasanze customary lands where they had practised individual farming before resettlement. The challenge that the leadership confronted was on how to get members sufficiently involved and integrated within fairly new social relations of production. In most instances smallholder farming systems are run along established routines but the same mode of decision making is not applicable to a large-scale enterprise such as the collective farming being introduced through Zhizha. The leadership was also not adequately prepared for mobilising sufficient momentum for such an initiative.

Salt Lakes is a group that approximates a marketing cooperative. Production remains individually organised and the group intervenes only through the mobilisation of specialised skills for the preparation of seed beds, handling of seedlings and their transfer, curing and marketing of the crop. The Salt Lakes approach seems to be more successful than the one being used by Zhizha. It nurtures individual enterprise and accountability unlike at Zhizha where those who have not made equal contributions of labour are still equally rewarded. The approach used by Salt Lakes encourages members to acquire productive assets. Through this gradual accumulation of assets a modest pool of small productive assets inclusive of spraying cans, hoes, wheelbarrows, water pumps and pipes is now available for use by members of the group. The purchase of productive assets has contributed to an understanding of the necessity of asset accumulation among the members of the group.

Farmer groups and labour pooling

Only one group, Salt Lakes in Goromonzi, was involved in the pooling together of labour, while both groups in Zvimba were actively involved in labour pooling. Labour pooling is a challenging form of collective action because of the large commitment of time that it involves, and it has a tendency to create conflict. Labour pooling remains poorly managed and female members especially at Chidziva and Zhizha see themselves as contributing more labour than their male counterparts.[45] The failure to coordinate this activity effectively has caused rifts in the groups and loss of production capacity. At Zhizha the failure of the leadership to mobilise adequate labour from the members has necessitated the hiring in of labour, which has harmed the profit margins on the marketed wheat.

Conclusion: challenges to collective action

Although there are obvious advantages of cooperation within local farmer groups, a number of challenges have been experienced by FTLRP beneficiaries in these cases. The case study material presented here suggests that local farmer groups have provided a platform for aggregating productive forces such as labour and farming assets, for building local processes of participation, and for defending land rights. They are important institutional arrangements for sustaining farm-based income generation within A1 resettlement schemes and there is a need to identify ways in which their potential can be realised. Local farmer groups in FTLRP contexts tend to be somewhat isolated, and most of them are not linked to the larger associations operating at district, provincial and national levels. By mid-2009, fast track resettlement areas still remained isolated from the national smallholders' farmers union (ZFU), national and international NGOs and other networks of civil society comprising a complex web of networks. The local farmer groups in fast track areas uneasily straddle civil society and the sphere of the state. The continued exclusion of fast track resettlement areas from important national civil society based networks has limited the possibilities of wider mobilisation beyond their local villages.

There is definitely a need to re-imagine how agricultural production as a whole can recover especially within a context where the smallholder sector has significantly increased since fast track. Although local farmer groups have emerged on almost every A1 settlement they are not necessarily an adequate response to the constraints faced by the newly resettled farmers. However it would be equally a barren methodology that fails to appreciate their significance and the clues these formations provide to understanding how local agency and social organisation in the newly resettled areas is evolving under austere economic circumstances. Their emergence has been a critical intervention in the survival of the newly resettled communities and there is a need to identify ways in which they can be nurtured in order to sustain farm production.

---

[45] Chidziva and Zhizha Focus Group Discussion Notes, September 2008.

References

Alexander, J. 2006. The Unsettled Land: State-Making and the Politics of Land in Zimbabwe 1893–2003. Oxford: James Currey Publishers.

Arnaiz, M.E.O. 1998. Coping with economic structural adjustment: Farmer groups in Shamva District. *In*: L. Masuko, ed. *Economic policy reforms and Meso-scale rural market changes in Zimbabwe: the case of Shamva*. Harare: Institute of Development Studies.

Barr, A. 2004. Forging effective new communities: the evolution of civil society in Zimbabwean resettlement villages. *World Development*, 32(10), 1753–1766.

Bratton, M. 1986. Farmer organisations and food production in Zimbabwe. *World Development*, 14(3), 367–384.

Bratton, M. 1994. Micro-democracy? The merger of farmer unions in Zimbabwe. *African Studies Review*, 37(1), 9–37.

Chaumba, J., I. Scoones and W. Wolmer. 2003. New politics, New livelihoods: changes in the Zimbabwean lowveld since the farm occupations of 2000. Sustainable Livelihoods in Southern Africa Research Paper 3. Brighton: Institute of Development Studies.

Cheater, A. 1984. *Idioms of Accumulation*. Gweru: Mambo Press.

Chidziva Farmers' Association Constitution 2004. Unpublished. Copy available at AIAS repository.

Dekker, M. 2004. Risk, resettlement and relations: social security in rural Zimbabwe. *Tinbergen Institute Research Series No. 331*. Amsterdam: Thela Thesis and Tinbergen Institute.

Goromonzi District Extension Department 2008. Register of A1 and A2 Farms in Bromley Ward (copy available at AIAS repository).

Government of Zimbabwe 2000. *Zimbabwe national census report*. Harare: Government Printers.

Government of Zimbabwe 2003. Ministry of Local Government's directive on local government procedures in newly resettled areas. Unpublished.

Government of Zimbabwe 2008. Report on macro-economic indicators. Unpublished report.

Helliker, K.D. 2006. A sociological analysis of intermediary non-governmental organisations and land reform in contemporary Zimbabwe, Thesis (DPhil). Rhodes University.

Helliker, K.D. 2008. Dancing on the same spot: NGOs. *In*: S. Moyo, K.D. Helliker, and T. Murisa, eds. *Contested terrain: land reform and civil society in contemporary Zimbabwe*. Pietermaritzburg: S&S Publishers.

Jiri, O. 2007. Goromonzi District field report on AIAS household survey. Unpublished, available at the AIAS repository.

Mail and Guardian. (South Africa), 18 January 2009.

Moyo, S. 1999. *Land and democracy in Zimbabwe*. Monograph Series, Harare: Sapes Books.

Moyo, S. 2000. *Land reform under structural adjustment in Zimbabwe: land use change in the Mashonaland provinces*. Uppsala: Nordiska Afrika Institutet.

Moyo, S. 2001. The Land occupation movement and democratization in Zimbabwe: contradictions of neoliberalism. *Millennium Journal of International Studies*, 30(2), 311–30.

Moyo, S. 2002. Peasant organisations and rural civil society: an introduction. *In*: M. Romdhane and S. Moyo, eds. *Peasant organisations and the democratisation process in Africa*. Dakar: CODESRIA.

Moyo, S. 2007. *Emerging land tenure issues in Zimbabwe. Monograph Series, 1/2007*. Harare: AIAS.

Moyo, S., I. Sunga and L. Masuko. 1991. *Agricultural collective cooperation; a case of the socio-economic viability of the Makoni district union of cooperatives, Part I: the socio-economic features of the collectives*. Consultancy Reports, Harare: Zimbabwe Institute of Development Studies.

Moyo, S. and P. Yeros. 2005. Land occupations and land reform in Zimbabwe: towards the national democratic revolution. *In*: S. Moyo and P. Yeros, eds. *Reclaiming the land: the resurgence of rural movements in Africa, Asia and Latin America*. London: Zed Books.

Moyo, S. and P. Yeros. 2007a. Intervention, the Zimbabwe question and the two lefts. *Historical Materialism*, 15(3), 171–204.

Moyo, S. and P. Yeros. 2007b. The radicalised state: Zimbabwe's interrupted revolution. *Review of African Political Economy*, 34(111), 103–21.

Moyo, S., W. Chambati, T. Murisa, K. Mujeyi, C. Dangwa, and N. Nyoni. 2009. Fast Track Land Reform survey: trends and tendencies, 2005/06, AIAS Research Monograph.

Muir, K. 1994. Agriculture in Zimbabwe. *In*: M.Rukuni and C.K Eicher, eds. *Zimbabwe's Agricultural Revolution*. Harare: University of Zimbabwe Publications.

Munemo, R.F. 2008. Evaluation of irrigation equipment at Lot 3 Buena Vista Farm in Goromonzi (unpublished).

Murisa, T. 2009. An analysis of emerging forms of social organisation and agency in the newly resettled areas of Zimbabwe, the Case of Goromonzi and Zvimba districts. Thesis (PhD). Rhodes University.

Murisa, T. and K. Mujeyi. 2008. Assessing the preliminary outcomes from 'Fast Track' and related agrarian reforms. Presented at the Africa Institute for South Africa (AISA) Conference on Zimbabwe held in Maputo, 17 July 2008.

Mutizwa-Mangiza, N.D. 1985. Community development in pre-Independence Zimbabwe. Supplement to *Zambezia*, 78–69.

PLRC. 2003. *Report of the presidential land review committee under the chairmanship of Dr C. M. M. Utete*, Volume 1 Main Report and Volume 2 Special Studies, August 2003. Harare: Government Printers.

Rahmato, D. 1991. *Peasant Organizations in Africa: Constraints and Potentials*. CODESRIA Working Paper Series 1/91. Dakar: CODESRIA.

Sadomba, W.Z. 2008. War veterans in Zimbabwe's land occupations: complexities of a liberation movement in an African post-colonial settler society. Thesis (PhD). Wageningen Agricultural University.

Scoones, I., N. Marongwe, B. Mavedzenge, J. Mahenehene, F. Murimbarimba, and C. Sukume. 2010. *Zimbabwe's land reform: myths and realities*. Suffolk: James Currey.

Sibanda, A.E. 2002. Voicing a peasant alternative: the organisation of rural associations for progress (ORAP) in Zimbabwe. *In*: B. Romdhane and S. Moyo, eds. *Peasant organisations and the democratisation process in Africa*. Dakar: CODESRIA.

Tagarika Irrigation Scheme Constitution 2005. (unpublished) Copy available at AIAS repository.

ZFU 2001. The ZFU position on land reform. Unpublished document.

Tendai Murisa recently completed his PhD at Rhodes University. He wrote this paper when he was a fulltime Research Fellow at the African Institute for Agrarian Studies in Harare, Zimbabwe. He is now a Programme Specialist at the Dakar based Trust Africa, a development foundation where he is responsible for the coordination of the Pan-African Agriculture Advocacy project. He still actively collaborates with the AIAS on specific research assignments.

Index